ORNITHOLOGY

From Aristotle to the Present

ERWIN STRESEMANN

1889–1972

ERWIN STRESEMANN

ORNITHOLOGY

From Aristotle to the Present

Translated by Hans J. and Cathleen Epstein
Edited by G. William Cottrell

With a Foreword and an Epilogue on American Ornithology
by Ernst Mayr

Harvard University Press
Cambridge, Massachusetts
and
London, England
1975

Library of Congress Catalog Card Number 74-25035
ISBN 0-674-64485-9
Printed in the United States of America

Originally published in German as *Die Entwicklung Der Ornithologie von Aristoteles bis zur Gegenwart* (Berlin: F. W. Peters)

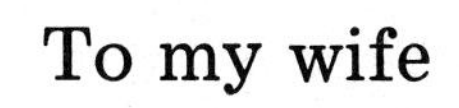

To my wife

Contents

Foreword

Ever since Stresemann's *Entwicklung der Ornithologie* was published in 1951 there have been demands for an English edition of this exciting volume. The merits of such a proposal are obvious. Ornithology, perhaps more than any other branch of zoology, resembles botany in its broad popular appeal. Yet in spite of the large number of volumes we have on the botany of every period from the Greeks to the present, there was nothing like it for ornithology prior to the appearance of Stresemann's volume. The fact that the author was perhaps the outstanding ornithologist of his era gives the volume great authority. But Stresemann was more than a specialist. He was a scholar of extraordinary breadth and erudition. Like a historian in the best sense of the word he understood how to capture the Zeitgeist of every period, the coming and going of fashions, the impact of other sciences and of contemporary conceptualizations, and, not least, he was fascinated with the personalities of the leading figures in the development of ornithology. This permitted him to escape the danger of a dry factual account and enabled him to present a thoroughly readable, indeed altogether fascinating, story. He made his intent quite clear by using (in the German title) the word *Entwicklung* ("development") rather than *Geschichte* ("history").

There are two reasons why a history of ornithology is of importance far beyond the boundaries of that field. Because birds are so accessible for study and research, ornithologists again and again have become leaders in certain new branches of biology. In zoogeography, in the theory of evolution, in the theory of geographic speciation, in environmental physiology, in population biology, in endocrinology, in the study of behavior (ethology), and in many other areas of biology some of the most important pioneering achievements were made by ornithologists. Indeed, no history of these disciplines can be written without a careful analysis of the input from ornithology. But there is a second reason for the importance of a historical treatment of ornithology. This is the excessive emphasis in recent histories of modern biology on experimental researches or, to put it negatively, the extraordinary neglect of the contributions made by all those branches of biology that might be conveniently bracketed under the name of natural history. There is no history of biology in which we can find a truly adequate presentation of the maturation of the species concept, of the study of geographic variation, of the theory of systematics (from the viewpoint of the zoologist), of animal ecology and ethology, and of the many other areas to which ornithology had made such significant contributions. Even those who are not in the slightest interested in ornithology will find Stresemann's accounts of these subjects fascinating and invaluable.

The volume was written in the immediate postwar years during the "hunger blockade" of Berlin, in a two-room apartment shared by Stresemann's family, always without hot water, electricity, or gas, often entirely without heat, and without free access to library facilities. Only a person of Stresemann's indomitable energy and extraordinary memory could have performed the miracle of completing the manuscript. The difficult conditions under which the book was written are reflected in the deficiency of footnotes and bibliographic detail, and it seemed desirable to fill this gap in the American edition. However, when the translation was begun, Professor Stresemann was nearing 80 and was no longer able to help. Most fortunately, G. William Cottrell came to the rescue and was able, thanks to his superb knowledge of the literature, to supply most of the missing information. Furthermore he painstakingly checked the entire translation, and either he or Margaret Mayr checked most of the citations in the original for accuracy and completeness.

Harvard University Press and the readers of this volume owe a great debt of gratitude also to the Epsteins for their superb translation. They have truly performed a labor of love.

Various readers of the draft translation expressed a desire for more information on the history of American ornithology. At the request of Harvard University Press, I prepared a short appendix which will permit better access to the literature on American ornithology. It is quite different in style from Stresemann's text and should be regarded, as stated in the title, as "Materials for a history of American ornithology."

Erwin Stresemann did not live to see the publication of this English edition. However, at the time of his death (20 November 1972) he knew of the completion of the translation and was looking forward with keen interest to the publication of this volume, which now stands as a fitting memorial to this great ornithologist.

Ernst Mayr

September 1, 1974

Preface

Birds are, at present, the best studied of any class in the animal kingdom. By now the number of bird species, and, for that matter, the number and distribution of the geographic races have been all but completely determined. Anatomy, physiology, ecology, and behavior of many species have been investigated in detail. As a consequence, the status of ornithology has lately been taken as both standard and guide for progress in zoology.

It is thus tempting to examine just this branch of zoology against its historical background. Current ornithological knowledge, opinions, and methods comprise a remarkably attractive image: a mosaic of new and old elements, some thriving, some atrophied, and some ossified. Even the last, unattractive but indispensable (for example, the symbolism of scientific nomenclature), may catch the imagination when the historian leads us back to the moment of birth, to the vital origin of such constructs.

No less rewarding is a closer acquaintance with the men whose work was essential to the advance of ornithology. In every age their number was small, but through them the torch ignited by the Greeks has been carried down the generations. Among them, we encounter wise philosophers, astute observers, and even some romantic enthusiasts. Many of them risked their lives to study the birds of unexplored lands; for ornithology has ever required the service not only of the intellect, but of the heart.

With this volume I have not endeavored to provide an exhaustive history of ornithology but rather to depict the field's development. To fulfill this goal I have, of necessity, selected material from an overabundance of sources. While attempting to follow the evolution of ornithology from modest origins to its present status, I have had to develop the story along several lines; for even in antiquity birds were studied in more than one aspect.

If I have met any success in offering something new even to the expert, I am indebted in no small part to favorable circumstances and willing assistance. Many important details were found in the ornithologists' letters preserved in the archives of the Deutsche Ornithologen-Gesellschaft, the Berlin Zoological Museum, and the Naumann Museum in Köthen. Of the many who helped me with their knowledge, Dr. J. Anker (Copenhagen), Prof. J. Berlioz (Paris), Prof. Dr. H. Engel (Amsterdam), Prof. A. Ghigi (Bologna), Prof. Dr. P. Thomsen (Dresden), and A. C. Townsend (London) deserve special mention.

The publisher, Mr. F. W. Peters, was most effective in putting my scheme into action and, busy as he was, in helping me with his lively interest and experienced advice, for which I wish to thank him particularly.

Prof. Dr. Erwin Stresemann

Berlin, April 1951

PART ONE

THE FOUNDATIONS

OF ORNITHOLOGY

1

From Classical Times to the Renaissance

The eventful history of ornithology is much more intimately interwoven with the growth of philosophy than many people today realize. In fact, our subject must be counted as a descendant of philosophy, which guided it for a long time before it began to find its own way. Eventually, together with the other biological sciences, it exchanged the role of a docile pupil for that of an experienced teacher.

The knowledge of birds achieved the rank of a methodical science only through its dependence on philosophy. A long time before that happened there had been people who knew about birds; indeed, the apparently instinctive desire which finds its satisfaction in hunting, trapping, observing, and admiring birds, and which can develop into a real obsession, may have existed from prehistoric times to our own day in roughly the same number of human beings.

Much has already been written about the prehistoric and early historical creations stimulated by interest in the world of birds — sculpture, wall paintings, poetry, and prose. These will not concern us here, however; we shall proceed at once to the beginnings of ornithology proper.

Aristotle (384–322 B.C.), who came from Stagira in Macedonia, was the first person to elevate knowledge of birds to the rank of a science. By the high Middle Ages, natural scientists justified their thorough absorption in the subject by the fact that he had declared the detailed study of birds to be a worthy occupation for the philosophic mind.

During the Renaissance, when the knowledge of his writings became the common property of scholars, they tried not merely to follow his suggestions but also to rediscover the species of birds he had mentioned by name, of which there may have been 140. Admittedly, Aristotle distinguished fairly clearly the appearance and habits of only some of them because often he knew so little of their characteristics that he either did not succeed at all, or succeeded only partially, in separating the groups ("genera") into species. A few chapters of his chief biological work, *The History of Animals* (περὶ τὰ ζῷα ἱστορίαι), contain his principal references to birds, but in general he did not treat them systematically. He interwove true and false accounts that he had found in older writers, like Aristophanes, with those that he had received directly from birdcatchers, fishermen, and farmers, which dealt with distribution and food, song and winter quarters, nestbuilding and breeding. To these he added his own personal observations, often deliberately, to illustrate general biological problems by example and to draw comparisons, as in the following paragraph:

> With most birds, as has been said of the pigeon, the hatching is carried on by the male and the female in turns: with some birds, however, the male only sits long enough to allow the female to provide herself with food. In the goose tribe the female alone incubates, and after once sitting on the eggs she continues brooding until they are hatched. The nests of all marsh-birds are built in districts fenny and well supplied with grass; consequently, the mother-bird while sitting quiet on her eggs can provide herself with food without having to submit to absolute fasting. With the crow also the female alone broods, and broods throughout the whole period; the male bird supports the female, bringing her food and feeding her. The female of the ring-dove begins to brood in the afternoon and broods through the entire night until breakfast-time of the following day; the male broods during the rest of the time. Partridges build a nest in two compartments; the male broods on the one and the female on the other. After hatching, each of the parent birds rears its brood.[1]

Aristotle did not intend to assign birds to rigid categories. Though later scholars read into his work a classification of birds (those with hooked claws; worm, thistle, and wood-beetle eaters; doves, terrestrial birds, fissipeds, and webfooted birds), he used these terms more descriptively than taxonomically. He also indicated another organizing principle: birds can be arranged in three classes according to whether they live on land, on the edges of rivers, lakes, and oceans, or on the water. For him such group characterization — which today would be called ecological — takes into account the essential differences in bird structure, because he considered that anatomy is shaped by basic needs, that is, by the demands of the environment. In order of importance, then, basic need ranks higher than structure. One of his examples states that when a hen has defeated a rooster, she begins to crow and imitates roosters by attempting to tread other hens. The result of these altered instincts is that the hen grows rooster's plumage. The same presupposition makes Aristotle ascribe to food and climate a direct influence on body size.

> In Egypt animals, as a rule, are larger than their congeners in Greece, as the cow and the sheep; but some are less, as the dog, the wolf, the hare, the fox, the raven, and the hawk; others are pretty much the same size, as the crow and the goat. The difference, where it exists, is attributed to the food, as being abundant in one case and insufficient in another . . . In many places the climate will account for peculiarities; thus in Illyria, Thrace, and Epirus the ass is small, and in Gaul and in Scythia the ass is not found at all owing to the coldness of the climate of these countries.[2]

The whole anatomic organization of birds is important for Aristotle because it is his aim to show which "parts" the "sanguineous" animals (those with blood as opposed to those without) have in common and which are peculiar to single groups. For this purpose he makes comparative examinations of the development of the esophagus, the stomach, and the cecum; he studies lung capacity; he mentions the location and the cycle of the gonads and describes the development of the chicken in the egg as exactly as was possible without the aid of optical instruments.

Even these brief examples indicate that Aristotle's ability to ask significant questions had made the study of the animal kingdom into a science, and had laid the groundwork for its eventual organization into morphology, systematics, physiology, embryology, biology, psychology, and so forth. In addition, he gave instructions on how a researcher should proceed. For him, the observation of animals should subserve teleology. Zoology must therefore establish the relation of a phenomenon to the *purpose* inherent in it. The study of animals is not ignoble, because

> if some have no graces to charm the sense, yet even these, by disclosing to intellectual perception the artistic spirit that designed them, give immense pleasure to all who can trace links of causation, and are inclined to philosophy . . . We therefore must not recoil with childish aversion from the examination of the humbler animals.

> Every realm of nature is marvellous; and as Heraclitus, when the strangers who came to visit found him warming himself at the furnace in the kitchen and hesitated to go in, is reported to have bidden them not to be afraid to enter, as even in that kitchen divinities were present, so we should venture on the study of every kind of animal without distaste; for each and all will reveal to us something natural and something beautiful. Absence of haphazard and conduciveness of everything to an end are to be found in Nature's works in the highest degree.[3]

For Aristotle everything living has been created to fulfill a purpose, because the two principles of Matter (ὕλη) and Idea (Form) operating together make up a unit in which the Idea, as the natural purpose (entelechy) working in matter, develops and shapes the material according to itself. The Idea is therefore the original and the creature its copy. The world is a hierarchy of entelechies, or purposive arrangements, which leads upward from the lowest to the highest; since ideas are eternal and immutable, so also must be the shapes created by them. Just as, in the hierarchy, apes occur between men and the other mammals, and water-lilies between plants and animals, the ostrich "has some of the characters of a bird, some of a quadruped . . . Further, it resembles a bird in being a biped, and a quadruped in having a cloven hoof; for it has hoofs and not toes."[4]

With his doctrine of entelechy Aristotle attempted to mitigate the conflict with empirical science into which Greek philosophy had been brought by the Eleatic and Platonic systems of thought. The Eleatics had maintained that the true Being of created things was not perceptible by the senses but was intelligible to the mind. They therefore posited the ontological equation, Being = Thinking (τό ὄν = λόγος). Later on we shall see what effect this fallacy had on ornithological systematics at the beginning of the nineteenth century, after the whole of the Middle Ages had already preferred to think ontologically. In such periods of intellectual history "concepts were valid instead of facts, and the imagined world of concepts penetrated so deeply into human flesh and blood that it needed the gigantic labors of natural scientists from Copernicus' time to ours to overthrow the colossus with feet of clay."[5]

Socrates (470?–399 B.C.) and Plato (427–347 B.C.) further developed the systems of the Eleatics. For them, too, only the content of universal concepts made up the true reality of things, and Plato finally designated Ideas, which were supposed to be established in the region beyond the fixed stars, as the only real creative and eternal causes of transitory phenomena. Thus the world for him was split between the immaterial (= the world of Ideas = heaven) and the material (= the world of phenomena = earth); the one is true reality, the other nothing.

As a natural scientist Aristotle attempted to overcome this Platonic dualism by establishing the location of Ideas within matter itself. By so doing he insistently called the attention of philosophers to the exact study of nature; but he was a great and lonely eminence in an epoch unprepared for such a comprehensive concept.

The following centuries produced nothing but epigones, who at best tried to encompass the master's knowledge and doctrine, without understanding their basic principles. Not propagators of firsthand knowledge but writers of fables were welcomed by those eager for enlightenment. In the first century after Christ, Alexander of Myndos collected miraculous stories about animals and endeavored to relate every bird to a given deity.[6] Like him, his younger contemporary, Gaius Plinius Secundus (A.D. 23–79), multifariously learned and perpetually active, collected the various products of decadent Hellenistic "zoology." Everything entertaining and paradoxical about birds that he could get hold of was arranged uncritically in the tenth of his 37 books of *Historia naturalis*, intermingled with excerpts from Aristotle. Pliny made no attempt to characterize the different groups of animals anatomically. He thought that he had discovered the most important distinguishing characteristic of birds in the structure of their feet, for according to him all birds of prey have hooked talons and all gallinaceous birds round claws, whereas all geese and similar water birds have webfeet with broad flat claws.

For nearly 1500 years this encyclopedia, almost unusable as a source of zoological knowledge, was held in the highest esteem, even by noted naturalists. A similar undeserved honor was bestowed on the *Collectanea rerum mirabilium*, mostly compiled from Pliny by a Roman, C. Julius Solinus (*c.* A.D. 275), and on the collection of animal stories made some 50 years earlier by the Roman litterateur Claudius Aelianus. This latter work, *On the Characteristics of Animals*, written in Greek by a quite uncritical man who never looked beyond the walls of his study, was compiled from more than a hundred sources; most of the bird stories came from Alexander of Myndos and Plutarch. Many of the "characteristics" that struck Aelian as remarkable had been freely invented by storytellers, but occasionally his reworkings contained a core of truth, an accurate observation, which he then interpreted for its moral. Gulls, which carry mussels high into the air in order to smash them hard against the rocks, behave "intelligently." The partridge and the Egyptian goose, which lure the hunter away from their nestlings by pretending to be crippled, "are distinguished among the animals by love for their young." "The mother swallow accustoms her young to the idea of justice by impartial distribution of food, training them to observe the law of equality." It never once occurs to this author that animals act from natural instinct; like man, they are good or evil, chaste or lustful, faithful or treacherous, reasonable or foolish. Here Aelian, an Epicurean, offers a contrast to the Stoics who, relying on Aristotle, believed they recognized in the purposive behavior of animals the working of Universal Reason, though they denied to animals themselves reason and its goal, virtue.

In the early Middle Ages, zoology degenerated still further. While Aristotle's writings fell into oblivion, the Neoplatonic and Neopythagorean doctrines absorbed by Alexandrian Christianity diverted intellectual studies more and more decidedly from the real toward the ideal and the abstract. Under the influence of Origen and his predecessors, men no longer regarded natural objects as existing in themselves, but tried to endow them with religious significance. This was the sole purpose of the animal creation, which was valuable only for its potential use in religious allegory. About A.D. 370, during the Patristic period, an "ecclesiastical natural history," known as the *Physiologus*, was compiled in Alexandria or Syria.[7] It was a collection of fables describing the mystical and magical properties of animals, which derived from Greek, Egyptian, and Jewish sources, interpreted afresh according to the principles of moral theology. It was the seedbed for a further crop of books about animals, the so-called bestiaries, as well as the *Etymologia* of St. Isidore of Seville (570–636), a work that organized under catchwords the entire knowledge of the church fathers, combined with as much classical learning as could serve the church's pedagogic purposes. The writings of Pliny, together with the *Physiologus*, the *Etymologia*, and similar moral treatises haunted by the supernatural, were repeatedly copied in monasteries and reworked into new encyclopedias. Even the Dominican Thomas de Cantimpré (*c.* 1210–1293), a man already influenced by recently rediscovered writings of Aristotle, compiled such a widely read work, roughly between 1233 and 1248, under the title *De natura rerum.* Over 100 years later the Regensburg canon Konrad von Megenberg (*c.* 1309–1374) undertook to translate selected chapters into German and to circulate them as *puch der natur* (book of nature). How well such a "natural history" suited the taste of the masses, brought up on a superstitious belief in miracles and authority, is clearly demonstrated by its six printings a century later, in the period between 1475 and 1500. A single sample of this late medieval zoological bestseller will amply suffice here:

> Of Storks. *Ciconia* signifies a stork or — to use another German term — an *Adebar.* This bird, according to Isidore, is ash-gray in color. Solinus says that it has no voice and can only clack with its beak . . . Whenever storks wish to fly across the sea, crows fly ahead of them and show them the way. Storks take many pains and much trouble over their young, whom they love tenderly, plucking out

> their own feathers and placing them in the nest at brooding time, so that the young will have a soft cushion. In return, the young storks retain great affection for their mothers, care for them as long as they themselves were cared for, and feed them also for as long. Therefore the stork is called the gentle bird. It hates snakes and pursues them diligently . . . In Asia there is a field in which the storks assemble and clack together, as if they were conversing. The last stork to arrive is torn to pieces and then they all fly away. Storks kill their wives when they have committed adultery and have not washed themselves with water after their misdeed. This has often been noted.[8]

Such distortions of natural fact and disdain for direct observation, barriers to a true knowledge of nature, were overcome at last by one of the most splendid embodiments of knighthood, Augustus Fredericus Secundus, Romanorum Imperator, Jerusalem et Siciliae Rex (1194–1250). He concentrated all his efforts on the here and now instead of the hereafter; in his poetry and his statesmanship he cultivated the worldly and sensual, because for him reality existed not in concepts but in individual beings. About 1230 he entrusted his court astrologer, the polyglot Michael Scot, with the task of translating into Latin Avicenna's Arabic version of Aristotle's works on animals. Frederick II thus channeled limpid Aristotelian thought through the muddy cesspool of contemporary learning. The chief beneficiary was zoology, in which the emperor was passionately interested.

> In various places in his kingdom he built parks for animals. The main part of his large menagerie, when not accompanying him on his journeys, was housed in Lucera . . . Symbolic for Frederick was his extensive "Vivarium" close to Foggia, a large area of marshes and ponds, fed by a well-regulated water supply from aqueducts, and populated by many species of water birds. What a fantastic picture — the place adorned with marble and *verdeantico* pillars and with bronze and marble statues, and in the midst of it all the German emperor, attended by Moorish slaves and noble pages, visiting his pond to observe the pelicans, cranes, herons, wild geese, and exotic waders![9]

In this way Frederick's knowledge of birds grew significantly; indeed, he became the first great ornithologist known to history. Endowed with rare natural gifts of keen observation, he was attracted while still a youth by the noble art of falconry, which the Saracens had brought with them from the Orient to Sicily and later had handed on to their conquerors, the Norman knights.

Between affairs of state and military expeditions, the emperor sought relaxation in hunting and became a master falconer, though only after great insistence by others did he finally decide to write down his experiences, a few years before his death. *De arte venandi cum avibus*, as he called his work, offers much more than the title suggests, even though it remains a fragment: the first of the seven books into which it is divided serves as an introduction to the entire field of ornithology. The classification of birds according to ecology and choice of food, their behavior throughout the day, their feeding, the many stages and varieties of migration, and avian anatomy are carefully presented and finally plumage, flight, and molt are described with painstaking exactness.

It would have been an incalculable advantage for ornithology if the succeeding centuries had been guided by the emperor's book, but the ecclesiastical naturalists and interpreters of nature took good care not to bestow such honor on an enemy of the spiritual authority, whom the pope had excommunicated and whom Dante had placed in the sixth circle of the Inferno, reserved for heretics. Thus neither Belon nor Gesner knew anything of the book. Although it was finally printed in Augsburg in 1596, it attracted the attention of ornithologists only in 1788. One of its two discoverers at that time was J. G. Schneider, whose new edition, with critical annotations, cannot be praised too highly.[10] The other discoverer was his friend, Blasius Merrem, who summed up his opinion as follows:

> The first book contains a general . . . description of all birds, which . . . far outstrips most of our more recent authors . . . He has most correctly described the bones, particularly those of wing and leg; he has most exactly distinguished the types of feathers and the way in which they are attached to the body; in a short summary of the internal organs he has noted so much of importance, and on the life and behavior of birds he has made such exact observations (which are wanting in all other writers), that I must confess that in no folio have I found so much that is good, new, and important for natural history as in these few leaves, and it would be easy for me, out of this old book, to astonish my readers with my gift of observation of important matters never before noticed, if only I possessed the genius of our age [that is, if I plundered the writings of others]. What a loss for natural history, that he was an emperor, and did not live in our time![11]

In conscious opposition to the philosophy of the church, Frederick pursued inductive research and — as a lonely precursor of the Renaissance mechanists — searched for natural causation. First and most important for him was the accurate examination and representation of natural objects. "Our intention in this book on hunting with birds is to show things as they are." His method is therefore to describe the objects, to compare them (thus establishing similarities and differences), and finally to investigate the causes of these relationships ("Modus agendi est . . . partim descriptivus, partim conventiarum et differentiarum assignativus, partim causarum inquisitivus"). For the latter purpose he also conducted experiments, for instance, to discover whether vultures find their food by sight or by smell. Sometimes he endeavors to explain shape functionally, according to the laws of mechanics or according to Aristotle's physiology. Thus, in Book II, chapter 5, he interprets variation in size among birds of prey as follows:

> In general, birds of prey that are born in the seventh climatic zone, or beyond, toward the North Pole, are larger, stronger, bolder, more beautiful, more able, and swifter, each with respect to its own species. They are larger because in the seventh zone and beyond there is intense cold, which tempers the intense heat that they have by their own nature, and through the tempering of the heat there is an increase of moisture, and through the increase of moisture the members are extended, which in turn affects the size of the body.[12]

The reason for migration he elucidates by comparing the behavior of tropical (Indian) with that of northern, aquatic with terrestrial, and insectivorous with graminivorous birds; migrations are set off by environmental conditions (cold or lack of food). The beginning of the migration period is signaled to the birds by "a natural sense, which reacts to change in heat and cold."

Because he was guided only by empirical reasoning in forming his conclusions, he was not even afraid to criticize Aristotle; for instance, Frederick did not believe the theory that birds migrating in wedge-shaped flight keep their hierarchic order throughout. In fact, he himself had noticed that the lead bird (*dux*) among cranes was relieved by another bird during passage, shifting back within the formation. The fable of the hibernation of birds, which had gained wide currency through Aristotle, Frederick found so incredible that he never mentioned it. In many places he also corrected or amplified Aristotelian notions of anatomy, exactly describing avian kidneys, for example, though Aristotle had maintained that "no bird . . . has kidneys."

The long chapter on flight ("De manieribus volatuum in avibus") testifies to the emperor's incomparable powers of observation. Although a great deal has been written on this subject since the eighteenth century, Konrad Lorenz in 1933[13] was the first direct observer among ornithologists to surpass him in variety of experience and acuteness of interpretation. The emperor gave exact accounts of the flight of the many species he had watched, and explained their differences in biological and physiological terms. He described rowing, gliding, and hopping modes of flight, mentioned the aerial roll of tumbler pigeons, exactly analyzed the relation of body weight and wing surface to the number of wing beats and the structure of feather shafts and plumes, established the correlation between flight capacity and the size of the pectoral muscles, allowed for the effect that wind direction has on flight, and realized that many birds flow low against a head wind "because the wind is more powerful in the upper regions." Raptors normally fly high around midday, especially in summer and on hot days, "to cool and air themselves aloft."

Another German authority also made a name for himself in ornithology at about the same time. This was Albertus Magnus (Albert von Bollstädt), Dominican and Master of Theology, who was born at Lauingen on the Danube in 1193, and who died in the Dominican monastery at Cologne in 1280.[14] In him, too, the new way of looking at nature is demonstrated, though by no means so impressively as in the Hohenstaufen emperor. Like his pupil, Thomas Aquinas, Albertus belonged to the group of Scholastic philosophers who, after becoming acquainted with Aristotle in Michael Scot's translation, attempted to harmonize ecclesiastical dogma both with the laws of logic and with Aristotle's erudition and philosophy. To this end, between about 1260 and 1270, Albertus wrote a commentary on many Aristotelian writings, including one on *The History of Animals*, which was called *De animalibus libri XXVI*, and which in its twenty-third book treats of birds. This gave him the opportunity to present, in addition to all the traditional "knowledge" that he gathered from Thomas de Cantimpré and similar sources, a great deal about German bird life that he himself had seen or heard about in his youth and on his many journeys. Studying these glosses, one comes to appreciate their author as a friend of nature who was not afraid to observe creatures attentively and to describe them affectionately for their own sake, although many of his factual contributions lack the accuracy and variety of Frederick II's, and Albert's theoretical commentaries reveal his ponderousness in detecting and understanding natural causation. For, like everyone who had grown up believing in supernatural causes and infallible authority, he found this kind of thinking difficult.

Although Albertus attempted to appropriate Aristotelian philosophy, he probably must be counted among those Scholastics who, defending the proposition "Universals are realities," exalted the idea above the concrete object, admittedly with the difference that for him universals exist "in the object" and not, as with the Platonists, "before the object." A great opponent of these Realists arose in the fourteenth century in the person of the English Scholastic, William of Occam (1270–1347). According to him, universal ideas are not, as in Plato, "outside the soul," but "in the soul"; they are simply abstractions, whereas true reality consists of concrete individual objects: "Universals are names existing after the object." In a few decades this Nominalism made great headway, so that even among the ecclesiastical theoreticians Nature was restored to her rights and Aristotle was raised to the dignity of a philosopher of the church.

In the fifteenth century the great movement of the Renaissance spread outward from the Italian city states over all civilized Europe. A new type of middle-class scholar came into existence, one who cultivated knowledge not for the service of church or state, but for its own sake and to acquire wisdom, as Aristotle and the ancient philosophers had done. At first, classical studies were significantly encouraged, especially through the rich contributions of the Greek refugees who sought asylum in Italy from 1430 on. Theodorus Gaza brought with him from Salonika the works of Aristotle, and in Rome translated *The History of Animals* from Greek into Latin. Printing, recently invented, helped the rapid spread of the newly discovered treasures. As early as 1476 Gaza's Latin *Libri de animalibus* appeared in Venice, to be followed in 1495 by a Greek edition. Horizons opened out wherever men looked. The whole system of concepts was radically altered; weary of abstract speculation, the new spirit demanded observation of the concrete. Artists like Pisanello or Dürer found their models in nature; Dürer's well-known comments may be recalled here: "Art indeed is inherent in Nature; whoever can tear her forth has captured her — do not ever imagine that you can or will make something better than the power which God has set to work in His natural creation. For intentions and faculties are powerless compared with God's creativity." In the same way, natural science again brought to mind its living source. Luther wrote at the time, with joyful enthusiasm, "We are now in the dawn of a new life, for we are beginning to acquire the knowledge of created things." It is symptomatic that the very first ornithologist of the new era was a zealous adherent of the Reformed doctrine.

William Turner, who seems to have set himself from his youth to study birds, was born about 1500 in Northumberland (reportedly as the son of a tanner).[15] A highly placed patron enabled him to study theology and medicine at Cambridge, where he took his B.A. in 1529–30. A member of his closest circle was Nicholas Ridley (1500–1555), later Bishop of London, who taught him Greek and converted him to Protestantism. Exiled from England as an outspoken apostle of the new faith, about 1539 Turner fled to Ferrara, where the Duchess Renata as a fellow supporter of the Reformation gave him asylum, and the University in nearby Bologna allowed him the opportunity to increase his knowledge of botany. After he had become a doctor of medicine at Ferrara, Turner went to Switzerland, made friends with Conrad Gesner in Zürich, and then about 1543 was taken up by the Archbishop of Cologne, Count Hermann von Wied, who as a friend of the Reformation attracted important humanists to his court. Later Turner became the personal physician of the newly Protestant Count von Emden, and took botanical journeys through East Friesland and Holland. He was permitted to return to England only in 1547, after the death of Henry VIII, where he was recompensed with lucrative positions for the hardships suffered under the previous regime. But Queen Mary's accession in 1553 forced him to flee again, and he could return only when Elizabeth became queen in 1558. This upright man died in 1568 after a life of vicissitudes as an unpopular Nonconformist.

The humanists in Turner's day, whose philological interests extended also to classical writings on natural history, demanded an accurate explanation of Greek and Latin names for plants and animals. Turner's friend Gilbertus Longolius, a Frisian, occupied himself with the bird names of the ancients. When he died in Cologne in 1543, he left behind a manuscript, *Dialogus de avibus et earum nominibus graecis, latinis, et germanicis*, which was published by Turner in 1544.[16] Though the compilation was unimportant for the natural sciences, this was not true of the little book of only 156 octavo pages that Turner himself wrote in less than two months on the same subject, and had printed in the same year (1544) at Cologne: *Avium praecipuarum, quarum apud Plinium et Aristotelem mentio est, brevis et succincta historia* [A short and succinct account of the principal birds mentioned by Pliny and Aristotle]. In his commentaries on the text of these two classical authorities, Turner found an opportunity to display at least a part of the rich ornithological knowledge that he had acquired in England and later on the Continent, especially in the neighborhood of Cologne. "He presents the life histories and the first clear descriptions of the appearance of individual birds, founded entirely on his own experiences and observations, with a verity and accuracy that secure for him with unquestioned justice one of the foremost places among ornithologists";[17] so Merrem judged him in 1788, and so we judge him today. Turner shows that he adheres to the same principles as those of Frederick II, by describing things, without any speculative additions, "as they are." An example is his commentary on the name Malakokraneus (Molliceps) in Aristotle, which he attaches to our great gray shrike (*Lanius excubitor*), for the following reasons:

> The Molliceps I think to be that little bird which Germans call

Nuinmurder [*Neuntöter*, red-backed shrike], not without a cause . . . In size it equals the least of the Thrushes, and to one observing from afar seems wholly grey. And yet, to one inspecting it more nearly, the chin, the breast and belly appear white, and from each eye there reaches to the neck, although somewhat oblique, a long black patch. It has so big a head that (were the bill stronger and larger), it assuredly would answer in proportion for a bird of thrice its size. The bill is black and moderately short, and hooked at the tip, but is the stoutest and strongest of all, so much so that the bird once wounded my hand, although protected by a double glove, and very speedily it crushes and breaks up the bones and skulls of birds. Each wing is wholly black, except that a white line of some size marks transversely the middle of the wing on either side. The tail is like that of a Pie, that is to say, longish and particoloured. Of all it has the shortest legs and feet proportionately to its body, and these parts are black. It has short wings, and flies as if by bounds upwards and downwards. It lives on beetles, butterflies, and biggish insects, and not only these, but also birds after the manner of a Hawk. For it kills Reguli and Finches and (as once I saw) Thrushes . . . It does not seize the birds it kills with its claws, after a swift flight, as Hawks do, but attacks them stealthily and soon (as I have often had experience) aims at the throat and with its beak squeezes and breaks the skull. Then it devours the crushed and bruised bones, and when anhungered crams into its gullet lumps of flesh as big as the gape's narrowness can take. Again, beyond the habit of the rest of birds, when prey happens to be more plentiful, it lays by some for future scarcity. For it impales and hangs the bigger flies and insects on the thorns and spines of shrubs, so soon as they are caught: of all birds it is tamed most easily, and when accustomed to the hand is fed on meat, and, should this happen to be somewhat dry or altogether bloodless, it requires drink. In England I have never seen it oftener than twice, although most frequently in Germany . . . Now if Aristotle's description of the Molliceps does not appear to any one in all points to agree with this let him ascribe it to the list of the Tyranni, or shew us a bird, which the description fits better than this.[18]

Even such a widely traveled naturalist as Turner could not solve the riddles posed by many of Aristotle's zoological and botanical names, for some referred apparently to species from the Mediterranean, which had not yet been visited by scholars of the modern period. Therefore, Turner, in his "Peroratio ad lectorem" in 1544, expressed the hope that another Alexander would arise to enable a new Aristotle to describe for us all the animals mentioned by the old Aristotle, together with their habits, their medicinal properties, and their modern names, in addition to many other creatures that he had not known. Though such princely favor certainly was denied to Turner, he did not hope in vain because it was shortly granted to another naturalist, much less capable, but one who worked to the best of his abilities.

In 1546 the French king, Francis I, who as an opponent of the emperor was concerned to improve his relations with the Sublime Porte, sent his confidant, d'Aramont, as ambassador to Constantinople. Two scholars, the humanist Pierre Gilles and the naturalist Pierre Belon, were graciously allowed to accompany the ambassador.

Pierre Belon, born about 1517 in a hamlet near Le Mans, at 18 attracted the attention of the bishop of that see, who undertook to educate him and in 1540 financed his studies at the University of Wittenberg.[19] The young Frenchman stayed there a year, and attached himself to the almost contemporary but already widely famous professor of botany, Valerius Cordus. On their journeys together into Pomerania and Saxony, as far as the mines of Joachimsthal, the two naturalists developed such a respect for each other that when Belon heard of Cordus's botanical trip through Italy with some of his students in 1544, he left Paris to join him. He found his friend in Liguria, but shortly thereafter, on September 25, 1544, Cordus succumbed to climatic fever [malaria?]. Thrown entirely on his own resources, Belon traveled to Padua to continue his studies. Even at that time he was particularly interested in fish and birds, as he says himself:

> While in Padua, we were accustomed to leave on Thursday evenings and travel all night along the Brenta in order to be in Venice from Friday morning through Saturday and Sunday, for the convenience of seeing both the birds and the fish, and, re-embarking Sunday evening, after conferring with the birdcatchers and fishermen, knowing that the boat would sail through the night, so as not to lose time, we returned to our studies on Monday morning. During that time, the aforementioned Fridays and Saturdays, there was not a single birdcatcher or fisherman who did not bring and show us anything rare that he had been able to catch.[20]

After returning to France from Padua, Belon found a new patron in Cardinal de Tournon, the king's powerful minister, who had already shown favor to the ichthyologist Rondelet; thanks to his advocacy, Belon was able to accompany d'Aramont to the East.

The ambassador arrived with his retinue in Venice on February 9, 1547. Three hired galleys took them to Ragusa, where d'Aramont and Gilles continued overland. When they made their entrance with great pomp into Constantinople on May 14, they found Belon already there. He had sailed by way of Corfu, Zante, and Cythera, and had stayed for a while on Crete, where, filled with scholarly curiosity, he had climbed to the top of Mount Ida. At the end of April he reached the Bosporus by way of Euboea. Soon his impetuous desire for action drove him on his travels again, and only in August was he heard of once more in Constantinople, after a major trip taking in Lemnos, Thasos, Mount Athos, Salonika, the Struma, Kavalla, and Maritza. In the same month he took ship for Egypt and made a quick journey to Memphis, Suez, El Tor, and the Sinai mountains. According to his own account, he then left Cairo on October 29 to travel over the old military road from Jerusalem to Damascus and return to Constantinople in the spring of 1548. He left there only in 1549 and in thirteen days arrived in Venice.

In Paris the cardinal received his adventurous "Aristotle" graciously, chose him as a companion when he went to Rome for the conclave in January 1550, and in the summer of the same year sent him to London and Oxford. For several years thereafter Belon worked on the written account of his travels. He chose to use his native language, an unusual decision for a scholar at that time, for which he was later given great credit, though Belon, as an autodidact with an inadequate basic education, was simply not capable of expressing himself freely in the language of scholars. Malicious reports maintained that he could not understand two lines of Pliny in the original. The story of his travels, *Les obseruations de plusieurs singularitez, trouées en Grèce, Asie, Judée, Egypte, Arabie et autres pays estranges* (Paris, 1553), is a colorful tapestry of personal experiences, reflections, and inventions, unimportant for the ornithologist because almost all he had to say about birds he saved for his work, *L'histoire de la nature des oyseaux, auec leurs descriptions, et naifs portraicts, retirez du naturel* (folio, Paris, 1555). The need to earn his bread forced on him a varied subsequent career, which ended violently when in 1564 he was stabbed to death by bandits in the Bois de Boulogne.

Belon's work on birds, in seven parts or books, was intended to be a real compendium of ornithology. In the first book the author provides a short general introduction to the subject (anatomy, physiology, biology), adorned with elaborate digressions into general natural history, in which he conforms to the teaching of Aristotle, Galen, and other classical writers, and barely ventures to offer any of his own opinions, except occasionally on birds. These he had studied closely during many dissections, in order to learn about their internal structure and to relate it to that of man. It is his comparison of human and avian skeletons that above all testifies to his well-developed capacity for judgment. But the more detailed discussions occur in the six remaining books, assigned to the six groups into which he had classified birds. Relying on Aristotle and Pliny, he proceeds according to ecological and morphological principles (without distinguishing them very clearly) to separate birds into raptors, waterfowl with webfeet, fissiped marsh birds (to which are also assigned kingfishers and bee-eaters), terrestrial birds (ostriches, gallinaceous birds, larks, etc.), large arboreal birds, and small arboreal birds (including swallows).

Like all contemporary scholars, Belon attaches the greatest importance to philological questions and to the interpretation of comments by Aristotle and Pliny, but this bare framework serves him almost always for the informal inclusion of his own experiences. Though he does not display the intellectual clarity, scientific method, and absolute love of truth emanating from William Turner's modest little book, the widely traveled and observant Belon has much to say that is new about the appearance, habits, and distribution of birds. The chough (*Pyrrhocorax pyrrhocorax*), which he calls "Chouca rouge," he had observed himself on the high peaks of the Cretan mountains, in Cornwall, in the Swiss Jura, on the Mont d'Or in Auvergne, "and in innumerable other places." The blue rock thrush (*Monticola solitarius*) he had encountered in Corfu, Zante, Cythera, Crete, and Negropont (Euboea). He came across pelicans in Rhodes, near Salonika, on the Nile and the Struma, and on the shores of the Hellespont, and more than once he saw stuffed pelicans hanging in entrances to houses in Germany and Bohemia. He succeeded in identifying several birds named by the ancients, such as *Merops* (bee-eater), which he found in Crete.

Belon's bird book was much praised in later periods; his contemporaries, however, did not consult it much, for in the very same year (1555) the third volume, "Qui est de avium natura," of Conrad Gesner's *Historia animalium* appeared in Zürich and absorbed almost the entire attention of readers eager to learn about birds. This is not surprising, for compared with Gesner's astonishingly erudite text, the French book seems like the entertaining product of a scientific amateur, who knows little more of his subject than what he himself has seen and investigated or what he has read up in a few authors. Gesner compensated overabundantly in another way for his lack of personal experience; not content with consulting the best specialists in the entire civilized world, he combed through classical and medieval literature with unparalleled thoroughness for apposite references, in order to produce a thesaurus of universal knowledge and satisfy the most exacting demands of the humanists. With all this preparation, he managed to fill 806 closely printed folio pages with the discussion of only 180 species of birds. Belon chattered informally about his subject, and mingled his own and others' observation in a way that cannot be disentangled. Gesner, on the other hand, knew how to control the mass of his material through precise organization. The text on each species is neatly divided into eight sections: synonymy; distribution; physical characteristics; behavioral peculiarities; how to hunt, capture, and tame the species; its use as food; its use as medicine; and philological and antiquarian notes on it. For the sixteenth century not only the abundant material and the lucid arrangement of Gesner's work, but also the 217 woodcuts, the majority by Lukas Schan, whose fidelity to nature completely outshone most of the 144 pictures by Belon's Parisian painter, Pierre Gourdelle, seemed to be a great advantage.

There have been many laudatory and devoted biographies of Conrad Gesner (1516–1565).[21] Starting in his early youth, his unceasing activity was motivated by the urge to assemble and organize facts. He collected all the titles of scientific works from classical times onward, in order to compile a *Bibliotheca universalis* in 21 books, arranged according to subject. With his *Mithridates* (Zürich, 1555), which contains the Lord's Prayer in 22 languages, he founded comparative linguistics. He was perhaps the first person to make a collection of objects for the study of natural history. All this and more he succeeded in combining with his self-sacrificing work as city physician (from 1554) and later as professor of natural history (from 1558) in Zürich.

His uninterrupted literary studies, a wasted physique, and weak eyesight kept Gesner from acquiring a practical knowledge of ornithology equal to his botanical knowledge. He did indeed describe some birds accurately from his own observation, but he owed the most important additions to older accounts to his correspondents who, each on his own, tried to find out the differences among native species and to learn their popular names. Three of the men who supplied Gesner with descriptions and pictures from Germany were outstanding: the Strassburg painter Lukas Schan (d. 1550), an experienced birdcatcher and bird fancier, already mentioned as the chief illustrator of the third volume of the *Historia animalium;* Georg Fabricius (1516–1571), the famous rector of the College of St. Afra in Meissen; and Fabricius' friend Johannes Kentmann (1518–1574), the city physician of Meissen. These two, and particularly Kentmann (who visited the famous Gesner in Zürich when returning home, during the winter of 1549–50, from a long period of study in Italy), knew a great deal about the waterfowl, waders, and marsh birds found in the Elbe district between Meissen and Torgau. Gesner had connections with England, too, for one of his correspondents there was William Turner, and another the queen's physician John Kaye or, as he himself wrote it, Caius (1511–1573), who published part of his ornithological letters to Gesner in *De rariorum animalium atque stirpium historia* (London, 1570). Gesner's friend Guillaume Rondelet (1507–1556) gave him information from Montpellier, and the young Ulisse Aldrovandi from Bologna.

Some idea of the spread of interest in native fauna during and after the Reformation may be gained from the fact that as early as 1585

Gesner's sumptuous four-volume *Historia animalium* appeared at Frankfurt am Main in a second Latin edition, with additions by the author and a fifth volume, "De serpentium natura." In the meantime a German edition (Zürich 1557, reprinted 1582 and 1600) of the particularly popular volume on birds had reached a wide public. This German *Vogelbuch* was addressed not to scholars but, as it states on the title page, to "all lovers of art / doctors / painters / goldsmiths / sculptors / silk embroiderers / huntsmen and cooks / not merely for agreeable diversion / but for practical and serviceable use." Therefore the translator, Rudolf Heusslin of Zürich, omitted all the scholarly apparatus and retained only what could be useful in practice. Though certainly no easy task, it was so splendidly performed by Heusslin that, up to the present day, a reader in a hurry would far rather consult the German bird book than the three times fatter Latin edition, which reproduces so much medieval pseudo-knowledge with historical accuracy. The chapter on *Jynx*, which in Gesner's Latin of 1555 occupied three closely printed folio pages, is shortened by Heusslin in 1557 as follows:

On the Wryneck, *Jynx*.

On the shape of this bird.

This bird has many sorts of names among the Germans / one group from its neck twisting, another from its resemblance to the Adder. Therefore it is called Wryneck / Turnneck / Adderneck / Adderturner / and Addertwist. It is somewhat larger than the Chaffinch / as is shown here in its true size. It is speckled in color. It has two toes pointed backward and two forward / a short and a long in each case side by side / like the Woodpecker and Nightingale / wherefore its body is less inclined forward than that of the other birds. The claws are large / and extend like those of jackdaws. It turns its neck about / while holding the rest of its body still / like the Adder. It also has a tongue like an Adder: this it thrusts forth more than four fingers' breadth / and then pulls it in again. The tongue is so strong and sharp at the tip / that it will go through / a man's skin / like a needle. It goes up through the throat / and curves over the skull into the nostrils. It is however double / it begins to curve at the neck: in front, however, both parts come together again on the head. The male may be told from the female by many points / especially by the yellower breast / says D. Gesner.

On the nature and charm of this bird

The Wryneck pierces ants very swiftly with its outthrust tongue / just as with us young boys pierce frogs with iron darts / shot from a bow / and swallows them: it never touches them with its beak / as do other

> birds their food. I once (says D. Gesner) kept a bird taken in April / alive for five days on ants / thereafter it died / and therefore it could not be known whether it might have lived longer: for it was very unhappy caged: and ate nothing but ants: as I learned from trying to feed it with little worms, flies, and the like. It pecks with its beak into wood / and makes thus a noise almost like a Woodpecker. They stick out their tongue / as fishermen do their lines / over which the ants go: when there are many, they pull the tongue back in / and eat them / says Oppian. If you hold one in your hands / it twists its neck about more than any other bird. It screams or cries with its voice. It nests in holes in trees like the Woodpecker. Sometimes also in farmhouses. In winter / as I understand / it is not seen / but especially in autumn. I have seen it caught in April and July: at which time it has young. It lays many eggs / eight or nine. As I have long observed / it does not fly away at once / when someone comes near: rather it becomes angry: it extends its neck / and thrusts with its beak: however, it does not bite: it pulls its beak back and then thrusts it forward again / thus threatening it shows its anger. At the same time its feathers / become erect on the neck / and the tail spread and raised.

For almost 200 years "Gesner" remained the German household book. Even in the seventeenth century it appeared in an enlarged Latin edition (Frankfurt, 1617) and two abbreviated German editions. The second of these appeared in Frankfurt in 1669: *Gesner redivivi aucti et emendati Tomus II oder Vollkommenes Vogel-Buch* [Volume II of Gesner revived, enlarged, and emended, or the complete bird book], edited by Georg Horst and by him "transformed from the old dark and incomprehensible phrases into modern, elegant, and lucid German speech / also adorned and enlarged with many new figures partly known but mostly of strange and foreign and even remarkable birds and their description." It is in every way inferior to the sixteenth-century German editions, particularly in the illustrations, which were copied from the skillful and attractive old woodcuts and came to grief completely in copperplate engraving.

The Bolognese scholar Ulisse Aldrovandi (1527–1605),[22] whose interest in natural science had been awakened by Rondelet in Rome in 1550, was fired to emulate Gesner. Though already an old man, he decided to outshine the master by writing an encyclopedia of the three realms of nature, which was to fill no fewer than 14 stout folios. Starting with birds, he published the three volumes of his *Ornithologia* in 1599, 1600, and 1603; they provided a far better collection of material than anything existing at the time. But only in his industry, not in his intelligence, could Aldrovandi compete with his model, Gesner.

Being largely uncritical, he had gone to work like the monk Thomas de Cantimpré, and had only rarely included his own discoveries among his borrowed matter, although in Bologna he was acquainted with Vesalius's pupils and, as shown by his careful description of the golden eagle's musculature, the green woodpecker's tongue, or the swan's or merganser's trachea, he was at least well qualified for zootomic research. (His famous pupil, the anatomist Volcher Coiter, called him "a man extraordinarily well endowed for the various arts and sciences, as well as philosophy, and truly among the first in natural science — my sponsor and preceptor.") In spite of these qualities he allowed anatomist friends to supply most of his new descriptions and illustrations (Marco Antonio Ulmo in Padua did the ear and eye of the long-eared owl and the oviduct of the hen, and the Silesian Johannes Bittner the skull and head muscles of *Psittacus*).

Aldrovandi's grouping of species was, however, new; he thought it especially meritorious because his predecessor Gesner (whom he almost always alludes to simply as "Ornithologus") had "made an unsystematic confusion of everything, merely using alphabetic order." But he, Aldrovandi, would work according to a systematic order, which Plato had called the soul of things. It was indeed true that continuing the medieval tradition of alphabetic arrangement was a weakness of the *Historia animalium*, but a weakness of which the author was fully aware. He explained it thus: "I have followed alphabetic order, since this work is almost as much a study in grammar as in philosophy." Aldrovandi was not so modest and reticent. Following certain suggestions in Aristotle, he classified birds into those that have a hard and powerful beak (raptors, parrots, ravens, woodpeckers, treecreepers, bee-eaters, and crossbills!); those that bathe only in dust or in dust and water (pigeons and buntings); songbirds (finches, larks, and canaries); waterfowl; and finally shore birds. The rash sacrifice of Aristotle's principles of classification was not even of any practical use.

After discounting material that had already been published by Gesner and Belon, or rejected by them as unreliable, the scientist finds nothing very useful left in Aldrovandi's *Ornithologia.* "By eliminating everything useless or irrelevant to the subject one could reduce it to a tenth of the original," was Buffon's subsequent judgment.

In this way Renaissance ornithology, which had blossomed so hopefully, withered within a short time in the stuffy atmosphere of the study. Going from bad to worse, after Aldrovandi came John Jonston (1603–1675), a hack writer of Scottish descent, born at Samter near Lissa but later resident in Silesia. His *Historiae naturalis de avibus libri VII*, appearing at Frankfurt am Main in 1650 as the second volume of a *Historia naturalis*, was mostly compiled from Aldrovandi and Nieremberg,[23] without the slightest firsthand knowledge. Nevertheless (probably because of its many good illustrations, engraved by Matthew Merian the Younger after the old woodcuts in Gesner and Aldrovandi and a few more recent originals), it was widely distributed and continued to be published far into the eighteenth century in Latin, German, English, Dutch, and French, its last appearances being in 1756 in Heilbronn and in 1773 in Paris. As always, the worst books on animals find the most buyers!

2

The Beginnings of Exotic Ornithology

For a long time the view apparently prevailed that Gesner and Aldrovandi in their mighty pandects had said everything that needed to be said about birds in the Latin and Germanic countries. Therefore intellectual curiosity turned from them toward the booty brought back by sailors after long and dangerous journeys from distant shores. There a new world of wonders had opened up, after the old world had surrendered all its secrets.

Even the first travelers to America had returned with remarkable birds. According to Oviedo, who was an eyewitness, when Christopher Columbus made his ceremonial entry into Barcelona in mid-April, 1493, he paraded "molti papagalli" (probably *Amazona leucocephala*) before the eyes of the astonished crowd.

In the following centuries Spanish ships provided Europe with two valuable domestic fowl of Indian origin, the turkey and the Muscovy duck. The turkey was introduced from Central America between 1525 and 1530, and by 1541 had already been bred so successfully that Archbishop Thomas Cranmer had to forbid his clergy to serve more than one such bird at a banquet. Soon afterward the Muscovy duck, probably brought from Cartagena by the Spaniards, appeared in literature. Caius sent a picture and description "of the Indian Drake" from England to Gesner before 1555; in the same year Belon reported that the newly introduced duck had already multiplied so much that it was beginning to be sold in the open market.

Live parrots also were good business. During the period of the great geographic discoveries, their possession was a sign of privilege for powerful lords and noble ladies. In antiquity Indian parakeets, probably *Psittacula krameri*, were kept as pets; later all trace of them disappears until the fifteenth century, when they were imported, though rarely, into Italy and even into the German-speaking countries. In the first decades of the sixteenth century the world discovered that there were many other kinds of parrot, and painters enlivened ideal landscapes with their brilliant plumage, which had never been seen before. In 1921 Killermann discovered a parrot from the New World with a yellow forehead, perhaps *Amazona ochrocephala*, in the Breviarium Grimani, which originated in the Low Countries about 1500.[1] The Cuban Amazon (*A. leucocephala*) first appeared in 1518, in Johann Burgkmaier's *Johannesaltar*, painted in Augsburg. Before 1564 Caius wrote to Gesner that he knew of "many kinds of Parrots" in England, and sent him the picture and exact description of a large and colorful specimen introduced from Brazil (probably Pernambuco), *Ara chloroptera*. One can understand Aldrovandi's delight when he saw live macaws for the first time at the Mantuan court in 1572. Duke William kept two species, *Ara ararauna* and *Ara macao*. The first of these two charming and valuable creatures had been brought back from the West Indies by Breton sailors a few years before 1569, and was given by Charles IX to Count Sforza, commander of the papal forces, in gratitude for his services during the Third Huguenot War.

Other species of parrot were imported during that period from West Africa. The Guinea Coast provided the gray parrot, *Psittacus erithacus*, which Lucas Cranach placed in Paradise in his picture of Adam and Eve, painted about 1520.

Long before the Portuguese had discovered the sea route to the East Indies, parrots from that area had occasionally been brought to Europe, but over land. About 1240 the sultan sent from Babylon to his friend Frederick II a white cockatoo, probably the species native to Celebes. Yet another cockatoo was included by Andrea Mantegna in his painting of the Madonna della Vittoria, about 1496. When the Portuguese began (after 1524) to send ships regularly to the East Indies, and after 1535 to the easternmost part of the Malay Archipelago, they seem hardly to have bothered with the shipment of animals because "East Indian" parrots were still a great rarity in the second half of the sixteenth century. Only in 1599 did Aldrovandi describe another cockatoo, owned by Prince Alessandro Farnese (d. 1592).

What could have been more credible evidence than these gaudy captives of the exuberant descriptions of tropical bird life with which sailors, priests, and conquerors spiced their reports? They returned not only with live birds but also, from time to time, with embalmed ones or with skins.

Even at the beginning of the sixteenth century no one in Europe had tried to preserve dead birds. In consequence, when Elcano, who had taken over the command of the Spanish voyage of discovery after Magellan's death, came back in 1522 with bird-of-paradise skins (*Paradisaea minor*) that had been given to him by the sultan of Batjan, he created a tremendous stir. People were astonished not only by the splendor of the plumage but also by the total absence of flesh and bones inside the skin. "When you look inside you will not find the little bird empty, for it is entirely filled with fat," wrote Melchior Guilandini (Wieland) to Gesner before 1555.[2] And in 1552 Francisco Lopez de Gomara said: "We think that they live on dew and spice-blossoms. But whatever the truth may be, it is at least certain that they never decay."[3]

Perhaps the example of the Moluccas hunters awakened European interest in methods of preserving bird skins. The oldest instructions that we know of come from Belon (1555):

> Whenever a bird fancier wishes to carry the bodies of birds from one region to another, he must proceed as follows: the skin of the bird should be cut in the anal region, all entrails removed, salt sprinkled inside and stuffed into body and throat cavities, and the bird hung up by its feet. This will ensure that it will be permanently preserved, along with its feathers, without being devoured by worms, and if one notices that the salt does not melt, then one must dampen it with a little strong vinegar, or else remove all the flesh; for every bird can easily be skinned, and when the skin has been salted (with the wings and legs left on) it can be kept as long as one wishes. May this also serve to guide all those who read this narrative and think of the general good, so that if they find a bird in their own country that is not described in this book or that I have not discussed, they may treat it according to my instructions and keep it in their collections. And if they should be pleased to send it to me, I should be much obliged to them.[4]

Belon seems already to have used this technique during his journey in the Levant (1547–1549), because he tells us that he observed a hawfinch ("Le Grosbec") in Greece "of which we have brought back the skin." In another place he writes, "The Fleming M. Antoine Martinellus showed us a dried and salted Chrysomitris [siskin or goldfinch] in Padua before our departure [1545], saying that his friend M. Turnerus, an English physician, had sent it to him."[5] Therefore it is not remarkable that around this period a few enterprising sailors should also have attempted to preserve exotic birds, for the first account of which we are again indebted to Belon. In 1555 he reported:

> Those who carry on commercial traffic with the new countries lose

> no opportunity to collect rarities, which they attempt to sell in the following way: since they are unable to bring back the birds of that country alive in their ships, they flay them in order to obtain the skins: chiefly those of the most beautiful colors, among which is the one we are now describing, and out of which the sailors have made their profit, having given it the name of the Brazilian blackbird. Many complete skins have been found.[6]

In this way he himself became acquainted with the fiery-red "Brazilian blackbird" (*Ramphocelus bresilius*), and the black-and-yellow icterid *Cacicus cela*, both of which probably came from Pernambuco. Also in 1555 Genser described a hummingbird from a skin.

These were the humble beginnings of exotic ornithology.[7] In the next 150 years many birds, dead or alive, must have been shipped to Europe, but since they vanished into the cabinets of curiosities and menageries belonging to rich dilettantes, it was only by chance that they could be seen by one of the few contemporary specialists. None of the birds lasted very long: the cage birds were attacked by maggots; the dried skins, which no one knew how to impregnate, were eaten by moths and dermestids, and they became known to future ages only when employed as subjects for a painting. Exotic birds never mentioned by contemporary writers were later discovered with astonishment in many a collection of pictures that amateurs of natural history had assembled between 1550 and 1650. Certainly anyone who wished to possess such rarities had also to possess a princely fortune. The splendid Villa Pratolino near Florence, built by Cosimo de Medici's successor, Francisco I, Grand Duke of Tuscany, was famous for its water displays and sumptuous aviaries. There in 1580 Michel Montaigne noticed particularly "little birds, like goldfinches, that have a tail with two long feathers like those of a large capon."[8] These were paradise widowbirds from the west coast of Africa. Francisco's brother Ferdinand I, who succeeded him in 1587, shared the same hobby; he even had the rarities in his aviaries painted by the superintendent of his galleries, Jacopo Ligozzi. When Aldrovandi was hunting up new material for his *Ornithologia*, the grand duke ordered the superintendent of his botanical gardens and natural history collections in Pisa, the Franciscan Francesco Malocchi, to send these pictures, with commentaries, to Bologna. (This was at some time after 1596.) Because Aldrovandi gratefully used them for his book, we know that in the Villa Pratolina before 1601 were kept such curiosities as both sexes of the Venezuelan helmeted curassow (*Crax pauxi*), which indeed had been seen by Herandez before 1570 in King Philip II's menagerie, the Mexican great curassow (*Crax rubra*), the Virginian cardinal (*Cardinalis cardinalis*), and three species of West African paradise widowbirds: *Vidua paradisea*, *V. macroura*, and *V. chalybeata*.

Fired to emulation by the example of the Italian princes, the Fuggers in Augsburg also acquired a splendid menagerie, described by Montaigne in 1580: "There is an aviary twenty paces square, twelve to fifteen feet high, surrounded everywhere with well-made brass wire nettings; inside are ten or twelve spruces and a fountain: this is all full of birds."[9] This hobby spread so wide, particularly in Italy, that in 1622 Olina could write in his *Uccelliera* that there were few princes who, in addition to their other diversions, did not also have an aviary.

In the last decades of the sixteenth century, when the introduction of rare objects from beyond the sea increased rapidly, rich amateurs occasionally had the idea of making a "museum" out of them. Here preserved tropical birds could often be admired, as in the apothecary Ferrante Imperato's natural history collection, founded about 1590 in Naples. According to Clusius, a Belgian called Jacob Plateau kept a few ornithological treasures "in suo instructissimo museo" in Tournai (which he sold in 1605 to Duke Charles of Croy and Aerschot in Beaumont [1560–1612]; they included a hummingbird ("Ourissia") from Brazil, a great auk ("Mergus americanus, ex America allatus"), a gannet, a great black-backed gull, and the heads of a crowned crane and a hornbill (*Rhyticeros undulatus?*) among others.

But far outshining all such collections were the animals belonging to the Emperor Rudolf II (1552–1612). Clusius reported in 1605: "I hear that his Majesty spares no expense to indulge himself in these trophies of travel, and in the knowledge of all the wonders of Nature."[10] Instead of concerning himself with the business of government in times of crisis, this monarch withdrew to study art and science, to the detriment of his subjects. In 1587 he built Schloss Neugebäu near Vienna, with a menagerie and a natural history collection which he added to at great expense. It seems that this eccentric recluse allowed no specialist a chance to see his treasures, which would have vanished without trace if he had not commissioned his Dutch court painters, Georg Hoefnagel (1542–1600) and his son Jacob (1575–1630), to portray in oils the zoological prizes from Schloss Ebersdorf and Schloss Neugebäu. These pictures, most of them admirable, including 90 of birds, are still in the Viennese Hofbibliothek. The earliest (*Crax*) is dated 1571, though most seem to have been painted between 1602 and 1610. Only when they were rediscovered by Georg von Frauenfeld in 1868 did anyone realize that the emperor had a dodo in his menagerie, in addition to the first cassowary brought to Europe, a great curassow (*Crax rubra?*), a blue macaw (*Ara hyacinthina?*), an *Ara ararauna*, a sulphur-crested and a rose-crested cockatoo (*Cacatua galerita* and *C. moluccensis*), a lory from the Moluccas (*Lorius garrulus*), two lovebirds

(*Agapornis pullaria*) from the Guinea coast, and other treasures never shown elsewhere. Further, one of these pictures provides the only source for our knowledge of the appearance of a large flightless rail with a long bill, *Aphanapteryx bonasia* Selys-Longchamps, that once lived on the island of Mauritius, but like the dodo was exterminated as early as the seventeenth century.*

Most of these rare birds came to Austria from Holland. Indeed, the Dutch had barely shaken off the Spanish yoke when they began to trade in their own ships with the inhabitants of the Malay Archipelago, bringing back rare and showy objects. While the Portuguese generally were content to bargain for bird-of-paradise skins in the Moluccas, the Dutch, more enterprising and efficient than they, saw business opportunities in everything. Their very first ship, returning from Java in 1597, carried a great zoological sensation — a young cassowary (*Casuarius casuarius*), a giant bird whose existence had not even been suspected in Europe. For several months an entrance fee was charged to view this miraculous creature in Amsterdam, until Count Georg Eberhard von Solms bought it and kept it for some time in his park at The Hague, later presenting it to the Prince Elector of Cologne (Duke Ernest of Bavaria). From Cologne the cassowary was finally sent as a gift to the imperial zoo in Neugebäu. Shortly after 1605 the dodo must have found its way there too and have been painted by Jacob Hoefnagel. This species also had been only recently discovered. In October 1598, while sailing to the Moluccas, Dutchmen landed on an island already known to the Portuguese, which they rechristened Mauritius, and found this ungainly flightless bird.

Dealing in these items was very profitable because the many amateurs immediately snatched animals, living or dead, from the hands of travelers returning to Amsterdam from the Indies. Clusius noted that the first lory from the Moluccas (*Lorius garrulus*) fetched the high price of 170 gulden, or 20 thalers. Along with parrots, the captains brought back bird-of-paradise skins; these were mostly *Paradisaea apoda* from the Aru Islands, which were much sought after for their plumes, but in 1603 Clusius discovered among them another and previously unknown species, the red king bird of paradise (*Cicinnurus regius*), whose skin went to the emperor for a large sum.

Strangely enough, in this period of great and startling novelties, Carolus Clusius (Charles de l'Escluse) was the one true scientist who devoted himself to exotic zoology. His work in botany earned him undying fame and he was, as Ernst H. F. Meyer says, "a man of eminent

*Along with other birds of paradise in the imperial collection, Hoefnagel painted a twelve-wired bird of paradise (*Seleucidis melanoleuca*), although ornithologists first learned of the existence of this striking creature through Valentijn in 1726.

ability, extraordinary memory, and universal education."[11] Born in Arras in 1526, when Flanders still belonged to the Hapsburgs, and a pupil of Rondelet at Montpellier, Clusius traveled through Spain and Portugal and then spent 14 years at the Viennese court, until the spirit of religious intolerance introduced by Rudolf II made life there so difficult for him, as a convinced Protestant, that in 1587 he removed to Frankfurt am Main. In 1593 he was called to be a professor at the University of Leiden (founded in 1575), where, intellectually vigorous to the last, he died in 1609.

Since he was favorably placed for the receipt of immediate news from Dutch ships of botanical or zoological discoveries, the almost 80-year-old Clusius decided to make an illustrated book out of the collected records of everything exotic that he had investigated or at least learned of. *Exoticorum libri decem* was published in 1605 in Leiden. As Clusius explained, although he was so advanced in years that he could hardly hope to reap the harvest of such a book, he had no doubt that others, having acquired sufficient material from it, would be able to write a natural history of these items more fully and completely. "I had hoped to obtain much more, because the Dutch voyages to far countries have developed so auspiciously; I gave the surgeons and apothecaries who sail on those ships lists of many exotic things that I knew occurred in the lands which they visited. Yes, I even implored some merchants of the company that finances these voyages to instruct their people to look out for and bring me the things that I had described. But up to now I have achieved nothing, although there were many promises."[12]

Clusius introduced the cassowary to scientific literature, for he had seen it himself in Count Solms's park, and described its egg from Dutch collections. He collected everything known about the dodo and discussed at length the dodo leg that his colleague the professor of medicine, Pieter Pauw (1564–1617), kept in his natural history cabinet. Using printed reports of voyagers, he provided a good description of the penguin (*Spheniscus magellanicus*) discovered by the Dutch in the Straits of Magellan in 1599. He went into detail on hummingbirds and the three species of birds of paradise (*Paradisaea apoda*, *P. minor*, and *Cicinnurus regius*) that he had studied. In his chapter on parrots he added three new species, *Deroptyus accipitrinus*, *Agapornis pullaria*, and *Lorius garrulus*, to those described from Rio de Janeiro by the Protestant preacher Jean de Léry (1578) and those treated by Aldrovandi. From the Duc de Croy's museum he received a picture and description of the scarlet ibis (*Guara rubra*). He was the first to work in the field of northern ornithology, describing two "aves resiccatae" from Professor Pauw's collection, the common loon and the puffin, which had been sent to him in 1603 by Henrik Højer, a doctor in Berg-en.* Still more important was his printing of a substantial report on the ornithology of the Faroes, which Clusius had received from Højer in 1604. Among the twelve clearly described species was included, as a rare visitor to the islands, the great auk.

What would not Clusius have given for the use of a manuscript on the natural resources of Mexico that had already been lying for decades in the royal library of the Escorial and was withheld from the public! Its author, Francisco Hernandez (1517-1578), chief physician to Philip II, was long dead. Sent to Mexico by his royal master to acquire exact information on plants, animals, and minerals, Hernandez lived there for seven years (1570-1577), and during his lengthy and dangerous journeys collected a mass of pictures and descriptions which, roughly arranged on his return, were supposed to have filled 17 volumes. He died before he could finish his work, whereupon Philip II entrusted the further exploitation of the Latin manuscript to his Italian physician, Nardi Antonio Recchi, who made a Latin summary of it and delivered it to the Escorial library. But he took a duplicate (with copies of the illustrations) to Italy and hid it in the Museolo M. Antonio Petilii. He too died without having made use of this treasure, but Prince Federico Cesi, an enthusiastic naturalist in Rome and founder of the Accademia dei Lincei, learned of it and bought the manuscript from Recchi's heirs, in order to have it printed, after rearrangement and provision of a commentary by three learned members of his Roman academy. But in 1630, when this work was finished, the prince died, aged only 45, and publication did not take place.

In the meantime a copy of Recchi's manuscript had been sent from Madrid to Mexico. There a Dominican, Francisco Ximenes, translated at least parts of it into Spanish and published them in 1615. In Europe, however, this Mexican version remained long unknown. The first person to acquaint scholars with Hernandez' animals and plants was Juan Eusebio Nieremberg, a Jesuit, who included a number of excerpts that he had made from Recchi's manuscript in the Escorial in his compilation, *Historia naturae, maxime peregrinae, libris XVI. distincta*, published at Antwerp in 1635. But even Prince Cesi's efforts bore fruit, for in 1651 the book prepared by him and almost printed finally appeared in Rome, as *Nova plantarum, animalium et mineralium mexicanorum historia a Francisco Hernandez medico . . . compilata.* Like Nieremberg's version, it was supplied with a number of illustrations of Mexican animals, the originals of which had been painted under Her-

*Fifty years earlier, the Catholic archbishop Olaf Storr (Olaus Magnus, 1490-1557) had included in his *Historia de gentibus septentrionalibus* (Rome, 1555) accounts of northern birds gathered in 1526 during his journeys out of Sweden, but these accounts were copiously strewn with elements of fable.

nandez' supervision in Mexico and preserved in the Escorial. They represented, however, only a small part of the supposedly 1200 colored sketches, which in 1671 were almost all lost in a fire in the Escorial.

For ornithologists these two editions became important sources. Hernandez conscientiously tried to describe the birds he had noted in Mexico (some 230 species) according to their most conspicuous characteristics, but because the pictures that should have illustrated the descriptions were burnt, an exact identification of many of the species, first attempted by Johann Wagler in 1831, has remained unsuccessful. Fortunately, Hernandez supplied them all with their Aztec names, some of which survive in Mexico to this day. Names like *quauhchochopitli* or *yxamatzcatltototl* are difficult for us to pronounce, so only the simpler ones, like *pauxi*, *hoatzin*, or *motmot* have been taken over from the Hernandez manuscript into the permanent vocabulary of ornithology.

For more than 200 years Hernandez remained the only authority on Mexican animals, because the Spaniards tried anxiously to prevent more exact reports on the natural resources of their colonies from becoming public knowledge. Therefore his descriptions were copied again and again in seventeenth- and eighteenth-century ornithological literature, such as the works of Jonston (1650), Willughby and Ray (1676), Brisson, and Buffon. Statius Müller and J. F. Gmelin named Hernandez' species of birds according to the Linnaean rule, and as late as about 1825 Georges Cuvier, in one of his famous lectures on the history of the natural sciences, signalized Hernandez' book as "almost the only work on the natural history of Mexico to have appeared up to now."[13] The barriers did not fall until Mexico became independent in 1821; two years later William Bullock brought back to Europe the first bird collection ever to have been made on the spot.

Nieremberg's excerpts from Hernandez' manuscript had only just been published when, at long last, another scientist appeared whose thirst for knowledge drove him across the ocean. The Dutch West Indies Company, whose troops had seized the important harbor of Olinda (near Pernambuco) from the Spaniards in 1630, six years later named Count Johann Moritz von Nassau-Siegen (1603–1679) as Stadtholder of their new Brazilian possessions and equipped him with a small military force to pursue the conquest.

It soon became apparent that this was a happy choice; he was equally outstanding as governor and general, and through far-sighted measures was highly successful in promoting the well-being of the captured territories, which stretched from Ceara in the north to Sergipe in the south. But his lively mind was not restricted to this task alone; before he had left for his post he had formed a staff of scientists for a thorough investigation of the territory, following the advice of Dr. Johannes de Laet (1593–1649), a well-known scholar who was one of the 21 directors of the West Indies Company. The versatile Amsterdam physician Dr. Willem Piso (1611–1678), whose name was originally Pies and who was of German descent, accompanied him as official doctor and natural science expert. He was to be assisted by two Germans, H. Cralitz and Georg Marcgraf, advanced students of medicine and mathematics, but Cralitz died in January 1637, soon after the expedition landed at Olinda, so that Marcgraf alone had to concern himself with natural history, geography, ethnology, and astronomy. This unusually gifted young man, born in 1611, the son of a schoolmaster in Liebstadt near Pirna, apparently fled from his home while still young, to escape the chaos of the Thirty Years' War. Dr. de Laet chose him because he had acquired much unfamiliar knowledge during a six-year journey in the Mediterranean countries. In fact, he achieved remarkable results in all the fields assigned to him in Brazil, but he died at 34, before being able to revise for publication his many notes on seven full years of work. Soon after crossing to the African coast, where he intended to carry on his investigations after the Brazilian project had been broken off, Marcgraf died of climatic fever [probably malaria] in São Paulo de Loanda, 1644.

With great devotion, Dr. de Laet undertook to decipher the secret writing used for the notes, arrange them as best he could, and see them through the press, because Piso at first could not spare the time. Thus as early as 1648 the *Historia naturalis Brasiliae* appeared, in folio, giving the results of the expedition and, in the fifth book, Marcgraf's materials on Brazilian birds. Piso reworked the entire natural science section after the death of de Laet, whose editorial capacities had disappointed him, and transformed and rearranged the whole as *De Indiae utriusque re naturali et medica*, 1658, which does not mention Marcgraf's name and omits much of his information.

Both works immediately created the greatest sensation in the learned world, to an extent that would hardly have been possible if they had not been illustrated. The Count of Nassau had helped here, too, because he had followed Marcgraf's activities with lively interest, and had even kept a number of animals in captivity in Brazil so that he could study their habits. A painter accompanying the expedition was detailed to make pictures of animals and plants, so that gradually a great number (425) of colored representations were amassed, which the Count lent from his private collection to be reproduced in de Laet's work. True, the woodcuts made after these attractive originals were so crude that they prevented ornithologists in many cases from making positive identifications of Marcgraf's descriptions, but the originals had fortunately been preserved. In 1652 the Count of Nassau entered the service of

Brandenburg and sold these pictures, along with all his collections, to the Great Elector. After J. G. Schneider, justly famous in antiquarian zoology, discovered them in 1786 in the court library, they were studied, first, tentatively, by Illiger, then, 1817-1819, by Hinrich Lichtenstein, and finally, in 1938, very thoroughly by Adolf Schneider, to remove the last nomenclatural doubts.[14] This last was just in time, because after 1945 they disappeared.

As a successful scientific explorer, Marcgraf acquired unrivaled fame among zoologists and botanists. About 1825 Cuvier said that his work "can still be considered a classic which can be consulted with absolute confidence on any topic that it contains."[15] Indeed, for more than 150 years Marcgraf's accounts remained the only source of information on the Brazilian fauna, because the Portuguese, who rebelled in 1648 against the Dutch occupation and in 1653 recaptured Pernambuco, the last Brazilian strong point, suppressed all further news from the country until, at the beginning of the nineteenth century, the spell was finally broken. Marcgraf enriched ornithology with 133 Brazilian species, mostly from the district of Pernambuco, among them such distinctive genera as *Rhea*, *Anhima*, *Mitu*, *Mycteria*, *Anhinga*, *Crotophaga*, *Guira*, and *Trogon*. Ray, Brisson, and Buffon worked Marcgraf's data into their books, and Linnaeus and Gmelin based no fewer than 45 specific names solely on his descriptions, which are often distinguished by great attention to detail, as may be seen in the following account of *Rhea americana:*

> Nhanduguacu in Brazilian, Ema in Portuguese. An ostrich. They are in this country very large but somewhat smaller than the African. Their legs are long, the foot [inferiora cura] measures about one and a half feet, the shank [superiora crura] one foot. They have three front toes with black, thick, completely blunt claws. There is one rear toe, but it is round and thick, so that on slippery ground or a plank surface they walk only with difficulty and fall easily. The neck is held curved like that of a swan or a stork; it is about two feet long. The head is gooselike; they have handsome black eyes and a not very broad flat beak, two and a half fingers long. The wings are small and useless for flight, but when they raise one like a sail they increase their running speed so much that they can rarely or never be overtaken by a hunting dog. The feathers on the whole body are gray as on a crane; they are longer and more beautiful on the back, and the whole feathered part of the body appears almost round. It does not have the tufted tail usually pictured [that is, for the African bird]; rather the feathers reach over the whole length of the back as far as the rump. Counters and other iron objects it swallows easily but cannot digest, passing them undamaged through the anus. It

lives on fruit and meat. It is common in the campos of Sergipe and Rio Grande, but it does not occur in Pernambuco. Its flesh makes good eating.[16]

This was the way in which Marcgraf gave mid-seventeenth-century science an insight into "West Indian" natural history. Soon afterward, in 1654, J. B. du Tertre, a Dominican who had arrived in the Antilles shortly after Guadeloupe and Martinique had become French possessions in 1635 and had spent many years there as a missionary, produced a little information about the local birds, though it was less exact or trustworthy than Marcgraf's and included borrowings from him. But no details of the fauna and flora of "the other Indies" were known at that time, even though the Dutch had already been exploiting large areas for 50 years and owned important trading centers. To fill this gap Piso, in 1658, included in his work the descriptions made by a Dutch doctor, Jacob Bontius, who had died in 1631 at Batavia (Djakarta) after staying four years in the tropics, but far from "breaking the ice," as Piso put it, these reports "de quadrupedibus, avibus et piscibus" only barely scratched it.

In the meantime the real father of East Indian natural history was already hard at work — Rumphius, "a great ornament to Germany."[17]

Georg Eberhard Rumpf, who later Latinized his surname to Rumphius, was born in 1628 in the county of Solms, probably at Münzenberg in the Wetterau.[18] His father, a builder, moved some years later to nearby Hanau, which meant that his son could receive a solid education at the Gymnasium. But the expense seemed to have been useless, for at 18 the boy was seized by an adventurous longing to visit foreign parts: "consumed by an insatiable desire to learn by traveling, I left my native land when I was grown up."* A Count of Solms-Greiffenstein amiably offered help. Under pretense of having enrolled the young man in the Venetian army, he lured him to Texel. There he and other Germans were conducted on board a ship which sailed, not to the Mediterranean, but toward Brazil, where, following the recall of the Count of Nassau-Siegen, the situation for the Dutch had begun to be critical. Luckily the ship, the *Swaerte Raef*, was captured during the voyage and taken to Portugal. There things seem not to have gone too badly for Rumpf, even though three years passed before he could get home (in 1648 or 1649). All his bad experiences had failed to dampen his love of adventure; now he was eager to see the wonders of the Far

*From Rumphius's letter, dated September 20, 1680, to the personal physician of the Great Elector, Dr. Christian Mentzel, living in Berlin. This was the same man who collected and arranged the illustrations for Marcgraf's *Historia naturalis Brasiliae*.

East, but with the fixed purpose of studying local plants and animals according to scientific principles.

Since the only possibility of this lay in working for the East Indies Company, toward the end of 1652 Rumpf enlisted in its service as an *Adelborst* (subensign), and arrived at Batavia in June 1653. Those were stormy times: in the Moluccas a menacing rebellion was raging, which Arnold de Vlamingh van Oudhoorn attempted to drown in blood. Rumpf, along with the fresh troops, was transshipped to Ambon, where he was assigned to technical jobs, with promotion to ensign in 1656. He made no secret of the fact that he disliked his military career. This helped; the governor met him halfway, making him an *Unterkaufmann* (junior merchant) in 1657, and transferring him to the town of Larike, on Ambon's peninsula of Hitoe. He proved so excellent in this capacity that he rose to the rank of *Oberhaupt* (chief administrative officer) for the whole of Hitoe. Now he could marry, and saw himself at the goal of his desires: the honorable position that he had achieved left him enough time to devote himself to scientific studies, for whose sake he had patiently endured all the previous hardships. In furtherance of his high endeavor, and abetted by the governor general, Joan Maetsuyker, he was even able to order books and instruments from Amsterdam. From then on his duties as a trader and administrative officer in Hitoe were simply a means to a higher end, "a mask," as he said later, "which I had to wear up to this time in order to earn the daily bread for myself and my family."[19]

But his luck did not last long. While he was still completely absorbed in surveying and investigating the luxuriant tropical world, in the spring of 1670 he was suddenly blinded by amaurosis. The indefatigable scientist bore it as a dispensation of Providence. In his own words, "The Divine will, doubtless understanding more than I of my salvation, suddenly withdrew from my sight the whole world and all the creatures in it."[20] He had already written notes on much of what he had seen and thought about in 15 years, while much else remained indelibly in his mind. With the help of secretaries and a draftsman supplied by the East India Company, and later with the assistance of his son Paulus Augustus, he could, "although blind, with the keen eyes of a lively mind,"* finish in his study at Ambon the great work that was closest to his heart and that made his name famous, the *Herbarium Amboinense*. Other works, all written by him in Latin, were carried on concurrently: the *Amboinsche Rariteitkamer* [Ambonian Cabinet of Rarities], the *Ambonsche Land-Beschrijving* [a description of the territory of Am-

*This is the first line of a poem written by N. Schlaghen, Governor of Ambon, to be attached to a portrait of Rumphius in 1695. (*Rumphius Gedenkboek 1702–1902*, p. ix.)

bon], and the *Ambonsche Historie*. He was still working on the *Ambonsch Dierboek* [zoology] when death took him, on June 15, 1702.

The *Ambonsch Dierboek* is what concerns us here — though it is precisely this manuscript that has been lost. Nothing would have been known of it beyond the fact that it was divided into three parts (dealing with animals of land, water, and air) and that it was illustrated with drawings, if Frans Valentijn (1666–1727), who was a rapacious plagiarist of Rumphius' material, had not steeped himself in this manuscript during his first stay in Ambon (1686–1694), and possibly also during his second (1707–1712), in order to plunder it for the third volume of his extensive encyclopedia *Oud en Nieuw Oost-Indië*n [The East Indies, Old and New] (Dordrecht and Amsterdam, 1726). Though this conceited windbag ("dominus loquacissimus," as Rouffaer called him) never admits anywhere that he used the zoology of his "bosom friend" Rumphius for his "Verhandeling der Vogelen van Amboina" [Treatise on the birds of Amboina], the expert soon detects that the core of his accounts can have originated only with an uncommonly gifted natural scientist, not a shallow scribbler. Now and then Valentijn gives himself away, as for instance on the cassowary: "I tried its flesh in 1668 and found the fat white and agreeable in taste" — apparently taken over word for word from Rumphius. Valentijn's procedure here must be the same as in other instances that can be analyzed, so that one feels justified in regarding what is specific and apt in zoological statements as being Rumphius, and what is vague and digressive as addenda by his garrulous plagiarist. The "Verhandeling der Vogelen" mentions more than 41 species occurring on Ambon and Ceram, and many from the surrounding islands. One can search in vain through the literature before 1670 for accounts that give such an accurate idea of the appearance and habits of birds.

Yet another German contemporary of Rumphius distinguished himself as an observer and commentator on distant countries, describing from his own experience "what he had found worthy of consideration on the ice, in the water, in the air, and on land." This was Friderich Martens, who must have been very young and gay when he made a trip to Spitsbergen on a whaler belonging to the Free Imperial City of Hamburg and "submitted to employment as ship's barber." "The ship was called *Jonah in the Whale* and the skipper was called Peter Petersen the Frisian." The little book, a genuine pearl of old travel literature, is entitled *Friderich Martens vom Hamburg Spitzbergische oder Groenlandische Reise Beschreibung gethan im Jahre 1671* (Hamburg, 1675).[21] The reader discovers in the "Fourth part, of the animals on Spitsbergen, and particularly of all the birds," the most delightful accounts of 15 species of birds from the far north.

What Clusius had foreseen had become reality before the century was out: the records of natural science enormously increased by "exotica."

PART TWO

THE DEVELOPMENT OF SYSTEMATICS AND THE STUDY OF EVOLUTION

3

The Influence of Methodology

Because the number of voyages overseas increased rapidly during the seventeenth century, true and false reports of foreign birds arrived more and more frequently, and previously unknown species were introduced, to the delight of collectors of "natural treasures" (the so-called "cimelia physica") or of live showpieces.

Aldrovandi, stimulated by the work of Aristotle and Pliny, had in his day attempted to organize the multiplicity of birds according to "philosophical" principles. Now, however, it became clearer that his methodology was impractical, because it demanded some knowledge of the habits of birds, and this often was exactly what was lacking, particularly for foreign species.

When Walter Charleton (1619–1707), physician to Charles II, set about classifying all birds systematically in his *Onomasticon zoicon* (London, 1668), including those known to him only from literature or from the museum of the Royal Society (founded 1662), he could think of nothing better than to reserve most of the exotics for the appendix, where he confined himself to arranging them by guesswork under "Terrestres" and "Aquaticae." On the other hand, in classifying species familiar to him he followed Aldrovandi, like his friend Merrett a little earlier (*Pinax rerum naturalium Britannicarum*, London, 1666), and thereby arrived at the following system:

There are two principal divisions: land birds and water birds.
Land birds are divided into:

1. Meat eaters (*Aquila*, *Accipiter*, *Collurio*, *Falco*, *Cuculus*, *Psittacus*, *Corvus*, *Bubo*, *Caprimulgus*, *Struthio*, *Emeu*, *Vespertilio*, etc.).

2. Seed eaters
   a. Dust-bathing (*Gallus*, *Otis*, *Oedicnemus*, *Ortygometra*, etc.);
   b. Dust- and water-bathing (*Columba*, *Passer*, *Troglodytes*, etc.);
   c. Singing (*Carduelis*, *Spinus*, *Fringilla*, *Chloris*, *Linaria*, *Alauda*, *Ficedula*, etc.).
3. Berry eaters (*Turdus*, *Sturnus*, *Coccothraustes*, *Caryocatactes*).
4. Insect eaters
   a. Non-singing (*Picus*, *Sitta*, *Jynx*, *Certhia*, *Merops*);
   b. Singing (*Regulus*, *Curruca*, *Hirundo*, *Parus*, *Motacilla*, *Luscinia*, *Upupa*, etc.).

Water birds are divided into:

1. Palmipeds
2. Fissipeds
   a. Fish eaters (*Ciconia*, *Phoenicopterus*, *Ardea*, *Porphyrio*, *Limosa*, *Haemantopus*, *Ispida*, etc.);
   b. Insect eaters (*Morinellus*, *Arquata*, *Crex*, *Totanus*, *Gallinula*, *Scolopax*, *Cinclus*, *Charadrius*, etc.);
   c. Plant eaters (*Grus*).

Nothing could more clearly demonstrate the uselessness of supposedly Aristotelian principles for avian classification than this attempt, which defies common sense.

The reaction was not long in coming. In 1676, also in London, appeared Francis Willughby's posthumous work, *Ornithologiae libri tres*, which finally jettisoned the 2000-year-old classical systematics and replaced them by one based not on function but on form. As points of reference the revolutionary used beak form, foot structure, and body size.

The result, much more satisfying than Charleton's, appeared as follows. Birds were divided into two main groups, land birds and water birds. Both of these were subdivided according to the Platonic principle of dichotomy (trichotomy only when size was a ground for distinction), until compact final groups remained. This was done without pedantry and only as far as necessary: for large flightless birds (*Struthio*, *Emeu*, *Dodo*) the final grouping was reached with the third subdivision, while many more complex groups required eight. In this way short diagnostic descriptions of the final groups were arrived at, an example being the account of the pigeons: terrestrial birds; with straighter or less curved beaks and claws; of medium size, with smaller and shorter beaks; those with black flesh larger. The gulls are: aquatic birds; those that swim on the water; palmipeds; equipped with shorter legs; four-toed; with a single rear claw; with narrow beaks; with a beak sharp at the tip and fairly straight; with long wings and skillful in flight. Further final groups are, for example, all the Fringillidae of modern systematics; the Corvinae, the Turdinae (including *Sturnus*); the Psittaci; the Picinae (*Picus, Sitta, Certhia, Aracari; Upupa, Passer arundinaceus*); the Gallinae (including *Otis*); the Limosugae (= Limicolae); the Steganopodes; the Anserinae and Anatinae of modern systematics; and so on.

For a long time this system remained unsurpassed, even though eager wiseacres soon tried to throw it out. It reflects the true relationships much more completely than the "natural system" thought up 60 years later by Linnaeus, because its creator was a true ornithological genius and a completely unconventional thinker. The real author of the new classification has never been completely established. Many later scholars attributed it to Willughby, others, perhaps more plausibly, to Ray. When Francis Willughby died on July 3, 1672, after a long illness, at the age of only 37, he left many loose papers containing scattered notes for the draft of a voluminous handbook of ornithology, which was far from ready for the printer. It was Ray who gave it its final form and, after four years of work, during which he was helped by other English ornithologists like Sir Thomas Browne and Ralph Johnson, had it printed in 1676.

John Ray, a brilliant scientist, and Francis Willughby, a painstaking researcher, passed much of their lives together.[1] At some time after 1653 they had met at Cambridge, where Ray, born the son of a village blacksmith in 1627 at Black Notley, Essex, had already acquired considerable knowledge, especially of "natural philosophy," along with his theological studies. He lectured on Greek and mathematics, and the young and well-born Willughby was attracted and held by the force of his mind and personality. Gradually the two friends developed a plan to describe the whole of nature according to their point of view, Ray dealing with plants and Willughby with animals. In the summer of 1662 they visited the west coast of England together and made an intensive study of the breeding habits of sea birds. That was the prelude to a greater project: in April 1663 they set forth for the Continent, accompanied by two young Englishmen, Nathaniel Bacon and Philip Skippon, in order to enlarge their scientific experience. From Calais they went to Holland, where they visited the great bird colonies near Zevenhuisen, a breeding ground for herons, spoonbills, and cormorants. Then they journeyed up the Rhine valley (through Cologne, Heidelberg, Strassburg, Basel) to Zürich, continued by way of Constance and Nuremberg to Vienna (September 15), and crossed the Semmering to Venice (October 6). They spent the winter there and in Padua. At the beginning of February 1664, they traveled from Bologna to Leghorn, and thence by ship to Naples. There Willughby and Bacon left the group and went quickly through Rome to Roussillon (August 31), in order to cross Spain. At the end of 1664 Willughby arrived back in

England. Ray and Skippon, however, left Naples at the end of April 1664, crossed to Sicily, and made a flying visit to Malta on May 13. A week later they climbed Etna as far as the snow line, and then proceeded by way of Leghorn to Rome, where they stayed from September 1664 till January 24, 1665. Then they traveled to Venice and Bolzano, and across the Ofen Pass to the Engadine. On March 25 they were in Chur. By way of Lucerne they reached Lake Geneva and then Montpellier. There they would have stayed until the next spring if an edict of Louis XIV, ordering all Englishmen to leave France within three months, had not reached them in February 1666. Ray and Skippon therefore returned to England, by way of Paris, on April 8, 1666, having been absent three years.

In those days, one wishing on such a journey to extend ornithological knowledge did not need to depart from the main routes; even in the city markets and the stalls of the birdcatchers the two friends saw many birds that were unknown to them in England. For example, hazel hens were for sale in Nuremberg, a black stork in Frankfurt, capercaillie in Venice and Padua, and a great bustard in Modena, and Ray saw cranes "often for sale in the Roman market." In this way they gradually obtained a quantity of widely different birds in fresh condition, so that they could make exact descriptions not merely of their external appearance but of their internal structure. They dissected each bird with the thoroughness of trained anatomists, and even watched for external and internal parasites and studied the contents of the stomach. For further information they sought to acquire pictures, buying in Strassburg a copy of the later famous *Vogel-, Fisch- und Thierbuch* by Leonhard Baldner, fisherman and gamekeeper. In Nuremberg they purchased a scrapbook with many paintings of birds, and they did not leave Bologna without visiting Aldrovandi's large collection of animal pictures. After returning, they worked through their zoological notes and souvenirs together in order to make systematic use of them, and they also visited museums and royal vivaria to look at exotic birds, which were as important to them as native ones. How ornithology would have benefited if Willughby's plan to visit North America in 1670 for research had not been dropped, because of the outbreak of the disease that was to kill him two years later!

Ornithologiae libri tres, in the final form bestowed on it by Ray, consists of several introductory chapters based on his own investigations and those of well-known authors (like Harvey), which provide a survey of the whole field: of the outer parts; of the inner parts; of generation; of age; of some of their [birds'] qualities (natural instincts, etc.); of the division of birds. After these follows a special section arranged according to the new system, with Hernandez' and Marcgraf's birds, as well as other exotics, fitted into appropriate places. Sometimes

Willughby's descriptions of plumage are so detailed that Ray cannot forbear commenting: "In addition (we must admit) he seems to us somewhat too painstaking in the description of the colors of individual feathers." If nest and egg are known to him, he describes them too, but in general his biological information is scanty.

In 1678 Ray, whose friend had willed him an income for life, published an English translation of the book, with some additions. After many years he reworked the special section once more, summarized what he considered particularly important, and enlarged it with addenda by other hands, concerning only exotic birds (drawn from the travels of Sir Hans Sloane in Jamaica from 1687 to 1689 and Paul Hermann in Ceylon from 1672 to 1680). The manuscript was ready in 1693 or 1694, but was published only after Ray's death in 1705 by his friend, the Reverend William Derham, under the title *Synopsis methodica avium et piscium* (London, 1713).*

With the Willughby-Ray *Ornithologia* ended the series of efforts to furnish a complete monograph on ornithology, a treatise that, after the fashion of Aristotle's writings, would collect everything worth knowing, not only about appearance, habits, and distribution, but also about anatomy and physiology. For subsequently, the mass of material increased so hugely that no one was capable of digesting all of it. Finally ornithology threatened to disintegrate among bands of particular specialists with very slight contact or ties — the anatomists, the physiologists, the "ornithographs," the behavioral scientists, the specialists in systematics. Each of them had his own aim, so that eighteenth-century ornithological handbooks, even when as voluminous as those of Brisson or Buffon, remain fragments.

Even in Willughby's book the different fields are not explored with the same attention. This is true, for example, of biology, and very much more so of anatomy, which, though not shuffled off, is not presented in such detail as was possible at that time. Aristotle and Frederick II had contributed a great deal to knowledge of the internal structure of birds; but their investigations were continued only in the sixteenth century, after Vesalius (1514–1565) had revivified anatomy according to Aristotelian principles and had trained a whole flock of pupils in Padua and Bologna. In this circle, which soon turned to zootomy, Belon and Aldrovandi probably acquired the anatomical knowledge that distinguishes their books on birds. Volcher Coiter (1534–1576), who studied the organic structure of birds with a thoroughness never before attained, also was a product of the Italian school. This Dutchman, born

*Derham added a manuscript on the birds of Madras, sent to James Petiver (see reference 3) by Edward Buckley, a correspondent of Georg Joseph Camel (1661–1706), a Jesuit active in Manila as a naturalist and particularly keenly interested in ornithology.

in Groningen, had completed his education with Fallopius, successor to Vesalius in Padua, and in 1564 had consulted Aldrovandi in Bologna. He worked in the Aristotelian spirit more completely than any other sixteenth- or seventeenth-century ornithologist, with the single exception of Harvey, though his contemporaries did not recognize this and completely disregarded his *Externarum et internarum principalium humanis corporis partium tabulae*, published in Nuremberg in 1573, while Coiter was city physician there. In a lengthy section entitled "De ovis et pullis gallinaceis," he describes more thoroughly than Aristotle the development of the chicken in the egg. A small chapter deals with the anatomy of birds and contains much good material on the development of the brain, the optic nerves, the olfactory nerves, the eyes, the hearing mechanism, the tongue, the trachea, lungs, and stomach, particularly those of the larger and smaller grebes, the corncrake, and a species of gull. In 1575 Coiter published a sequel (also in Nuremberg), *De avium sceletis et praecipuis musculis*, with very exact diagrams and descriptions.

In the second chapter of this sequel, "De differentiis avium," Coiter endeavors to follow Aristotelian guidelines. He looks at the arrangement of the skeletal bones to find connections between form and function, "for nature has fitted the instruments to their offices," and introduces his attempt to classify birds with a quotation from Pliny: "The first and most important distinguishing characteristic among birds is the feet." Therefore he divides them into fissipeds and palmipeds; as in Pliny, the former are subdivided according to claw shape, and further subdivided according to the placing of the toes; the latter, according to the spread of the web. Thus, long before Linnaeus, Coiter had developed a bird classification that relied only on morphology, and was not much further from the ideal of the "natural" system than that of the Archiater.*

Scientific research was most successfully continued by Hieronymus Fabricius ab Aquapendente, for 54 years (1565–1619) professor of anatomy in Padua, who in 1615 and 1618 provided important contributions to knowledge of the anatomy and physiology of birds, in which the crop, glands and gizzard, intestines, skeletal structure of wing and leg, flight, walk, respiration, and other matters were considered. His most important pupil was William Harvey (1578–1657), who introduced the methods of Fabricius to England and continued his researches there. As a starting point he chose, among others, a posthumous work by Fabricius, *De formatione ovi et pulli* (1621), inspired by Aristotle; Harvey's result, published in 1651 as *Exercitationes de generatione animalium*, added enormously to knowledge of embryonic development

*In 1747 Linnaeus was appointed by the king of Sweden his personal physician, or archiater. In the following decades many scientists referred to him by this title.

and of sexual organs in the chicken. Another pupil of Fabricius, Caspar Bartholin, introduced "comparative" anatomy to Denmark, where later his son Thomas and the latter's pupil, Nicolas Steno (Niels Stensen), also worked on birds. This branch of research was cultivated about 1670 in many European universities, among them Paris, where Claude Perrault and G. J. Duverney dissected exotic birds that had died in the Jardin du Roi, and from 1671 published detailed reports on the results. It is true that at that time nobody considered using anatomic characteristics as bases for classification; the "ornithotomes," like the "ornithographs," contented themselves with collecting facts and exploiting them, if at all, only for physiology.

Even in Willughby's day the microscope was already trained on the structure of birds. Robert Hooke concerned himself with the finer details of feathers, and employed excellent illustrations to elucidate the way in which the distal barbules and the proximal barbules of a goose's feather work together, and how the iridescent barbules of the peacock feather are constructed (*Micrographia*, London, 1665). However, Willughby's book makes no mention of this interesting discovery.

Of all the stimulating suggestions that ornithology received from Willughby, none was more eagerly taken up than the idea that birds should be classified according to morphological and not biological considerations. Anyone who set up a "cabinet of birds" for the "delectation of eye and mind" now saw a possibility of organizing it methodically.

In Willughby's lifetime the keeping of birds "in natura" was certainly not common. Though some people, as in the previous century, attempted the difficult art of making durable preserved specimens, the old, easier, and cleaner process of drawing or painting a picture of the freshly killed animal, and making a collection like Gesner's or Aldrovandi's, was more popular. There were only a few places in England where one could admire "stuffed" birds; the inquisitive could find them, scattered among a colorful miscellany of other curiosities, in the Museum Tradescantium in South Lambeth, where there were even a dodo and a great auk, and in the Royal Society's London museum, whose "rariora" were enriched shortly before 1668 with a hawfinch (*Coccothraustes*) presented by Charleton, who says: "I arranged not so long ago to have a female of this kind, which had been killed by my own pistol and preserved in powdered spices, placed among the rarer objects in the gallery of the illustrious Royal Society."[2] Toward the end of the century the preserving of birds became commoner. To make specimens more durable, most of the collectors must have used the process described in 1698 by James Petiver (1663–1718), a London apothecary:

As to Fowls, those that are large, if we cannot have their Cases

> whole, their Head, Legs, and Wings will be acceptable; but smaller Birds are easily preserved Entire, by opening their Bodies which is best done, by cutting them under the Wing, and take out their Entrals, and then stuff them with Occam or Tow mixt with Pitch, or Tar, and being throughly dried in the Sun, wrap them up close and keep them from Moisture.[3]

If such preserved specimens were simply placed on shelves or hung from the ceiling, as shown in the frontispiece to Ole Worms's *Museum Wormainum* (1655), they could not have lasted long. Therefore people soon became more careful. The famous Amsterdam collector Albert Seba (1665–1736) kept most of the smaller birds sent to him from both the Indies, just like his fish, snakes, and frogs, "in liquore," since alcohol and *Malzwein* (half-matured malt or beer vinegar) had come into use as good preservatives. On the other hand, Paul Hermann (1640–1695), the professor of natural history at Leiden, kept shut up in cupboards the many "dried" birds that he had brought back from his travels in Ceylon and India (1672–1680). He was outdone by his colleague, the Leiden anatomist Frederik Ruysch, who kept each "dried" bird in a wide-necked bottle, with its cork sealed by a fish bladder and red velvet. "Also one comes across dry phials, and in them rarer dried small birds, particularly 180 sent from both the Indies, and holding between their claws twigs of rare plants."[4] A similar method was used by Johann Leonhard Frisch (1666–1743) in Berlin; he kept "each bird in a separate wooden or glass case. But one must take good care that these cases shut extremely tightly."[5]

The progress in techniques of preservation, which made bird collecting a pleasure, not only assisted comparative research but also provided a basis for the large color-plate books that now, in the rococo period, were much sought after as ornaments to the library. Interest in nature, indeed, became fashionable, was cultivated in the best circles, and was sublimated in art, where naturalism became playful. The eye demanded to be refreshed by colorful or rare shapes, if possible exotic; uncolored pictures were too common. Thus Mark Catesby (1682–1749), a protégé of Sir Hans Sloane's, could venture to publish the accounts of his discoveries in Carolina, Georgia, Florida, and the Bahamas in two sumptuous illustrated folios (London, 1731, 1743), which met with great success, especially because of their hundred or more colored plates of birds. Catering to the current demand, the illustrator Eleazar Albin in London decided, in the same period, to bring out a *Natural History of Birds* with colored pictures, chiefly of British birds (three quartos, 1731, 1734, and 1738). Because he was as clumsy with the pen as with the brush, and understood nothing of his subject, he borrowed most of the text for his first two volumes from Willughby and Ray. Much more able was the skillful English animal painter George Edwards (1694–1773), who soon afterward compiled a new *Natural History of Uncommon Birds*, "most of which have not been figur'd or describ'd, and others very little known from obscure, or too brief descriptions, without figures, or from figures very ill design'd." This folio edition swelled into four volumes (1743–1751) with 210 colored plates (189 of birds) and was continued from 1758 until 1764 as *Gleanings of Natural History* (three volumes with 128 plates of birds). Both instructive and agreeable, untrammeled by any rigid system of arrangement and yet accurate, clearly it was just what the public wanted. The Nuremberg engraver J. M. Seligmann recognized this promptly, made new colored engravings of all 474 of Catesby's and Edwards's ornithological plates, and published them between 1749 and 1776 with German, French, and Dutch texts, so that many recently discovered exotic birds became subjects of drawing-room conversation before the native birds had all been described and portrayed in color.

It is true that all this was completely disorganized sciolism. But that age of buoyant frivolity was hardly able to enjoy any meticulous order; it entertained itself with chinoiserie and the asymmetric ornamentation of rococo. Everyone who sought to gain even a little public favor followed the fashion, and not only artists but also some scholars toyed with it.

Pierre Barrère (1690–1755), whose *Ornithologiae specimen novum . . . in classes, genera et species, nova methoda, digesta* appeared in 1745 in Perpignan, reveled in paradoxes, which 15 years later appalled his fellow countryman, Brisson. He commented that, in Barrère's system, the bustard is placed between swallow and bunting, the latter between bustard and pelican, the bird of paradise between cassowary and ostrich, and so on, exactly as if the author wanted to put all the birds into one cupboard, and, in order to save space, filled the inevitable gaps between large birds by fitting in small ones. Buffon, who published the first volume of his monumental *Histoire naturelle, génèrale et particulière* in Paris in 1749, was too clever to indulge in such folly, but he too sacrificed to fashionable and frivolous caprice by opposing the "methodists." The usual systematic arrangement of natural objects and the procedure followed therein are, he maintained, simply arbitrary. Everyone is therefore free to choose the sequence he considers most convenient and most commonly used. The most natural arrangement, which Buffon believed that he ought to follow, was, in his own words, the one that a completely impartial man would choose. Thus he dealt first with the animals most interesting and important to man, and proceeded gradually outward to those entirely unfamiliar to him. But the "methodists" were advancing steadily, and even the versatile mind of a Buffon could not hold out permanently against them.

When the 27-year-old botanist Carl Linnaeus,[6] from the small Puritan town of Uppsala, arrived in Holland in 1735 to take his medical degree in that land overflowing with gold, he carried in his knapsack the outline of a new system of classification, and his visits to the cabinets of natural curiosities must have strengthened his sense of mission. These showy collections made by rich amateurs needed a system, the clear arrangement of innumerable objects according to a strict plan such as he had already thought out in Uppsala with his friend and fellow student Peter Artedi (Arctaedius). He presented his ideas and had the satisfaction of realizing that they had fallen on fruitful ground; Dr. J. F. Gronovius of Leiden was so enchanted that he paid for the publication of the *Systema naturae*, with its new hierarchy of concepts (Classis, Ordo, Genus, Species), and it appeared in Leiden in the same year — printed in the form of placards on just a few sheets.

The fame of this foreign organizational genius spread so rapidly that two attractive proposals were made to him. The apothecary Albert Seba, rich in honors and in possessions, attempted to engage the young Swedish doctor for scientific research on his treasures.[7] He had just published in Amsterdam, at great expense, the first two volumes of his "Accurate description of the most richly endowed treasury of Nature, and an illustration with the most skillful pictures, for a universal history of the physical world," an ostentatious folio from his own pen, reflecting exactly the fussy confusion of his "treasury of Nature." There in handsome plates are to be seen plants, snakes, monsters, birds, butterflies, mammals, and grasshoppers in wild disorder, to which the superficial descriptions of an equally baroque text are supposed to supply a commentary. Seba wished Linnaeus now to work over the fish for the third volume. But fish, and all cold and clammy animals, did not please him in the least ("Terrible are Thy works, O Lord!" he wrote later, in 1766, as a heading for the class of Amphibia), and the chaotic Seba material cannot have been more to his mind. It was therefore most fortunate that just then his friend Artedi arrived from England and, as an ichthyologist, gladly accepted Seba's offer, which helped him out of severe financial difficulties. Linnaeus, on the other hand, offered his services to the banker Georg Clifford, as overseer of his magnificent gardens, and from that time on was treated "like a prince," to use his own words.

A very productive period now began for both men. The tranquil and critical Artedi, like his enthusiastic fellow student, was a fanatic for method and a devil for work. In a few months he completed his task and prefaced his catalogue of fish with a *Philosophia ichthyologica*, in which the scientific knowledge of fish is thus defined:

> Ichthyology is a science that first names all the parts of a fish, then

> lists the true names of genera and species, and finally calls attention from time to time to valid characteristics. This last section, however, must be kept short and succinct, because long and digressive descriptions of habits and characteristics are pointless in ichthyology and the rest of natural history, since the employment of the true and natural method for the identification of creatures according to genus and species is the prime and sole aim of natural history.[8]

This program deserves to be quoted here at length because, until the end of the century and even longer, it was esteemed and followed by most zoologists as the embodiment of wisdom. Artedi himself was able to pursue it in only this one book, because in the autumn of 1735, as he was coming home late in the evening after visiting Seba, he fell into a canal and was drowned. From then on, Linnaeus followed the guidelines laid down by his unforgettable friend, as well as a botanist with modest zoological knowledge was able to do.

The completion of his method of nomenclature had to wait 20 years. In 1735 Linnaeus, like his models Tournefort and Artedi, considered it most important to establish genera, which Willughby and Ray in 1676 had not done; they used the term *genus* not only to describe abstract groups of the kinds we call classes, orders, families, or genera (as Aristotle had done), but sometimes also in the sense of species. Linnaeus accompanied each generic name with a short diagnostic description, whereas he gave species only as examples. In this way he could subscribe to the concept of species defined by John Ray in his *Historia plantarum*, I (1686):

> Among plants there is no other positive sign of conformity to a species than the growth from seed of specifically or individually identical plants. The forms that are differentiated by belonging to different species preserve their specific nature constantly, and one species does not grow from the seed of another or vice versa.

The idea of the genus *Numenius* is elucidated by reference to its species: *Gallinago*, *Arquata*, *Limosa*, *Recurvirostra;* that of the genus *Columba* by *Columba*, *Palumbus*, *Turtur*, *Oenas*, etc. In the *Fauna Svecica* (Stockholm, 1746), Linnaeus gives a diagnostic description also of species; for example, he distinguishes the following species in the genus *Motacilla:* Motacilla pectore nigro, Motacilla pectore abdomineque flavo, Motacilla pectore albo, corpore nigro, etc. The two techniques are combined in the sixth edition of the *Systema naturae* (Stockholm, 1748). Here the author arranges all the birds known to him in 85 genera which, as in 1735, he groups into 6 orders. In so doing he by no means shares the point of view of Willughby, who clearly had chosen the overall appearance of a bird as the criterion of affinity and had not hesitated to split off small groups if their appearance demanded it. Linnaeus, on the other hand, chiefly valued the establishment of a system whose clear and balanced divisions would impress it on the mind. Anyone who, like him, took as distinguishing marks first of all the beak and then the foot, could create as many little groups ("orders") as he liked. He decided on six, divided into:

I. Accipitres (*Falco*, *Strix*, *Psittacus*);
II. Picae (*Picus*, *Buceros*, *Cuculus*, *Paradisaea*, *Corvus*, *Upupa*);
III. Anseres (all birds that swim);
IV. Scolopaces (a collection of fissipid waterfowl);
V. Gallinae (*Gallus*, *Struthio*, *Otis*, *Fulica*);
VI. Passeres (*Columba*, *Turdus*, *Alauda*, *Trochilus*, *Sitta*, *Loxia*, *Motacilla*, *Parus*, *Procellaria*, etc.).

Even the last vestige of the "classical" system, the primary differentiation between land and water birds, had been eliminated from Linnaeus's "natural system."

The Swedish botanist's fame grew rapidly; however, it was not the classification but the proposal of a strict methodology that found a welcome here and there among zoologists, and because the methodology could be varied at will it was soon deliberately played with. Pierre Barrère in Perpignan (*Ornithologiae specimen novum*, 1745), Jacob Theodor Klein in Danzig (*Historiae avium prodromus*, Lübeck, 1750), and Heinrich Gerhard Möhring in Jever (*Avium genera*, Bremen, 1752) all joined in the game; though they all, like Linnaeus, took beak and foot as points of reference, their results differed widely. So now there were four new ornithological systems, though none had been designed by a real expert on the subject. As Buffon remarked, "They are only arbitrary reports and different points of view from which natural objects can be studied."[9] For the task that these classifiers had set themselves their material was far too limited.

By 1756 two private collections had outstripped all others. Sir Hans Sloane (1660–1753) owned a most important one ("incomparable," according to Linnaeus, who had admired it in 1736). A disciple and friend of John Ray's, as a young doctor he had accompanied the Duke of Albemarle to Jamaica (a British possession since 1655), and had collected all kinds of natural history specimens there from 1687 to 1689. In order to develop this basic collection for a museum (which had induced him to publish a much admired, richly illustrated folio, *A Voyage to . . . Jamaica*, in two volumes, 1707 and 1725),* he bought Petiver's remarkable "Gazophylacium" after its owner's death (1718),

*The plates of Jamaican birds are extremely primitive, because Sloane understood nothing about preparing specimens and had a clergyman draw on the spot all birds that had been shot.

and subsidized Mark Catesby's important scientific journey to Carolina, Georgia, Florida, and the Bahamas (1722–1726). A Maecenas to naturalists and painters (like Edwards), highly regarded as physician and botanist, secretary of the Royal Society and for many years editor of its *Philosophical Transactions*, he occupied a position in English intellectual life corresponding to that of Sir Joseph Banks at the end of the century. He willed his important library and his large collections to the nation, under certain conditions, making it possible for the British Museum in Montague House to open to the public in 1759. Sloane's ornithological collection had grown to 1172 items, including, however, "partes avium" (beaks, bones, eggs, nests, and so on). For that period it was enormous.

The animal kingdom was even more completely represented in the cabinet of the French naturalist René Antoine Ferchauld de Réaumur (1683–1757), which he had set up at his country seat in Bas Poitou. Using his instructions on how to prepare specimens, the most adventurous French travelers (including Michel Adanson, who had carried out scientific research in Senegambia from 1748 to 1753, and the botanist Pierre Poivre, who had been in Cochin China, Madagascar, and Manila) had brought him animals from all over the world. Réaumur's private means enabled him to employ a scientist as curator of his museum, which contained, besides stuffed birds, a notable collection of nests and eggs. This "démonstrateur" was Mathurin Jacques Brisson (1723–1806), an able man who, stimulated by the publications of Klein and Linnaeus, worked extremely hard on the methodical arrangement of the highly varied collection. In the process he soon recognized the deficiencies of the *Systema naturae*, for anyone who tried to classify a newly discovered species according to Linnaeus's short diagnostic descriptions of genera found out that "there are not enough characteristics given." Brisson's own freshly developed system for mammals appeared in his *Règne animal* (1756), followed in 1760 by a similarly arranged *Ornithologie*, a six-volume standard work of ornithography. It contained all known species, together with 320 new ones that Brisson had found partly in Réaumur's museum and partly in the extensive collections of the Abbé Aubry (rector of Saint-Louis) and of the physician Mauduyt de la Varenne. There were some 1500 in all, described in as much detail as possible, whereas Linnaeus had taught that descriptions should be as brief as possible. Linnaeus was accustomed to stretch a genus; Brisson, defining it more sharply, reached a total of 115 genera of birds. Unlike Linnaeus, he grouped them not in 6, but in 26, orders. Whenever their size seemed to require it, he subdivided the orders into sections. Though Brisson too was guided only by easily recognizable external characteristics, chiefly beak and claw ("as far as possible I have kept to obvious, simple characteristics, easy to comprehend, so that too much complexity should not lead to error"[10]), he used these taxonomic indications so prudently and with such a fine feeling for essentials that he made an astonishingly small number of major errors. In general, his heavily subdivided system approaches our own more closely than any attempt at classification made in the following 80 years. It did not please his contemporaries, however; it was difficult to grasp, and difficult to learn by heart; above all, the author, who detested anything doctrinaire, had neglected to assign names to his orders and sections. Thus no one followed this modest expert; even he himself did not uphold his theories again, for soon after 1760, to Buffon's satisfaction, the museum of his hated rival Réaumur was absorbed into the Cabinet du Roi. Brisson lost his job, abandoned zoology, and in 1762 succeeded the Abbé Nollet as professor of physics in the Collège de Navarre. But even in this position he found opportunity to perform a great service for his former favorite subject — by converting his pupil Étienne Geoffroy Saint-Hilaire to the study of zoology.

Shortly before Brisson entered the arena with his *Ornithologie*, only to leave it again immediately, Linnaeus the persistent organizer had an idea that finally assured victory to his system and made his name immortal: he evolved binomial nomenclature, or rather, he employed it with strict logic in his writings. From then on only one Latin word served him to designate a species. By his edict the names of genus and species were coupled in a "natural" unit ("class and order are the work of wisdom, genus and species the work of nature"). He employed his new technique first in 1753 for plants, in 1754 (in the *Museum Adolphi Friderici Regis*) also for birds and other groups of animals, and only in 1758, in the tenth edition of the *Systema naturae*, for all organisms known to him. Some of his combined names had long been in use; Gesner in 1555 and all later writers had divided European tits into *Parus cristatus*, *P. major*, *P. ater*, *P. palustris*, *P. caeruleus*, and *P. caudatus*.

Many people immediately recognized the great advantages of the new, clear nomenclature, stabilized by a law of priority (1751), though whoever settled for it had to take over the Linnaean system as well, which even in its new guise (1758) was not much more "natural" than before. Linnaeus had not abandoned his six orders of birds, but he had shifted some of the genera. *Psittacus*, *Sitta*, and *Trochilus* were now under Picae, *Otis*, *Struthio*, and *Fulica* under Grallae (= Scolopaces 1748), and *Procellaria* under Anseres; a Frenchman who despised the method sneered that the cards had been shuffled again. At any rate, the zoographs had at least acquired a solid foundation for international understanding and therefore had every reason to revere the Archiater. He, however, overestimating the value of such acceptance, thought in his good-natured conceit that "God had granted him the greatest insight into natural science, greater than anyone before had achieved."

"The hideous laughter of the grinning satyrs, and the gloating of the little monkeys who jump on my shoulders, I have borne philosophically. I have gone my way and have kept the course set for me by Fate."[11]

Among those who applauded such persistence were several who made important contributions to ornithology: first of all Pontoppidan (1763) and Brünnich (1764), friends from Copenhagen; then Pallas (1764), the most important ally; and a little later also J. R. Forster (1767) and J. A. Scopoli (1769). In naming new species of birds they accurately followed the Linnaean prescriptions. And when in 1773 Statius Müller, a Dutch professor of natural history in Erlangen, brought out in several volumes *Des Ritters Carl von Linné vollständige Naturgeschichte*, "according to the twelfth Latin edition, and according to the guidelines provided by Houttuyn's Dutch book," presenting the German public with a zoological reference work for the household, constructed on the Linnaean system, this Latin nomenclature found general acceptance in Germany. No one who rejected it was considered a scientist.

This was not true everywhere. The English ornithologists had become so accustomed to Ray's classification that for a long time they hesitated to give it up for a foreign one; even when Tunstall in 1771 (*Ornithologia Britannica*) and Latham in 1781 (*A General Synopsis of Birds*) took over the general outlines of Linnaean grouping, they retained the old major division into land and water birds. The Linnaean system was not very precisely followed in other ways in England. For international communication the nomenclature selected was in Latin, a consideration that might seem self-explanatory to the nations on the Continent, but that did not suit the English leaning toward individualism. The two leading English ornithologists of the time, Thomas Pennant and John Latham, considered English names quite sufficient for their large descriptive books on birds. That was shortsighted and arrogant, but when Latham noticed that he had acted against his own interests, and in his *Index ornithologicus* of 1790 Latinized all his earlier names according to the Linnaean prescription, he was too late; the sedulous compiler Johann Friedrich Gmelin, in Göttingen, had beaten him by a short head with the thirteenth edition of Linnaeus's *Systema naturae* (Leipzig, 1788–1789), edited by himself, and was recognized henceforth, according to the Linnaean law now accepted also by the English scientists, as the author of the new species instead of Latham, although all he had done was to translate the English text accurately into Latin. We learn by example; under the influence of the Linnean Society, founded in London in 1788, the new species described by the French "anti-methodists" were treated in the same way, as a cheap ticket to immortality.

The long-standing French opposition to the Linnaean system had quite different motives. France was the home of the little monkeys who, when Linnaeus forced his way into Nature's dark forest, "jumped onto his shoulders and bared their teeth at him with horrid laughter." In 1758 everyone recognized Linnaeus's grim caricature of the famous Buffon and his followers.

It was fortunate for the development of ornithology that it was not abandoned solely to the influence of the Linnaean spirit, but at the same time received help from quite a different, in fact an opposite, direction. Linnaeus's greatest antithesis, Buffon, born in the same year (1707), was by taste and intellectual inclination a man of the world, and was convinced that the "nomenclateurs," far from doing the natural sciences a service, had harmed them seriously by placing classification in a false light, distracting the attention of scientists from their true tasks with dry systematics, and turning it toward unimportant and even quite arbitrarily chosen matters. Buffon was concerned to survey the great relationships in nature in a bold intellectual flight, and to fit his purpose he made use of everything that philosophy and astronomy, physics and chemistry, geography and meteorology, anatomy and physiology, had to offer. Unhampered by any of the childish piety that hindered Linnaeus during his entire life, he was skeptical of all traditional doctrines and worked at a new theory of the development of the earth and of life. For him species were not in the least shaped by God's hand according to His design for creation, and therefore clearly distinguished and unalterable. He was much more inclined to suspect the existence of a number of fundamental types and their gradual multiplication through cross-breeding, "degeneration," and the local influences of food and climate. The smaller the species of bird (as he says in the introduction to his *Histoire naturelle des oiseaux*, 1770), the larger the number of its closely related species, so that one can assume that they all came from the same ancestor.

> A sparrow or a warbler has perhaps twenty times as many relatives as an ostrich or a turkey; for by the number of relatives I understand the number of related species that are sufficiently alike among themselves to be considered side branches of the same stem, or at least ramifications of stems that grow so closely together that one can suspect they have a common root, and can assume that originally they all sprang from this root, of which one is reminded by the large number of their shared similarities; and these related species probably have separated only through the influence of climate, food, and the procession of years, which brings into being every realizable combination and allows every possibility of variation, perfection, alteration, and degeneration to become manifest.[12]

The faster diversification of small birds is favored because they breed more rapidly than the large ones and have more young; and the larger the number of closely related species, the better their chances, according to the laws of mathematics, for a further increase through crossbreeding.

For this reason, says Buffon, considerable difficulties are met with in the differentiation of species, especially among small birds; many seem to him to be simply "varieties." He illustrates this with the example of the shrikes (*Oiseaux* I, 1770). Some scientists distinguish the German great gray shrike (*Lanius excubitor* L.) as a separate species from the French lesser gray shrike (*Lanius minor* Gmelin), although the difference is almost exclusively a matter of size, which could well arise simply from its feeding habits or, more accurately, from plenty or dearth in its habitats. Therefore it is not surprising that the European gray shrike is even more varied in more distant regions such as America, Africa, and India. The Louisiana gray shrike (*Lanius ludovicianus* L.) is almost identical with the European; the gray shrike from the Cape of Good Hope (*Lanius collaris* L.), that from Senegal (*Tchagra senegala* L.), and the blue vanga from Madagascar (*Leptopterus madagascarinus* L.) are three further very closely related varieties which all belong to the same species, that of the European gray shrike. Even the barred antshrike from Cayenne (*Thamnophilus doliatus* L.) can "on solid grounds" be regarded as a climatic variety of this widely distributed species.

Such a grotesque application of an Aristotelian theory revealed to every thorough researcher that the impressive verbiage concealed a brilliant superficiality. As early as 1776 Sonnerat ventured a criticism. Every island in the Philippine archipelago possesses one or more species of parrots peculiar to it. But these islands are evidently pieces broken off from a continent and separated during the rotation of the earth.

> Are these species now held to have died out on some islands and survived on others? If so, on what plausible ground can this assertion be maintained? Must one take refuge in the theory of climatic and dietary influences? In places so near each other, under the same sky, where the same fruits are found, these influences are insufficient to produce the changes that one would like to assign to them. Therefore another explanation must be sought, which I leave to the discussions of the natural scientists.[13]

Anyone who believed Buffon's assertion without testing it would form absolutely false zoogeographic notions and would misunderstand the Linnaean concept of species. An early victim of this hypothesis was the Italian Jesuit G. J. Molina, who in his *Saggio sulla storia naturale del Chili* (Bologna, 1782) seriously considered many Chilean birds to be merely varieties of European species.

Such an irresponsibly mounted attack on the doctrine of the stability of species was doomed to collapse harmlessly. As in other fields, Buffon's impatience had here charged much too far beyond the limits set by experience. But his raw material was very meager; when he began to publish his *Histoire naturelle* in 1749, the Cabinet du Roi, whose curator he had been for ten years, contained only about 60 to 80 birds, most of which had suffered badly from insect pests because no one knew how to protect them.[14] Only after the accession of Réaumur's treasures did the King's Cabinet become far the largest collection in the world. In 1770 it contained 700 to 800 species of birds, though most of these were represented only by a single specimen.

As Buffon wrote at the time, one really ought to collect a male, a female, and some nestlings in order to become properly acquainted with the species. Such an ideal was, however, unattainable. One need only imagine that, if there are 2000 species of birds, 8000 individuals would have to be collected, to realize the impossibility of establishing such a vast collection. Further, it would have to be more than doubled if all the varieties were also to be taken into account.

To share Buffon's belief that he had observed transitions in many parts of the animal kingdom, not only among groups but among species, meant that one was forced to condemn Linnaeus's hierarchical nomenclature. Exaggerating a good idea, Buffon maintained: "In general the proposition must be accepted that the more one multiplies subdivisions among objects in natural history, the closer one comes to reality, for in fact only individuals exist in Nature, whereas genera, orders, and classes occur only in our imagination."[15] Therefore, in his horror at "the confusion with which the mania for methods has filled Natural History," he went so far that, without seriously trying to arrange the groups according to their probable relationships, he placed them in his first volumes of the *Histoire naturelle des oiseaux* in order of interest to the general public.* At the beginning came birds of prey, ostriches, bustards, and gallinaceous birds. He casually dropped "L'Agami" (*Psophia*) and "Les Tinamous" in between the antbirds (Formicariidae) and flycatchers in the fourth volume. But that was not enough; as an enemy of formalism in nomenclature he chose his specific names as unsystematically as possible — they were intended to be not instructive, but amusing. He called the species belonging to his "Fourmiliers" (antbirds) Le Roi des Fourmiliers, L'Azurin, le Béfroi, la Palikour, le Colma, le Tetema, and so on, and did not bother about Linnaean names, even emphasizing that the description of behavior ("l'histoire") is much more important than nomenclature. Before anyone can hope to solve the problems of relationship it will be necessary "to write a separate natural history for the birds of each country." The strict methodists

*As he had already done in his *Histoire naturelle* (1749 ff.)

might well find such words distasteful; they still agreed with Artedi that the search for external diagnostic characteristics is "the prime and sole aim of natural history."

But what a stimulus this was for alert observers of bird behavior, whom Buffon now encouraged in their activity and who gladly made their observations available for his great work! Four of his contributors were particularly distinguished for their accurate information: Dr. A. G. Lottinger in Saarburg, Emmanuel Baillon (d. 1802) in Montreuil-sur-Mer, L. Deshayes in Santo Domingo, and the young Charles Nicolas Sigisbert Sonnini de Manoncourt (1751–1812), who at Buffon's urging cheerfully took a job on the island of Cayenne where, like Pierre Barrère 50 years earlier, he could study tropical animals. The disciples of Linnaeus traveled to distant countries in order to add to his museums or herbaria, but Sonnini, as a pupil of Buffon, returned after three years (1772–1775) with a great deal more than the skins of 160 mostly undescribed species of birds; he had taken careful notes on the habits of many of them and was able to make valuable contributions to the last part of the *Histoire naturelle.* In fact, he was a model for later scientific travelers.

There was also a foreigner who considered it an honor to gratify the famous Buffon with souvenirs of his journeys. He was the great Scottish adventurer James Bruce, who, after staying several years at the Abyssinian court in Gondar, returned home by way of Paris in 1773 in order to present the Cabinet du Roi with pictures in color of rare Abyssinian birds, which he had managed to preserve among all the dangers encountered in crossing Nubia.

The ornithological volumes of the *Histoire*, supplemented by 973 colored plates, "Planches enluminées" (1764–1783), which F. N. Martinet had painted under the supervision of the younger Daubenton, make an impressive array; the first appeared in 1770, and the tenth and last in 1783. All species then known are included. Buffon's share remained small; it was good that he had entrusted his friend Guéneau de Montbeillard (1720–1785) with carrying out his plan, because the latter was less given to speculation and more to analysis and study of the literature. After completing the sixth volume (1779) he retired in favor of a still abler editor, the Abbé Gabriel Leopold Bexon (1748–1784), to whom the highly renowned master gladly resigned the "boring" waterfowl, "these melancholy water birds — one does not know what to say about them, and their number is overwhelming!"

For a long time "Buffon," either in the original or in translation,[16] exerted an unparalleled influence and was devoured by young and old. The elegance of its language is enchanting, but its greatest value lay in the revival of an appreciation of living things in their natural environment. This spirit spread rapidly everywhere, above all in Italy, where the Jesuit Francisco Cetti (1726–1780), an admirable naturalist and able writer, filled his little book on *Gli Uccelli di Sardegna* (Sassari, 1776) with accurate and detailed biological observations. And even though Buffon's work is not free from considerable defects, "Sunt delicta tamen, quibus ignovisse debemus" [They are nevertheless failings that we ought to overlook], a Horatian tag that seemed apt to Boddaert as early as 1783.

Many ornithologists at the time thought that they had to choose between Linnaeus and Buffon, because the contradictions seemed impossible to resolve. For a while the flock of Linnaeans, the orthodox "nomenclators," were more influential than the opposition which, after first finding a famous champion in François Levaillant (the main subject of Chapter 5), later wearied of the noisy arguments, its members quietly contenting themselves with the results of their biological research, and not troubling the ranks of the systematizers, whose means of communication they took over without protest.

Pallas was the first zoologist who tried to synthesize the two schools, but in 1788 (the year of Buffon's death) a farsighted author appeared who, influenced by Pallas's work, foresaw their final amalgamation in the true and proper form of natural history, and urged it for the sake of scientific progress. Blasius Merrem, a zoologist and pupil of Blumenbach from Göttingen, was an important thinker. Zoology would have made great strides if it had received his undivided services, but difficult circumstances and the need of money forced him to scatter his considerable gifts. He was born on February 4, 1761, in Bremen, the son of a merchant. At first, wishing to become a theologian, he studied Arabic as well as Greek and Latin. But when he was 15 his father forced him into commerce. After enduring this for a year he could not bear it any longer, and enrolled as a student in the *Gymnasium illustre* in Bremen. Here he concentrated on natural history, after his uncle, J. P. Berg, professor of theology at Duisburg, had bought him Müller's book on Linnaeus, and he himself had acquired Linnaeus's own works. It is characteristic of Merrem that his critical sense was immediately aroused; while he was still in Bremen he drew up a new system for mammals, which in his own words was "at least no more unnatural than the old one." In the autumn of 1778 he entered the University of Göttingen, undecided on his choice of faculty. At first he tried medicine, but gave it up because he could not bear to watch operations. Young Blumenbach, who at the age of 24 had become professor of medicine at Göttingen (1776), converted him completely to zoology, and helped him so much that in 1780 he was already lecturing at Göttingen on Leske's zoology. In July of the same year he received his doctorate. The admirable *Vermischte Abhandlungen aus der Thiergeschichte* [Various essays on zoology] (Göttingen, 1781) had been written while Merrem was still a student. He was by this time specially interested in ornithology, chiefly anatomy. Stimulated by Pieter Camper's and John Hunter's publications on the pneumaticity of avian bones, he injected wax through the trachea and discovered the air-sac system ("Über die Luftwerkzeuge der Vogel" [On the air-apparatus in birds], *Leipziger Magazin*, 1783).

The fragments of his planned large-scale monograph, *Versuch eines Grundrisses zur allgemeinen Geschichte und natürlichen Eintheilung der Vögel* [An attempt at an outline of the general history and natural classification of birds], though published at Leipzig in 1788, had been written at Göttingen, because only with the famous university library at hand could he have essayed such a comprehensive work, prefaced by a history of ornithology. Merrem intended "to provide Germany with something so complete on one class of animals as no other nation could boast of," but "German naturalists and their friends did not support my scheme, my overmastering passion, on which I would have spent all my powers, employed all my energy, and spared no effort . . . For most readers of natural history books want them as wall decorations, because of the colored pictures." So the book got no further than some highly promising sample installments that reached only a few people. In financial straits, Merrem had to look for a different job and, to add to his chances, began in 1783 to study law in Göttingen. Finally, in 1785, he was awarded the professorship of mathematics, physics, and political economy at Duisburg, for a miserable salary. This forced him to take up subjects that had hitherto lain beyond his field of interest, and left him hardly any time for the prosecution of his zoological studies. When he was at last called to Marburg, in 1804, as "Public Lecturer in Economics, [and] Political and Fiscal Economy," he was already worn out by his money troubles and broken by illness, so that his later achievements did not nearly fulfill the promise of his youth. He died in Marburg on February 23, 1824, as director of the Zoological Institute that he had founded.

In the preface to his *Versuch*, he argued:

> What path must the true student of nature choose in order to acquire a thorough scientific knowledge of natural bodies? Natural history, as its name teaches, should be a branch of historical knowledge. Observation and experience, therefore, not ratiocination, must be the foundation on which it rests. The manner in which the history of single persons, taken individually and in association with others, the history of a state, and the history of all states, territories, and peoples, in association with one another constitute a general history

of the world, must be the manner also in which knowledge of single kinds of creatures constitutes the foundation of all natural history, and these histories in association with one another must constitute a general natural history. Just as it is unavoidably necessary for the true and genuine student of history to omit nothing that may give the character of a single person, so that no trait will be missing that could influence the progress of history, because he dare not overlook even the smallest event, for often the things that seem least noticeable produce the most important alterations in states: so little also dare the student of nature leave unnoticed the smallest and seemingly most unimportant properties of the creatures that are the object of his examination, for only a more precise knowledge of nature than we yet possess must teach us what properties are more essential, more important than others, and more intimately associated with the nature of the objects. As the historian must labor to explain the actions of men from the formation of their spirit and the directions that it takes, so the naturalist must seek to deduce from the entire inner and outer structure of bodies the properties thereof. Thus it is not sufficient to enumerate colors, and to describe in only a few words the form of some few outer parts, and the numbers of teeth, rays, stamens, and pistils; plants, animals, and minerals must be examined and described not from their outer parts alone but with the greatest possible exactitude from their entire formation, yes, even from the entire inner structure of their body. Everything about their habitats, growth, increase, and characteristics, and in addition for animals everything known about their behavior, habits, and instincts must be most carefully noted, their bodies most accurately observed in various situations, with their behavior in them, and any gaps in knowledge must be filled. That is the only way in which we can assemble material for a preliminary outline of a general science of nature, in which by comparing many objects the more important characteristics are separated from the unimportant, and thereby the nature of closer resemblances and wider differences among the multifarious species of creation can be demonstrated. Until then I make not the slightest reproach against mere description of the exterior of an object, whenever nothing more thorough is possible, so long as it is really complete, and represents the entire constitution of the outward form according to its structure and the relations of the parts to one another. In my contributions I have done this myself; I consider it useful because it extends our knowledge, though usually only a little; but it should not be regarded as the essential part of natural history, and one should not, instead of representing the object as a whole, whenever possible, with all its inner and outer

> parts and its characteristics and relations, merely trifle with the description of its colors . . .
>
> Certainly every naturalist must agree with me, that this science would profit extraordinarily from such a method as I have here recommended. But will this ever be possible? I truly believe so, but the work must not be undertaken by one person alone. Each one must choose a single class, or a single order, and through such a compilation we shall obtain the most complete account of all objects in natural history. A true connoisseur and judge of texts in physiology and natural history, who is at the same time himself a great physiologist and naturalist, ought finally to extract the results from all these individual labors, and in this way, I believe, we might easily obtain a general physiology and history of all known natural objects.

Exactly 100 years later the "great physiologist and naturalist" appeared, to carry through the "natural classification of birds" according to these principles. He was Max Fürbringer. However, the labor of generations was needed before this could be carried through to permanent success.

4

Scientific Expeditions

Between 1767 and 1795

Though Linnaeus, Brisson, and Buffon differed in their concepts of the goals and methods of natural science, they agreed in their endeavor to make an inventory of animals. Their lists enabled collectors to put their cabinets in order and to separate known from as yet unnamed species. According to the catalogues, over 1500 species of birds had now been discovered. But as the sum of knowledge grows, so does the fascination of novelty. Encouraged by learned and ignorant alike, an outdoor, instead of merely an indoor, hunt for *novae species* started off, with romantic expectations. No longer did explorers take only astronomers, cartographers, and surgeons with them; from now on the naturalist, as collector and describer of plants and animals, joined them on an equal footing.

This new and most profitable period for ornithology began about 1767, with an enterprise launched by Louis XV (to be discussed later). The French Revolution and its subsequent wars held up further development for more than 20 years, until immediately after Napoleon's overthrow it proceeded all the more vigorously.

Carried away by the spirit of the times, and spurred to emulation by the French example, the English and Russian rulers decided to equip expensive expeditions, promising themselves profit not only for trade and commerce but also for the natural sciences. The dates of the expeditions were fixed by the transit of Venus, expected on June 3, 1769, because observations from different parts of the earth were intended to help in making new determinations of the parallax of the sun. Venus therefore decided the careers of such men as James Cook, Joseph Banks, Reinhold Forster, and Pallas.

The first of this illustrious company to be celebrated should be he who contributed so greatly to the development of ornithology, as to other branches of zoology: Peter Simon Pallas.

Pallas was born on September 22, 1741, in Berlin, the son of a professor at the Collegium Medicochirurgicum and chief surgeon of the Charité Hospital. Growing up at home in happy circumstances, and taught by intelligently chosen tutors, he soon revealed his gifts by unusual proficiency in Latin, French, and English. At 13 he showed such promise as a scholar that from 1754 onward his father allowed him to attend the lectures at the Collegium, those on surgery being given by his father, on botany by Gleditsch, and on anatomy by J. F. Meckel, the elder. He was already interested chiefly in animals; he experimented on insects, and in 1756 or thereabouts he tried his hand at a system of his own for classifying birds. In 1758, before he was 17, he had finished his anatomy course in Berlin and went to the universities of Halle and Göttingen for further study. In 1760 he proceeded to Leiden, where he defended a thesis on internal parasites, which he had written in Göttingen. This was a pioneering study in helminthology, in which among other matters he demonstrated that Linnaeus had wrongly organized the entire class of worms. In July 1761 he went from Leiden to London for nine months, and then returned home, because Professor Pallas wanted his son finally to start practice. His father's plan to send him to the front as an army doctor fell through when Prussia and Austria ceased hostilities in November 1762, so that the young doctor was allowed to go back to Holland. In August 1763 he arrived in The Hague, where he stayed for almost four years, supporting himself by scientific writing and technical advice to rich collectors of natural history specimens (like Abraham Gevers and Dr. Voet). He was almost the first zoologist to recognize the methodological significance of Linnaeus's binomial nomenclature, which he employed consistently from 1764. This evidently pleased Linnaeus greatly, because in 1765 he called Pallas "a most acute young man and remarkably well versed in entomology, ornithology, and the whole of natural history."[1] Pallas also created this impression in person. "Doctor Pallas, a very ingenious young man from Berlin," wrote Pennant in his diary, after meeting him on July 30, 1765, in The Hague.[2] In his autobiography (1793) he noted:

> I esteem my meeting with Doctor Pallas, at the Hague, a momentous affair, for it gave rise to my synopsis of quadrupeds . . . From con-

> geniality of disposition we soon became strongly attached. Our conversation rolled chiefly on natural history, and, as we were both enthusiastic admirers of our great Ray, I proposed his undertaking a history of quadrupeds on the system of our illustrious countryman, a little reformed.[3]

In reality Pallas was then chiefly concerned with mammals, especially the new and rare species kept in the Stadtholder's zoo. What attracted him to investigate them more closely was more than their outward appearance and the search for "diagnostic" features that had generally satisfied Linnaeus and his school. For Pallas, the study of an animal included the results of careful dissection and the conclusions drawn from comparison of its structure with that of other animals. In his own words,

> the natural genera, if they are to be well founded, are to be judged according to the sum of resemblances and the clue provided by the relationships among the species. One should not hope to make a consistent, infallible description of generic characteristics in one or two sentences, which will almost never remain without exceptions . . . Nature . . . does not tie herself down to an artificial school system and diagnosis of species; she wants to be observed, not hemmed in[4]

—a criticism that could not have been very agreeable to the "General" of the naturalists. In 1767 Pallas started publication of a collection of monographs on mammals and birds which he called *Spicilegia zoologica*, or "Gleanings in zoology," as if he wanted to make clear that Brisson and Daubenton had not left much for him. There was no reason for such modesty; K. A. Rudolphi in 1812 praised the writer of the *Spicilegia* by saying: "We have no author who gives better and more lucid descriptions of animals . . . He ignores everything superfluous, includes no foreign subject matter, and is precise without being discursive."[5]

Although he was now widely known among naturalists through his publications, Pallas did not succeed in finding a post in Holland. And a plan with which the young Prince of Orange tantalized him, a project for a scientific expedition to the East Indies, came to nothing. Finally his father lost patience, and in the spring of 1767 demanded that he return to Berlin to start medical practice. In the midst of this anxiety he received an important letter: Catherine the Great called him to St. Petersburg for a large and honorific task. After a little indecision he accepted, and in July 1767 traveled from Berlin to Kronstadt by way of Lübeck.

At the St. Petersburg Academy of Sciences, of which he had been appointed a member, he found an entire staff of scholars already collected to advise on important matters, because the empress had decided to take the opportunity provided by the transit of Venus to promote many branches of science, and to send out not only astronomers but all the naturalists in the Academy on scientific journeys to different parts of her great empire. There were Professor Johann Peter Falk, born in Sweden in 1727, Ivan Lepechin from St. Petersburg, Johann Anton Güldenstädt from Riga, and the 23-year-old botanist Samuel Gottlieb Gmelin from Tübingen, who in 1765 had stayed briefly with Pallas in The Hague and had arrived at the Academy only a few months before. Pallas had to make all kinds of preparations for the journey, which was scheduled to begin in the summer of 1768, but even so he could not resist undertaking a detailed study of the Academy's zoological collections. Among them was the remnant of the spoils brought back by his famous compatriot, Georg Wilhelm Steller, from Kamchatka and Bering Island (1741–42). At last they were subjected to expert study, and illustrated descriptions of the five most striking new species of birds were prepared for publication in the *Spicilegia zoologica.*

In June 1768 Pallas and Gmelin followed Lepechin and Güldenstädt, who had been sent off with only a short interval between them. Later Johann Gottlieb Georgi, from Stettin, and Carl Ludwig Hablizl, a student from Königsberg, joined Güldenstädt. Gmelin, like Güldenstädt, was supposed to cover the southern districts, Pallas those farther east. Before winter began Pallas had reached Simbirsk (Ulyanovsk); in 1769 and 1770 he explored the slopes of the Urals and the valley of the river Ural, and in 1771, sometimes in the company of the students Suyev and Sokolov, he got as far as Krasnoyarsk. In 1772 he went through Kyakhta as far as northern Mongolia and the Amur River area. In 1773 he turned west again from Krasnoyarsk and wintered in Tsaritsyn (Stalingrad now Volgograd), on the Volga. "Exhausted in body and gray-haired at thirty-three,"[6] he reached St. Petersburg again on July 30, 1774. The Academicians whose expeditions had been studying geography and the natural sciences had been ordered back in great haste by the empress, because the Pugachev rebellion was spreading on all sides. But not everyone who had gone out in 1768 managed to get back. Falk had shot himself in Kazan in 1774; Gmelin had met a pathetic end on June 27, 1774, as a prisoner of the khan of the Kaitaks in Dagestan; and Hablizl was still in distant Persia, in the Gilan mountains, absorbed in zoological research.

Only Pallas possessed the diversified scientific training and enormous energy for work demanded for the evaluation of all the various results, which was the chief concern of the highly cultivated empress. To avoid the danger of her scientific explorers' meeting an early death and taking the wealth of their experiences to the grave with them, she had ordered them at the beginning to prepare their manuscripts during their stay in winter quarters, and to send them on immediately to St. Petersburg. When Pallas reappeared there he had already become an international celebrity, because the empress had had his reports printed, as *Reisen durch verschiedene Provinzen des Russischen Reiches* [Travels through various provinces of the Russian empire]. The first volume, closing with October 1769, came out as early as 1771, and the second, written during the winter of 1771–72 in (Novi) Selenginsk, came out in 1773, both issued by the Press of the St. Petersburg Academy. When he brought back the manuscript of the third volume (published in 1776), it was almost ready for publication. This account of his travels was extraordinarily important for all the sciences, including zoology and zoogeography. Already in the first volume Pallas, without any literature available, had named six new species of birds and had described them with his usual care; to these new discoveries he added others in the second (Selenginsk) volume, even furnished with admirable illustrations, which must have been made by an artist traveling with him. That these species really were all previously unnamed proves the thoroughness of his ornithological knowledge.

As early as 1770, when he discovered surprising animals in the lower Volga region, Pallas resolved to write a "circumstantial account of the animals of Russia and northern Asia." Beyond all the tasks heaped on him soon after his return to St. Petersburg (concerned not only with botanical, paleontological, and ethnological subjects, but also with the compilation of a polyglot dictionary of the languages in the Russian empire, which was of particular interest to the empress), this remained his favorite project. In addition to his own material he had at his disposal the collections, descriptions, and publications of the other men who had set out with him in 1768. He could also use the writings of the eccentric Daniel Gottlieb Messerschmidt, who at Peter the Great's command had traveled through Siberia for seven years, from 1720 to 1727, as far as the river Lena; added to these was the legacy of the unfortunate Steller, who had died on his return journey to St. Petersburg. This was a great deal, but it was not enough for Pallas. In particular, the animal population of the most easterly part of the empire, where Steller had been a pioneering naturalist, had been insufficiently investigated. When the geographic discoveries made by Cook and his men in 1778 and 1779 in the Bering Sea area (well within the Russian sphere of influence) became known, Pallas was chief among those who urged the empress to quick action. He recommended Joseph Billings (an Englishman who had been one of Cook's companions on his last voyage) as leader of a Russian expedition to explore the sea between

Asia and America and to map its lands and coasts. He also saw to it that a naturalist was finally found for the party, a young doctor born in Darmstadt, Carl Heinrich Merck. He was not well suited to the job, but he did his best to meet the expectations of his great model, Pallas, and he collected and described for him everything that seemed important during his eight years' journey, 1786–1794.[7] In that way many birds from the Far East reached Pallas — over 200 specimens of 120 species. Among these were 20 new species, mostly from Okhotsk, Kamchatka, Unalaska, Kodiak, and Prince William Sound.

By now matters had advanced far enough for Pallas to begin his great work, *Fauna Rossica*, a title that he later changed to *Zoographia Rosso-Asiatica.* He went on one more scientific expedition, 1793–94, in the southern provinces that had been incorporated in the empire ten years before. Then in August 1795, taking part of his ornithological collection with him,* he retired with his family to the idyllic country seat in the Crimea that the grateful empress had presented to him. There he worked for the next 15 years almost exclusively on his *Zoographia.* He wished to produce in this labor of his later years a real natural history of the animal kingdom, "not a meager treatise, like others, with a thin skeleton of names and synonyms and a thickly encrusted hide of observations and descriptions, but a full and copious work, so constituted as to serve as a universal zoology."[8] In addition, it was supposed to be illustrated with 210 plates. These plates, however, worked a curse upon publication: Pallas, constantly put off and finally shamefully left in the lurch by the Leipzig engraver C. G. H. Geissler, did not live to see his work appear. Accompanied by his widowed daughter, he left the Crimea in 1810 and moved back to his native Berlin, in order to expedite from near at hand the completion of the plates. But shortly after receiving from St. Petersburg the final proofs of the first volume and the beginning of the second for correction, he died, on September 8, 1811. Another 16 years had to elapse before the first copies of the work were made public and could contribute to the progress of ornithology.†

For decades the *Zoographia Rosso-Asiatica* remained a much admired and much used book. It not only provided zoologists with a vast quantity of facts about morphology, variation, distribution, and habits, but also influenced their theoretical views. For at the head of all his

*Pallas had expressly reserved these specimens for himself when he sold his collection to the empress. See *Pallasia,* 3 (1925), p. 11.

†Not till 1827 did the St. Petersburg Academy of Sciences release the three volumes (printed 1811–1813) for sale, thus finally effecting publication; see Erwin Stresemann, "Date of Publication of Pallas's 'Zoographia Rosso-Asiatica'," *Ibis,* 93 (1951): 316–318. The plates, only 48 in number, were issued in six fascicles, 1831–1842.

investigations Pallas set the search for general principles: "Thus the complexity of the whole of nature is to be so set forth as to elucidate the order of things and the general laws of creation." Following these precepts, he worked out a skillful synthesis of Linnaean and Buffonian tenets; from Linnaeus he took over the general systematics, including binomial nomenclature, and from Buffon the dynamic outlook and the idea that species are mutable, particularly under climatic influence — a conception of which the birds in the vast open spaces of Eurasia apparently produced many striking proofs. In contrast to most of his contemporaries and many who used his systematics for decades after him, Pallas realized the necessity of distinguishing between true species and their climatic variations; like Cuvier (1798) he used two names only for the former, and decided to distinguish the latter as varieties by calling attention to their differences and their geographic distribution, without giving them names, even when their distinguishing characteristics were extremely noticeable. ("I have not neglected any varieties, which are of the greatest importance in zoology.") Therefore he, along with Cuvier, should count as one of the first to form ideas capable of future development about the nature of species and their "climatic races." Very soon after the *Zoographia* finally appeared, C. L. Gloger appropriated these ideas of the great scientist and, as author of a very popular little book on the mutation of birds under climatic influence (Breslau, 1833), diverted them into a blind alley.

While Pallas was acquiring new laurels by his discoveries on the Mongolian borders, the botanist Joseph Banks (1743–1820) decided to travel round the world on board the *Resolution*, with the great navigator James Cook. Banks had joined Cook for the first time (1768–1771) at his own expense and had observed the transit of Venus on Tahiti. Now, in May 1772, he considered himself ready for a still more ambitious project. His friend Dr. Daniel Solander, a pupil of Linnaeus's and a member of the British Museum staff, was to accompany him again, but Banks intended to take not only Solander but also two other naturalists and three draftsmen. However, he had formed a completely mistaken idea of the size of the *Resolution*, and when he entered the narrow cabins reserved for the naturalists, which were hardly large enough to sling the hammocks in, he was appalled. The immediate sequel was a quarrel with the Admiralty and with the king's powerful minister, the Earl of Sandwich, who was more than ready to flout the pretentious herb gatherer, and brought all further discussion to an abrupt end. Using a fund of £4000 at the Admiralty's disposal, he quickly engaged two Germans for the expedition, one as naturalist and the other as draftsman, and left the dumbfounded Banks out in the cold.

The two unexpectedly lucky men were Johann Reinhold Forster (1728–1798), the learned pastor of the little village of Nassenhuben near Danzig, and his eldest son Georg (1754–1794), who had always been inseparable. As early as the summer of 1765, when Forster had been sent by Count Orlov from St. Petersburg to the lower Volga area, to survey the land intended for German colonization by Saratov's government, and to report on conditions among the existing settlements, he had been accompanied by Georg. The 11-year-old's passionate enthusiasm for natural history collecting, stimulated by the eleventh edition of Linnaeus's *Systema naturae* (1760) which his father had earlier brought him from Danzig, infected the father too, so that now in southern Russia they liked nothing better than to collect together and identify plants and animals, But disappointment followed disappointment. In St. Petersburg Forster waited in vain for the expected recognition of his careful report, and when he finally went home in disgust he found that his parish had been assigned to someone else. To escape impending ruin he set off with Georg in June 1766, to try his luck in the British Isles, the land of his forebears (his ancestor George Forster had emigrated in 1630 from Scotland to a part of Prussia then Polish and had married a German woman).

His extraordinary knowledge of classical and many modern languages, whose groundwork was laid when he was a pupil at the Joachimsthal Gymnasium in Berlin, helped Forster greatly during his emigration, as did his practical experience in the natural sciences. A substantial treatise entitled "Specimen historiae naturalis Volgensis," published in 1767 in the fifty-seventh volume of the *Philosophical Transactions* of the Royal Society of London, soon won him the goodwill of leading scientists; in fact, it shed the first light on the fauna and flora of southern Russia, treated according to Linnaean systematics, and it identified 74 species of birds. One of them, the black lark (*Alauda yeltoniensis* Forster), was new; father and son had discovered it at Lake Elton (or Yelton). He was made a Fellow of the Royal Society, and his stimulating company was sought by men like Dalrymple the geographer, Pennant the naturalist, and Tunstall the collector. He reviewed the fauna of India, China, and the American colonies, and it was not long before he had compiled and arranged all facts known about them according to the Linnaean system, because he hoped that he would at some time be able to make an expedition himself to one of these regions, still so little known. Finally, he was even entrusted with the identification of the specimens in the collection of birds that Andrew Graham, at the suggestion of Daines Barrington, had sent from Hudson Bay to the Royal Society's museum.* It contained nine new species, which Forster described in 1772 with all the care of a Pallas.

*"This collection was entirely neglected, and was falling fast to decay, till of late years, when it was restored, or rather refounded, by the indefatigable pains of the Hon. Daines Barrington," Peter Brown, *New Illustrations of Zoology* (London, 1776), p. 2. In 1781 the Royal Society's collection was transferred to the British Museum.

He had to accomplish all this in the spare time left after his hard day's work. His linguistic knowledge helped him to eke out a living by translating into English, together with his son, French, German, or Swedish accounts of travels undertaken for the study of geography and natural history, such as Peter Kalm's *Travels into North America* (1770–71), J. B. Bossu's *Travels through . . . Louisiana* (1771), and Per Osbeck's *Voyage to China and the East Indies* (1771). He can hardly have expected, when he was translating Bougainville's *Voyage autour du monde, 1766–1769*, from the Paris edition of 1771, that he would ever follow this famous discoverer to Tahiti. Yet before a year was out the incredible happened, and the two Forsters were Cook's companions on board the *Resolution*.

According to Georg Forster, Cook had received orders from the Admiralty to use the summer months for exploring in the direction of the South Pole, but as soon as the season became too cold, stormy, foggy, or dangerous to return to the Tropic of Capricorn and determine more exactly the position of the islands already discovered, by means of modern astronomical instruments and new calculations. If he found no large land area, then he was to sail east, always as close as possible to the South Pole, until he had circumnavigated the globe.

On July 12, 1772, the *Resolution* and her smaller companion vessel, the *Adventure*, under Captain Furneaux, weighed anchor in Plymouth harbor, and three months later reached the Cape of Good Hope. Here, at Alphen (an estate near Cape Town), the Forsters met an enterprising and skilled Swedish naturalist, the young Dr. Anders Sparrman (1748–1820), who had traveled to the Cape on a Swedish merchantman in January 1772, with letters of recommendation from his teacher Linnaeus, and had already been there for several months, studying natural history and ethnology. Forster, who had realized in the meantime that he could not manage to perform alone all the duties assigned to him, welcomed this as a heaven-sent opportunity and immediately proposed to Sparrman that he should accompany the expedition with all expenses paid — a generosity made possible by the grant of £4000. No wonder that the Swede was almost speechless with surprise and delight at such an honor.

The further progress of this voyage round the world has often been described and can be only lightly touched on here. After a four-month journey through the southern Indian Ocean, keeping a course near 60° S. for a distance of 600 nautical miles, the *Resolution* reached New Zealand. In August 1773 Cook made Tahiti. Two months later he set out again for New Zealand in order to make another survey of Antarctic waters, and on the way discovered the Tonga Islands. Having searched in vain for the "large southern continent" in the South Pacific, he returned to his friends the Tahitians by way of Easter Island and the Marquesas. In June 1774 he set course again for New Zealand, visited the Tongas for the second time, discovered the New Hebrides, New Caledonia, and Norfolk Island, and finally set sail from New Zealand for Tierra del Fuego. In April 1775, in Table Bay, Sparrman took leave of his good friends and he spent almost another year studying South African natural history and traveling through the Hottentot and Bushmen regions as far as the Boer settlement of Bruintjes Hoogte, near the Kaffir border.

Those universal explorers, the Forsters, returned to London on July 29, 1775, with an incomparable haul of scientific specimens. Banks and Solander, Cook's companions from 1768 to 1771, had confined themselves almost entirely to botanical material. Because they were unskilled at preparation, they had given a number of birds and other animals to their draftsman, Sydney Parkinson, who had made watercolors of them, and therefore the little that is known of the birds of New Zealand, Tahiti, and New Holland (Australia) observed on this voyage rests almost entirely on pictures. J. R. Forster, on the other hand, whose genius led him to pay attention to everything of consequence, was an experienced preparator and dissector, and what could not be recognized from a dried skin or described in words was conveyed in many pictures by Georg's skillful brush. The enthusiastic explorer's dogged energy had achieved prodigious results:

> When we had crawled over a large part of the island from early morning to dusk, collecting plants, birds, fish, shellfish, and lithophytes, and had finally returned to the ship with our load, we often had no time after our hastily snatched meal to wash; for frequently we had to write down quickly the most important details until far into the night, so that the experiences that were still fresh in our minds should not be crowded out by new impressions. But even then it was a most irksome task to select the most important from all the plants and animals that there were to describe. Our limbs were weary from the strenuous expedition, our minds were dulled with fatigue, and our exhausted powers demanded the sweetness of sleep. Often I had to wash my eyes, which were closing in sleep, with cold water, or I even rinsed my face and my whole body in a cask of sea water, and after such a bath went back refreshed in body to my night's work, in order to make notes on everything that could not be postponed till the next day.[9]

If anyone possessed the gifts and experience to make an enthralling account of everything of scientific interest encountered on this voyage, it was the elder Forster. "Fully resolved," wrote his son,

> to complete the purpose of his mission and to communicate his discoveries to the public, and not allowing himself any time to rest

> from the fatigues which he had undergone, he inscribed and presented the first specimen of his labors to His Majesty within four months of his return.[10]

This was *Characteres generum plantarum, quae in itinere ad insulas maris Australis, collegerunt, descripserunt, delinearunt, annis MDCCLXXII–MDCCLXXV. Johannes Reinholdus Forster . . . et Georgius Forster* [Characteristics of the genera of plants collected, described, and illustrated by Johann Reinhold Forster . . . and Georg Forster on a journey to the islands of the South Seas in the years 1772–1775], with 78 plates, published in London in 1776. The narrative of the journey, the chief work, remained Forster's most particular concern, but the samples of it that he submitted to the Lords of the Admiralty did not meet with their approval. They preferred to have Cook's and Forster's diaries printed separately. Forster set to work again, but his second and third versions were, like the first, rejected out of hand by the tyrannical and capricious Earl of Sandwich.

> Perhaps through this opposition he was to be made to feel a foreigner, perhaps his mode of thought, in the few comments that he risked when making comparisons [between the customs of the Europeans and those of the Polynesian natives] was considered too philosophical and free, perhaps also the interest of a third person was concerned.*

In short, Forster was finally deprived of all reward and his right to over £1000 that was his by contract. Since this contract had mentioned only the father by name, the son now felt himself free to describe the journey himself, in *A Voyage Round the World* (2 vols., London, 1777).† Never before was a scientific expedition described more thoroughly, comprehensively, or thoughtfully than in this work, in which the literary talent of the 22-year-old Georg was joined to his father's scientific seriousness and knowledge of humanity. But the book's success could not rescue the great scientist; the quarrel with the Admiralty, the expenses of a large family, and his lack of forethought in money matters had driven him deep into debt, which in due course reached the sum of £1600. He fell into the hands of hardhearted creditors, who succeeded in restricting him to London, threatening him with debtors'

*This passage exists only in the German edition of Georg Forster's work, *Reise um die Welt*, I (Berlin, 1778), Vorrede.

†The following year the father succeeded in publishing in London miscellaneous notes relating to the voyage under the title *Observations Made during a Voyage Round the World, on Physical Geography, Natural History, and Ethic Philosophy.* More than two thirds of the work, which was dedicated to his fellow members of the Royal Society, was given over to "Remarks on the Human Species."

prison (in which he was apparently confined for a time in 1779), and no one knows how the matter might have ended if Georg, entirely devoted to rescuing his beloved father, had not gone to the Continent in October 1778, in order to raise a ransom. In February 1779, he succeeded in getting Zedlitz, the Prussian minister of education, to invite his father to Halle as professor of natural history, but only a collection of contributions, including some from princes of small German states (particularly Duke Ferdinand of Brunswick), finally released him from his London liabilities, and it was not until 1780 that Forster received permission to leave his unfriendly foreign abode to enter on his professorship at Halle. He was active there until his death, 18 years later, but with a deep-seated resentment that was to torment him for the rest of his life.

Removed from the close intellectual companionship that had so stimulated him in London, deprived of a respectable library and even of natural history collections, Forster was not able to develop his former vigorous intellectual flights within the narrow limits of a German provincial town. He had not managed to save and bring to Halle anything but his scientific notes; his collections were scattered in England, and his son's watercolors, made for the diary, had gone to the creditors, who sold them to Sir Joseph Banks. Without these important illustrations Forster undoubtedly did not wish to publish his *Descriptiones animalium*, which he had completed while still on his travels. Though it would have been of the greatest use to zoology, it remained a hidden treasure until Forster's books were bought after his death in 1798 for the Royal Library in Berlin. Long before this, though only superficially and with many mistakes, John Latham had described Forster's ornithological discoveries from stuffed or illustrated specimens, without mentioning the name of the man to whom they were due, and although every specialist who had looked at the manuscript in the Berlin library, starting with Illiger in 1812, had commented on its lasting significance, Lichtenstein waited till 1844 before printing it, with a bungled commentary.

Forster's contemporaries, and even posterity, had little enough gratitude for a scientific explorer who would have reached the same heights as a Pallas if he had had similar opportunities. This judgment is confirmed by the only two longer ornithological treatises published while Forster was in Halle (besides various proofs of his skill in translation), though they were probably written while he was in London. His "Historia Aptenodytae generis avium orbi australi proprii" [History of Aptenodytes, a genus of birds inhabiting the southern hemisphere] (*Commentationes* of the Royal Society of Sciences, Göttingen, vol. 3, 1781) and his "Mémoire sur les Albatros" (published in 1785 in the *Mémoires* of the Académie des Sciences, Paris) are two monographs that stand comparison with Pallas's *Spicilegia* and can rank even today as models of brilliant, thorough, and comprehensive presentation.

Though an adverse fate deprived Forster of the fruits of his labors, the work he had had to abandon proved profitable to others. At the end of the eighteenth century, his collection of Polynesian and New Zealand birds provided no small part of the original impulse for the zealous English efforts to develop exotic ornithology. Forster had given only a small number of the birds to the Royal Society, and the majority found their way into private collections, particularly the cabinets of Ashton Lever and Sir Joseph Banks, both in London.

Ashton Lever, born at Alkrington near Manchester in 1729, was one of those who, thanks to an inheritance from a wealthy father, are able to indulge to the full their passion for collecting. He started in 1760 with a cabinet of shells. Soon he plunged with continuing zeal into acquiring every kind of object, and when he moved to London in 1774 he took his museum of curiosities with him and housed it in its own building, Leicester House in Leicester Square, which was open to the public for an entrance fee. Nothing was more welcome to this eccentric than increasing the drawing power of his museum with zoological and ethnological items brought back from the travels of the famous Cook. Forster's souvenirs were the beginning. They were exhibited in a "Sandwich Room," as a compliment to the powerful minister, who rewarded Lever in 1778 with a knighthood.

In the meantime Cook had set out for the third time in July 1776 with two ships, the *Resolution* and the *Discovery*, with the particular aim of finding a passage beyond the Bering Strait, from the Pacific Ocean to Europe, either westward or eastward. The Earl of Sandwich would not allow any more naturalists, after he had had so much trouble with those unruly cranks, but luckily the ship's doctor, William Anderson, took over the scientific duties. He had been the doctor on Cook's second voyage and had learned some zoology, botany, and ethnography from Forster and Sparrman. Like his great commander, Anderson was not to return. Nevertheless, when the ships reached England again in August 1780, they brought back a notable scientific booty, including many unknown species of birds from the Kerguelen Islands, Van Diemen's Land (Tasmania), the South Seas, the newly discovered Sandwich (Hawaiian) Islands, the Pacific coast of North America, Unalaska, the Bering Strait area, and Kamchatka.[11] These treasures, too, were promptly divided between Lever and Banks. By 1784 the Museum Leverianum possessed 28,000 items and had become world-famous. But now it became evident that its owner had been too prodigal. To make some money, he received permission from Parliament to hold a lottery, with 36,000 tickets, at a guinea each — all blanks except one, which entitled the holder to the entire museum! On the day of the draw Lever had sold only 8,000 tickets and still had 28,000, but as ill luck would have it, the winning ticket was in the smaller pile, and was one of only two bought by a dentist, Mr. James Parkinson, The unhappy loser was apparently all too thorough in his efforts to drown his sorrows, and shortly after the event he died, on January 24, 1788, at the Bull's Head Inn in Manchester.[12]

Parkinson immediately tried to give new luster to the cheaply acquired Museum Leverianum. He transferred it in 1790 to a new and handsomer building, the Rotunda, near Blackfriars Bridge, and concentrated chiefly on enlarging the bird collection. In so doing he came into contact with English and foreign collectors and was advised by Dr. George Shaw, a zoologist at the British Museum and an enthusiastic Linnaean. That explains why the Museum Leverianum continued for a while to improve. H. F. Link, a much traveled professor of natural history from Rostock, visiting the English capital in 1799, was full of praise:

> The British museum contains amid a vast quantity of insignificant trifles, a few important specimens, but in the present state of science it is no longer instructive. On certain days it is shewn to those who had previously procured tickets. The Leverian museum may be seen for a trifle, and the collection there of stuffed birds, and viviparous animals, exceeds every thing of the kind I have seen. It is well arranged, and to each specimen is affixed the Linnaean name.[13]

But in the meantime the London public's curiosity had shifted to other objects, so that Parkinson's enthusiasm flagged more and more; he finally decided to auction off the entire contents of the museum in May and June 1806. One of the most persistent bidders was Leopold von Fichtel, on behalf of Karl von Schreibers (1775–1852), the new director of the fast-developing Imperial Court Cabinet in Vienna. He acquired many great rarities, including some of Forster's birds, a Mascarene parrot (*Mascarinus mascarinus*) and a white gallinule (*Porphyrio albus*) from Lord Howe Island, (both species now extinct), which are still in the collection. Other historic items went to the goldsmith William Bullock (the owner of the "Liverpool," a rival museum opened in Piccadilly in 1805), to Lord Stanley, John Latham, Thomas Pennant's son David, Edward Donovan, and other private collectors.

The history of the museum would not have been so thoroughly detailed here if it had not been so important for the development of exotic ornithology. In fact, when Pennant, Latham, and Shaw were trying to compile a list of known species of birds, they found their richest quarry in the Museum Leverianum, where the number of types of previously undescribed species quickly passed the hundred mark.

Thomas Pennant (1726–1798), the eldest of the three, brought out his richly illustrated folio, *The British Zoology*, between 1761 and 1766, and revived his countrymen's languishing interest in their native birds.[14] Soon afterward he turned his attention to birds from foreign countries, and the first results of his interest appeared in 1769 in his *Indian Zoology*, a slim folio volume with eleven colored plates of Sinhalese and Javanese birds. These plates and their text have a story of their own: they are the oldest documents remaining from a systematic study of the birds of Ceylon and Java. The groundwork was laid by Joan Gideon Loten (1710–1789), who had been the Dutch governor of Ceylon from 1752 to 1757, and in his last year of office had undertaken a journey to Batavia and Macassar, where he continued to work at ornithology. A Christian Sinhalese and skillful animal painter, Pieter Cornelis de Bevere, was commissioned to make watercolors of his specimens.[15] Loten then came to England in 1759 with 144 pictures (103 of birds), and showed them to George Edwards, who immediately used several for the illustrated book he had just begun, *Gleanings of Natural History*. Pennant was also briefly interested in them, in a dilettantish way, though for his *Indian Zoology* he used only a small selection made by Sir Joseph Banks and accompanied by a superficial text, before he turned his attention to other matters. He was now chiefly interested in the higher animals of North America and, following in J. R. Forster's footsteps, he collected the material for an American zoology. He had already made great progress when the astonishing spoils brought back in 1780 by Cook's companions from the northern shores of America and Asia changed his plan. As Pennant wrote in a foreword, an "American Zoology" would certainly be produced at some time by the new nation, whose independence had finally been recognized by England in 1782. His book now became *Arctic Zoology* (vol. I, 1784; vol. II [Birds], 1785).

Most of the unknown species that Pennant, thanks to the generosity of Lever and Banks, was able to describe in this significant work had been treated in the meantime by John Latham (1740–1837), who used the same sources. Latham, a successful doctor in the little Kentish town of Dartford, not far from London, had, like Lever, begun as a collector. They were already exchanging speciments in 1772. But whereas Lever was interested only in quantity, Latham channeled his curiosity into the study of Linnaean systematics. The large number of new species from J. R. Forster's collection that had reached Lever and Banks must have stimulated the scholarly Dartford physician to compile a list of all known species of birds, with brief descriptions. Although the latest edition of Linnaeus (1766) was only a dozen years old, it was already completely out of date. Latham used it as a framework and increased the number of orders from six to eight by adding doves and ostriches, though otherwise he made only minor changes in the Linnaean arrangement. However, the conservative Englishman thought the Latin names just as superfluous as had the rebellious Buffon. Latham's dogged industry produced *A General Synopsis of Birds* (1781–1785), in three volumes, which is chiefly important because Latham was the first to describe the 200 species that Cook's naturalists had brought back from his second and third voyages. It is, however, a mass-production job with every consequent failing; the descriptions are superficial and often so disfigured by errors as to be completely unrecognizable, and the locality data are unreliable. Later it was established that, when Latham identified the countries of origin of the birds "from our great navigators," he got a third of them wrong. It was not his fault; such errors continued to hinder the progress of science for a long time to come, until it became standard practice to attach a locality label to every specimen. Even at that time, however, there was a recognized procedure, as recommended by J. R. Forster in his instructions for field collecting in the *Catalogue of the Animals of North America:*

> To the quadrupeds, birds, reptiles, fish, and in general to all the specimens, must be fixed lead tickets by means of a wire, and a number on the lead scratched in; which must be referred to, in a paper, where under the same number the collector would be pleased to write the name by which the animal goes in his country, or among the various tribes of Indian nations, with the food, age, growth, nature, manners, haunts, how many young or eggs it brings forth, in what manner it is caught, what it is used for, etc., etc.[16]

But how often the purchasers, paying no attention to anything except the showpiece, must have thrown away such accompanying lists without thinking!

We have already learned (Chapter 3) of Latham's motives for issuing a two-volume summary of his *General Synopsis of Birds* in 1790, under the title of *Index ornithologicus*, in which he Latinized all his earlier names according to the Linnaean system. The 2921 species of birds appearing there were not all contained in the *General Synopsis* and its *Supplement* (1787), because it was only after 1787 that science opened up a new continent, New Holland (Australia), the first samples of whose birds enchanted the ornithologists.

The European settlement of this quarter of the globe began with 750 transported convicts, who were landed in Botany Bay on February 20, 1788. It had been so named by its discover, Cook, because his botanists Banks and Solander had made such a rich haul there in 1770. But soon afterward it proved so unhealthy that it was abandoned in favor of the large neighboring bay, Port Jackson (renamed Sydney Cove), and another group of convicts was sent to Norfolk Island, discovered by Cook in 1774. Barely a year had passed before the versatile chief of this enterprise, Governor Arthur Phillip, published a book with 54 plates of illustrations, wherein natural history had its place (19 plates of birds): *The Voyage of Governor Phillip to Botany Bay: with an Account of the Establishment of the Colonies of Port Jackson and Norfolk Island* (London, 1789). Latham, who had been engaged to work on the newly discovered birds, was able to make here his first contribution to Australian ornithology and, as his greatest sensation, to describe the emu (*Casuarius novaehollandiae* Latham). More novelties followed immediately, because the surgeon general of the new settlement, John White, was also interested in zoology, and illustrated his *Journal of a Voyage to New South Wales* (London, 1790) with 29 plates of birds. They were nearly all engraved from the stuffed specimens that the Leverian Museum had received through White from Sydney or Lord Howe Island, as the foundation of Australian ornithology. This time not Latham, but Parkinson's friend, George Shaw (1751–1813), described and named them, but ten years later his rival was able to outstrip him again. In the meantime White had encouraged several convicts (especially Thomas Watling, who had been sentenced in Scotland for coining) to paint for him pictures of birds accompanied by scientific data. These (or sometimes copies of them) he sent to A. B. Lambert and other correspondents of his in England, from whom they were borrowed by Latham as he was engaged in adding a second *Supplement* (London, 1801) to his *General Synopsis of Birds*, bringing it completely up to date. He discovered other unknown species of Australian birds in the possession of a passionate collector, Major General Thomas Davies, of Blackheath, who had just received them from the new governor, Captain Gidley King. Latham's new supplement was almost printed when he saw the most remarkable of all the Australian species, the lyrebird, and was just able to insert two leaves describing and illustrating it. This latest surprise, which Davies and Parkinson received almost at the same time in 1799, also came from the immediate environs of Port Jackson, for as yet no zoological collector had penetrated farther into New South Wales.

On the other hand, the zoological exploration of the southern coast of Australia had already begun. The British Government had provided Captain George Vancouver, on his great voyage of discovery in the North Pacific (1791–1795), with an able naturalist, Archibald Menzies. In the autumn of 1791, while the *Discovery* and *Chatham* were at anchor for 12 days in King George Sound, near the southwestern corner of Australia, he collected several new species, including the rare musk duck (*Biziura lobata*). From California, too, where just as little had been collected, Menzies brought back to London remarkable birds, including the California condor and the California quail (*Lophortyx californicus*), to the delight of George Shaw and John Latham, who soon set to work on this small bird collection in the British Museum.

During these years when Russia and England had harnessed natural history to trade, France was no mere idle spectator. As early as November 1766, Louis Antoine de Bougainville, a sailor with experience both in war and in peace, set out at the command of Louis XV in the frigate *La Boudeuse* for a voyage of discovery round the world. In June 1767, off Rio de Janeiro, his ship met the flute *L'Étoile*, whose doctor was the naturalist Philibert Commerson (1727–1773). After a stay in Montevideo, the two vessels sailed together through the Strait of Magellan to Tahiti (April 1768) and on by way of Port Praslin in New Ireland, Kajeli on Buru, and Batavia to the Île-de-France (Mauritius). There Commerson was left in November 1768 because he intended, with the encouragement of the island's governor, Pierre Poivre, to study the flora and fauna of the Mascarene islands and Madagascar. In the history of successful voyages of discovery this first French circumnavigation of the globe ranks high, but it yielded little for zoology. The versatile and conscientious Commerson would have made a name for himself if he had returned to France with Bougainville in 1768, instead of hazarding his life further. But after he had died in the Île-de-France in 1773, a victim of the tropical climate, his executor Buffon bothered very little with his collections and descriptions. Most of them decayed unused. One can count on the fingers of both hands all the new species, collected by Commerson in Montevideo, Tahiti, and Buru, that were illustrated by the younger Daubenton and soon afterward described by Guéneau de Montbeillard, in Buffon's large *Histoire naturelle*.

Pierre Poivre (1719–1786), who intervened so fatefully in Commerson's life, was noted earlier in these pages as a contributor to Réaumur's museum, to which he had sent many new species of birds from Cochin China (Annam), Manila, and Madagascar. He was originally a missionary in Canton, and then in Cochin China. When the ship on which he had embarked in 1745 for his homeward voyage, after four years of work, was captured by an English ship off the island of Bangka, he lost his right arm and became a prisoner. After he was freed he went to Batavia, and there learned of the lucrative trade in Moluccan spices that the Dutch had been vigilantly preserving as their monopoly for more than a hundred years. From that time on Poivre endeavored to obtain the same advantages for France. First, as an agent of the French East India Company, he returned to Cochin China in 1749,* and brought back to the Île-de-France various spice plants from Fai-Fo, in the bay of Tourane (Da Nang). To acquire nutmeg trees as well he went on a secret mission to Manila. He had so little success that he was forced to make the voyage again in 1754 in a small vessel. This time he was more fortunate, but the new "Directeur des Jardins" on the Île-de-France opposed his bold plans, and, after spending the winter of 1755–56 in Madagascar, Poivre finally went home to France in a huff. There, after some time, the importance of his undertaking was recognized, and he was sent out again as intendant to the Île-de-France and Île-de-Bourbon (Réunion) in 1767. Immediately he prepared to acquire clove and nutmeg trees in large quantities. The first expedition, which he sent in 1769 to the northern Moluccas by way of the Philippines, was not able to accomplish much. So in June 1771 a second expedition was dispatched, consisting of the royal flute *Île-de-France* and the corvette *Nécessaire.* A young nephew of Poivre's who was also a naturalist, Pierre Sonnerat (1749–1814), was allowed to join it. He had accompanied his uncle in 1767 to the Île-de-France and then had visited Madagascar in 1770 as Commerson's companion and factotum. Again the ships chose the clandestine detour by way of the Philippines (Manila, Antigua on Panay, and Zamboanga, from December 1771 to February 1772) in order to reach the northern Moluccas. Here, from March 9 to April 6, 1772, on the coast of the island of Gebe, where a quantity of the desired seedlings was obtained, Sonnerat was able to add some great prizes to the zoological collections that he had already begun on the Philippines, because the Rajah of Patani, and the envoy of the Rajah of Salawati, who visited the French ships, brought him the skins of birds of paradise, including those of some very remarkable and then unknown species, *Parotia sefilata*, *Manucodia chalybata*, *Lophorina superba*, *Epimachus fastuosus*, and *Diphyllodes magnificus.* On June 4, 1772, the expedition returned successfully to the Île-de-France with its valuable cargo. Soon thereafter Sonnerat accompanied his uncle back to France, where the new birds of paradise were illustrated by the younger Daubenton in his *Planches enluminées* and discussed by Guéneau de Montbeillard in Buffon's *Histoire des oiseaux* (1775), arousing general surprise and admiration.

Sonnerat gave a vivid account of this journey in his book, *Un Voyage à la Nouvelle Guinée* (Paris, 1776), describing and illustrating a whole series of birds that he had come across in Manila, Antigua (Panay), and the northern Moluccas. Although he ensured himself a place among classical ornithologists of the Far East, his present-day reputation is

*At that time, Annam and lower Cochin China were united in the Kingdom of Cochin China.

none too good, because Sonnerat (like his countryman Levaillant soon afterward) was not content with the simple truth but embroidered it a good deal. To swell the numbers of the birds, he presented South African, South American, and other species as Philippine or Papuan, and even maintained that three species of penguins he had illustrated were observed by him personally in New Guinea — an outrageous invention perpetuated in J. R. Forster's name for the subantarctic gentoo penguin, *Pygoscelis papua.*

When Sonnerat's book appeared, he had already set out again, this time to the coast of Coromandel and through the Strait of Malacca to Canton. On the return voyage he stayed for some time in India, where in 1778 he helped to defend Pondicherry against the English, and again visited the Île-de-France and Madagascar. His report of the journey, from which he had returned to Paris with "more than 300 birds of different species,"[17] acquainted his contemporaries with a large number of surprising new species; indeed, for 70 years it remained the most important contribution to the ornithology of China. Sonnerat continued to follow Buffon's example by giving his species only French names. Giovanni Antonio Scopoli (1723–1788), at that time professor of botany in Padua, found this intolerable. In his *Deliciae florae et faunae Insubricae* (three parts, Pavia, 1786–1788), he arranged Sonnerat's species from both voyages in their "proper" genera and equipped them with legal specific names,[18] just as the Dutchman Pieter Boddaert had done in 1783 with the descriptions illustrated in Daubenton's *Planches enluminées.* A blind adherent to Linnaean systematics, who did not dare add anything to the master's 81 genera, he found no other method than to stuff everything described by Sonnerat into the traditional straitjacket. But, lacking the necessary experience, he heaped error on error, putting a button-quail from Luzon (Sonnerat's "caille de la Nouvelle Guinée") in the genus *Oriolus*, a duck into *Sterna*, the secretary bird into *Otis*, and a sandgrouse into *Tringa*, thus unintentionally revealing not only his own inadequacies but also the deficiencies of the Linnaean principles of classification.

In the meantime the tremendous success of the English voyages under Cook's command had aroused the French to a similar undertaking. It was entrusted to the Comte de La Pérouse, who sailed to the Pacific in 1785 with two frigates. Not only geographers but also zoologists had high hopes of him, because he was accompanied by two naturalists, but after the ships had sent home a report from Botany Bay in February 1788, they disappeared. In 1791 the National Assembly sent out the *Recherche* and the *Espérance*, under d'Entrecasteaux, to look for the vanished ships east of Australia. They cruised for a long time in the western Pacific, using New Caledonia as a base, and made important geographic discoveries, without finding any trace of the missing ships.* When d'Entrecasteaux died of illness on board his vessel in July 1793, his deputy gave up the search and sailed by way of Butung to a Javanese harbor. There the expedition broke up because the devisive issues of revolution had destroyed discipline. In the meantime news arrived that a revolutionary army had invaded the Netherlands; in retaliation the crews of both French ships were interned in Java and all the scientific collections confiscated. Riche, the zoologist, and Labillardière, the botanist, were indeed allowed to return to France in 1795, but Riche died two years later of chagrin at the loss of his notes and material.

*According to remnants discovered 30 years later, La Perouse's vessels were wrecked off Vanikoro, in the Santa Cruz Islands.

5

François Levaillant (1753–1824)

France had spared no expense on her expeditions sent out from 1766 onward to explore distant seas and lands — but, as we have noted, ornithology gained little from them. Instead, an impecunious adventurer managed to achieve what the national undertakings could not. His accounts of his travels caught the attention of the world and made François Levaillant the idol of a whole generation.

Almost everything we know about the first part of this remarkable man's life comes from his own writings. An important source is the foreword to his *Voyage de Monsieur de Le Vaillant dans l'intérieur de l'Afrique* (Paris, 1790), a literary gem that has always enchanted the true ornithologist, because it magically brings to life again his own youthful adventures.

According to this account, François Levaillant was born in 1753 to French parents living in Paramaribo, the chief Dutch town in Surinam (Dutch Guiana), where he grew up in absolute freedom until he was 10. His father and mother, enthusiastic naturalists, often took their small boy with them when they went into the jungle to collect; soon he himself began to shoot birds with a blowgun and to start his own "cabinet." In 1763 the family returned to Europe and lived in Metz, the birthplace of the elder Levaillant. Here François attached himself to an apothecary, Jean-Baptiste Bécoeur (1718–1777), the owner of the biggest and best collection of European birds that, says Levaillant, he had ever seen. Bécoeur, the inventor of arsenic soap, was not only a master at preparing and exhibiting, but also an exceptionally good field collector. His teaching made young Levaillant into a real hunter and field ornithologist. The autobiography says nothing about anything else that he may have learned; one can only gather that he spent two years on the left bank of the Rhine in Germany and seven years in Alsace and Lorraine, birdwatching and collecting. He seems to have spent the longest time in Lunéville, where he became acquainted with Richard, a collector of native birds (who 40 years later sent Vieillot a new species of pipit, *Anthus richardi* Vieillot). Levaillant arrived in Paris for the first time in 1777. Here during the next 3 years he learned still more by studying in varous natural history cabinets, particularly in the large bird collection belonging to the physician Mauduyt de la Varenne, which had already been useful to Brisson (1760), and he sharpened his critical judgment in the company of specialists. Then the urge for great adventures drove him abroad. In July 1780 he went to Holland and became acquainted with owners of large ornithological collections and menageries. None was so impressed with the intelligent young Frenchman's burning enthusiasm as Jacob Temminck. Convinced that Levaillant was the right man to add new and rare species of birds to his collection, Temminck provided him with money and recommendations for a journey to Cape Colony. Levaillant took ship from Texel on December 19, 1780, and stayed in South Africa for over 3 years, from April 1781 to July 1784; during this time he made two long journeys to the farthest limits of the European sphere of influence.

For the first of these he set out from Cape Town on December 18, 1781. By way of Swellendam he reached the lands of the Auteniqua (Houtniqua) Hottentots, who lived in the area of present-day Knysna, at the beginning of February 1782; later he proceeded to Plettenberg Bay and, after crossing the Sondag River, he arrived at the beginning of October in the Boer settlement of Bruintjes Hoogte, where Sparrman had been 7 years before. On October 10 he stood on the bank of the Great Fish River, alive with hippopotamus, and pledged friendship to the Gonaqua tribe of Hottentots. This was on the border of the mysterious Kaffir country, which he entered only briefly when he went on a three-days' march to the east. On December 4 he said goodbye to his black friends and his beautiful Narina, whose name he gave to the first known of the three African trogons, and went toward the Sneeuwberg mountains so that, after crossing the upper Sondag River, he could follow the southern border of the Great Karroo on his return to Cape Town, by way of the Boer settlement of Cambdebo. He arrived back at the end of March 1783.

After this, following a series of excursions in the Cape district, he set out, on June 15, 1783, with a small band of porters for the Orange River, 700 kilometers away. He was furnished with advice and recommendations from the commander of the forces in Cape Town, Colonel Gordon, who knew this area from his own experience and who in 1779 had taken an Englishman, William Paterson, to the lower reaches of the Orange River. Traveling mostly near the coast and finally through

Little Namaqualand, Levaillant reached his objective only with great difficulty. Here he collected for several weeks in the gallery forests, and found himself rewarded for all his hardships with many ornithological discoveries. At the end of October he even crossed the river, made the acquaintance of Great Namaqualand, and was lucky enough to shoot a giraffe, which later became the pride of the Paris Museum. But he certainly did not go much farther north, and everything narrated in his *Second Voyage* about his journey from the Orange River to the tribal lands of the "Houssouanas" belongs to the realm of fable. Therefore no one knows when he returned to Cape Town, before embarking on July 14, 1784, for Europe.

Four months later he landed at Flushing, bringing with him over 2000 skins of exotic birds — by the standards of that time, a tremendous number. His friends Boers and Temminck welcomed him back and he spent several days with them, leaving part of his treasures with Temminck, and then returned to his beloved family in Paris.

The burning desire to become rich and famous through his travels spurred him on unceasingly. But if he had expected to find any of the Paris scholars able to take an expert's pleasure in the new and rare Cape birds, he was bitterly disappointed. The great Buffon had grown old and had retired in 1785; in the same year Guéneau de Montbeillard, his former ornithological collaborator, died, and among the younger generation at the Cabinet du Roi there was not one whose practical experience could touch Levaillant's. In these circles he could learn nothing more and, what was still more painful, he could not even find a Parisian buyer for his unique collection, although the highly regarded Cabinet du Roi needed it acutely; the birds there, described by Sonnerat in 1773 as an "immense collection," numbered only 463 on June 10, 1793, after its rebaptism as the "Museum d'Histoire Naturelle." (In fact, during the last years of the Buffon era, on Mauduyt's advice, the birds on exhibition had been disinfected with sulfur fumes, which destroyed a large number. "The shafts of their feathers were completely burnt up by the acid, so that whenever they were lightly touched they collapsed into a black and oily dust."[1] In addition, Buffon had made only a small selection from the many bird skins sent him by his admirers, and had exhibited in the galleries only those that would make the strongest impression on the public. The rest remained packed in boxes, where they soon fell victim to insect pests.[2]

So Levaillant found himself compelled to sell many of his South African birds, after he had mounted them splendidly himself, to rich amateurs in Holland, especially Temminck, Raye van Breukelerwaerd, and Holthuyzen. Then he set to work to write a book about his journey to the land of the Kaffirs, and happened upon a publisher who advised him to have it reworked by a skilled professional writer, Casimir Varon. It had hardly appeared as *Voyage de Monsieur le Vaillant dans l'intérièur de l'Afrique par le Cap de Bonne Espérance dans les années 1780, 81, 82, 83, 84 et 85* (Paris, 1790, in both quarto and octavo editions) before the traveler's name was on everyone's lips. The book's style and content both hit the public taste exactly. Even such an exigent critic as Georg Forster was full of praise. At the beginning of a detailed review in 1790, he said, "This is yet another foreign production which makes one wish that the feeling for the importance of form in a good book, a feeling so admirably widespread in France, could be inoculated into a certain numerous class of native writers."[3] Several further printings followed. In the year of publication an English translation appeared, and in 1791 a German translation by J. R. Forster, and everyone was impatient for the promised sequel. The chaos caused by the Revolution delayed it until 1795, when it came out in two quarto volumes entitled *Second Voyage.* A German translation by J. R. Forster followed in 1796. (Later biographers have asserted that, during the Reign of Terror, Levaillant languished in a dungeon and was rescued from the guillotine only by the fall of Robespierre, but I can find no reliable source for this and therefore doubt its accuracy.) Unfortunately, Levaillant, in his eagerness for fame, had allowed an imaginative and well-read author (P. S. B. Le Grand d'Aussy, according to Cuvier) to ornament and even add to his experiences, whose climax was a freely invented journey from the Orange River "as far as the Tropic of Capricorn"!

Levaillant had now progressed so far as to think seriously of seeking scientific laurels by publishing a book that would portray the South African birds in lively style and attractive pictures. He could not resist the temptation to use other collections of exotic birds, newly arrived in Paris, to eke out his material; among them was one that a physician, René Geoffroy de Villeneuve, had brought back from Senegal about 1795. After Levaillant's wealthy patron Temminck had provided him with the necessary funds, he published in 1796 the first volume of his *Histoire naturelle des oiseaux d'Afrique* simultaneously in three editions, folio, quarto, and duodecimo! The charming plates, painted by J. L. Reinold, were printed by Audebert (d. 1800) by a color process introduced into France by Barraband. In spite of its high price this volume seems to have had a wide sale, and even in Germany publishers were found brave enough to bring out editions in translation (quarto by J. M. Bechstein, 1797–1802, and duodecimo by J. R. Forster, 1798).

In this way Levaillant rapidly became famous among cultivated people throughout the world. One of those who made a pilgrimage to see him was Count von Hoffmannsegg. On his way to Portugal he wrote from Paris to his sister on December 2, 1797:

My greatest impressions came from the natural history collections. What will certainly interest you is Vaillant, to whom for so long we have owed many agreeable hours. One of my acquaintances introduced me to him first at a breakfast, at which only naturalists were assembled. He is a man of 40, of middle height, with charming regular features and very penetrating, intelligent eyes; his genius is soon recognizable, but not in external mannerisms. He converses easily, is obliging and without pretensions, except for those that one can easily concede in him. It is quite possible that often in his tales he improves on nature as he observed it, but certainly not so vastly as many here believe, and I do not know why all his stories should not basically be true. I was never able to understand how he could preserve so many birds in a single day, and asked him about it. He assured me that he did not need more than three minutes to prepare a small bird completely for preserving. That sounded exaggerated. But he had a small bird brought immediately and prepared it completely right in front of us, before three minutes had passed by the clock. And he said that many a time he had shot 30 or more himself and then prepared them, all in the same day. And now I see indeed that it was possible. Since then he has demonstrated the method to me several times, and it will help me greatly. A few days later I visited him. He has an agreeable competence, having earned a great deal from his collections and through his books. Now he is issuing copperplate engravings of all his African birds; they are, however, very expensive, although they are very beautiful; the drawings for them, which a German, Reinold, makes for him, are still more beautiful; he already has 1500 on hand. The larger part of his collections is now in the public Cabinet of Natural History; private collectors have acquired another part. But he himself still has a great deal, and things more beautiful than one can imagine. Many birds are not even stuffed, but are simply skins, just as he brought them back. He also has a lot of insects still. He lives in fine style, and everything to do with him looks affluent. As soon as his book is finished, he greatly desires to travel again, but I doubt that it will be possible in his present family circumstances. He would be very useful to command a dangerous enterprise.[4]

In 1799 the second, and in 1802 the third, volume of *Oiseaux d'Afrique* came out in the same splendid format, but that was not enough. Relying on the luster of his name and the inexhaustible buying power of his compatriots, who were spoiled by their military successes, Levaillant joined the popular illustrator Jacques Barraband (1767–1809), and in 1801 launched no fewer than three magnificent illustrated works at the same time, all of them in large folio (and also in quarto):

1. *Histoire naturelle d'une partie d'oiseaux nouveaux et rares de l'Amérique et des Indes . . . ouvrage destiné par l'auteur à faire partie de son Ornithologie d'Afrique* [The natural history of some of the new and rare birds from America and the Indies, a work intended by the author to form part of his African ornithology]; 2. *Histoire naturelle des perroquets* [parrots]; 3. *Histoire naturelle des oiseaux de paradis et des rolliers et des promerops* [birds of paradise, rollers, and sugar birds]. No ornithological writer before or since has ever attempted anything so presumptuous. But everything Levaillant touched seemed to turn out well, even though at the same time other French publishers were lightening the purses of their bibliophile public with bird books at unheard-of cost. From 1800 to 1802 Desray published in Paris Audebert and Vieillot's *Oiseaux dorés, ou à reflects métalliques* [Golden birds, or those with metallic reflections], in 32 issues at 30 francs each. The same publisher followed it, in 1805–1809, with Vieillot's *Histoire naturelle des plus beaux oiseaux chanteurs de la zone torride*, and in 1807–08 with the same author's *Histoire naturelle des oiseaux de l'Amérique septentrionale*, "containing a large number of species described or illustrated for the first time," both of them with plates by Prêtre. In addition, A. G. Demarest's *Histoire naturelle des tangaras* [tanagers], *des manakins et des todiers* [todies], "with illustrations printed in color, after drawings by Mademoiselle Pauline de Courcelles, a pupil of Barraband," was published between 1805 and 1807 by Garnéry, in Paris.

Levaillant, outsider among scientists, promised scholars and laymen alike a detailed and fully illustrated presentation of the most colorful and bizarre birds from all over the world, not merely from Africa. In fact, he was not undertaking too much, because for a long time he had been gradually becoming more expert. He had built up his own collection and, at the Abbé Aubry's auction, had been able to obtain birds received by Aubry from Poivre; from his travels in Holland, Levaillant knew better than anyone else the splendor and richness of the cabinets belonging to Jacob Temminck, Gevers Arnz, Holthuyzen, and Raye van Breukelerwaerd, and the former museum of the Prince of Orange; years before in France he had studied the carefully chosen collections of Aubry and Mauduyt; he knew the menagerie in Cape Town, full of "East Indian" birds; and he was supposed to have been in Madrid and Lisbon. He was not unacquainted with the latest English discoveries, because in Paris and London there were collectors of pictures (like Colonel Woodford) who exchanged watercolors of birds. Above all, the Paris collector Gigot d'Orcy (1733–1793), *receveur générale des finances*, had obtained many rarities from London and (from 1789) had commissioned his draftsman Audebert to paint them. In addition, Levaillant could consult the Muséum d'Histoire Naturelle as often as he wished, because the *aide-naturaliste*, L. Dufresne, who was himself an enthusiastic bird collector, gladly granted him access. From 1794 to 1818 Dufresne, as "aide-naturaliste" and "Chef du Laboratoire du Muséum d'Histoire Naturelle," headed the Bird Department. According to Vieillot, 1807, he possessed "une précieuse et nombreuse collection," which was bought in 1819 for the Royal Scottish Museum in Edinburgh. His successor was Desmoulins, who was in turn (1832) followed by the bird collector Florent Prévost (died 1870).

Thanks to the untiring energy of the young director, Etienne Geoffroy Saint-Hilaire (1772–1844), the Paris Museum had again become on the whole the richest in the world. The Revolutionary armies had brought in many remarkable and important rarities when they carried off the Prince of Orange's famous collection as booty from The Hague to Paris in 1795. Its condition in 1797 was described by H. F. Link:

> The museum of natural history in the botanic garden at Paris is far more interesting than the British museum, and contains a great number of specimens, and very extraordinary productions. London possesses nothing that can be compared with it, and the Leverian museum exceeds only in two branches [birds and mammals]; the arrangement, however, of that at Paris is not very good: the names of the birds and viviparous animals are taken from Buffon, very many natural productions have no names at all, and the mode of placing them requires great improvement. The magazines of this museum are full of unarranged treasures, which require both money and time to be properly placed; mean while it is much to be lamented, that in their present situation many of them are going to decay.[5]

Thus it is certain that no one in England or France had seen or critically examined so many species of birds as Levaillant had. His great color-plate books, especially the one on parrots, are therefore more complete than anything else published at the time and display excellent judgment. But literary production was not Levaillant's only means of support. After he had sold his own African birds to the Dutch,* and at long last also to the Paris Museum, he continued to procure rare and splendid birds for private collectors and dealers, which he converted with unparalleled skill from unsightly skins into real museum specimens.

At this time, in 1802, another German naturalist, K. A. Rudolphi, an anatomist who later became famous, visited Levaillant in Paris. In the memoirs of his travels he writes (1804):

> I will say a few words about the famous Le Vaillant. I had an intro-

*A large number of birds collected by Levaillant landed in the Leiden Museum, along with Temminck's collection, and are still to be found there.

duction to him [from Raye van Breukelerwaerd in Amsterdam] and was received in a most friendly manner. He is no longer young, but still as enthusiastic as a boy, has a finely made, lively face, and is really very interesting. From what I saw of him I can hardly believe that he intentionally deceives, but his liveliness must often have caused him to lay his colors on too thick, and his imagination perhaps added a good deal without his noticing. He had only a few stuffed birds with him, but they were extremely beautifully prepared; he said, however, that he had many unstuffed ones . . . He spoke in a most lively fashion of the inclination to natural history that he had from his earliest youth; how often, during the night, a good idea for preparing a bird would come to him and he would immediately rise to try it out. He was very sorry that I had not seen Teiners' [a mistake for Temminck's] collection in Amsterdam, because his and Raye's contain every bird that one could possibly want to see in Europe. He had provided almost everything for these two cabinets, and Sonnenberg,* who is at the moment the best preparator in Holland, is his pupil. In the Paris Museum collection, in the Jardin des Plantes, there are also many things of his . . . His wife seems still young, and talked with us too; she also knew the manucode, the sixfilet [two birds of paradise], and so on, in fact she took part quite freely, and in his company certainly everyone must enjoy natural history; thus of his three pretty children (two boys and a girl) the older boy† loves it already, and may perhaps some day follow his father.[6]

But the Goddess of Fortune deserted her favorite Levaillant a little while later. The interest of the newly rich in luxurious bird books had been only a flash in the pan; now his bold and adventurous spirit found no public support for its enterprises and brought him into financial difficulties. After the sixth volume of *Oiseaux d'Afrique* appeared in 1808, the series could not be continued, although text and plates were ready for further installments. Then a deeper and deeper silence settled around the once idolized traveler. To Dr. W. E. Leach (successor to George Shaw at the British Museum in 1813), who found him in Janu-

*Sonnenberg-Gallant, a painter in Amsterdam, who prepared birds at a charge of 1 to 6 ducats each. (H. Engel, "Alphabetical List of Dutch Zoological Cabinets and Menageries, "*Bijdragen tot de Dierkunde*, 27 (1939): 318).

†As Rudolphi anticipated at the time, this son of the second marriage, baptized with the highly auspicious name of Jean-Jacques-Rousseau, added to our knowledge of the African fauna forty years later. From 1840 to 1842 he was assigned as a battalion commander to the French "Mission d'Exploration scientifique de l'Algérie," and supervised the collection of mammals and birds for the Paris Museum. In 1850 he described several of these as new species. He wrote an *Introduction à l'histoire des mammifères et des oiseaux du nord de l'Afrique* (Philippeville, 1851). (Information *in litteris* from Professor J. Berlioz, Paris.)

ary 1818 living in an attic apartment up many flights of stairs, he made the bitter joke that the longer he lived, the higher he rose in the world. In the same year he attracted attention again with a supplement to the *Histoire naturelle des oiseaux de paradis.* Soon thereafter he left the capital with his numerous family and retired to his little country house in La Nomé, near Sézanne, where he died, extremely poor, at the age of 71, on November 22, 1824. Among his papers was found the manuscript for a book on African ducks and waders; according to Johann Wagler, it was offered in Paris for not a very high price, without finding a single buyer.

Though the general public had long forgotten Levaillant, professional ornithologists still prized his writings as the best, indeed the only substantial, accounts of African birds. He was admired as the discoverer of a large number of South African birds (there must in fact have been about 50 species) and even more as the writer of elegant descriptions, which many sought in vain to imitate. Even though the author's veracity was openly doubted soon after the appearance of the *Second Voyage* in 1795, this could not seriously damage the respect in which he was held. Indeed, as late as 1830, one of the pioneers of South African ornithology, Andrew Smith, himself a model of precision, called him "the accurate Levaillant, an authority I so much admire."[7]

The author of the *Oiseaux d'Afrique* was so highly valued because, as one of the imaginative Buffon's most gifted pupils, he had the knack of touching up and beautifying, with Nature's own colors, the dreary areas where scientific knowledge was meager. None of his contemporaries gave him his due so fully as Georg Forster, writing in 1791 in a second long review of Levaillant's *Voyage:*

> Without over-anxiously keeping to diary form, the author produces an uninterrupted story of the course of events in which he was involved during this expedition, and weaves together his observations not only of peoples but of natural history so artistically that one is carried away and can listen to the end without fatigue. His enthusiasm for his branch of science gives even the descriptions of new material (which are, however, only rarely included) a peculiar fire, and when one has finally realized how much pains and skill it costs him before he can shoot this or that rare bird, how often, in his eagerness to pick up the game that he has killed, he either nearly loses his life by falling into a pit dug for elephants, or barely escapes drowning in deep rivers, or crawls under the trunk of a fallen tree so that a raging elephant will jump over him, one is even more anxious to learn the distinguishing characteristics of the creatures for whose capture he was willing to pay such a price . . .

> One cannot say that there is anything much new in his description of these objects; its sole advantage is that it brings everything clearly before the eye, and the reader of this remarkable work is fascinated by the lively feelings of the author, the individuality of his way of thinking, the youthful and innocent enthusiasm with which he declaims against the misuse of over-civilized refinement and pours himself out in praise of nature untouched by art. It may well be that this aesthetic accomplishment sometimes draws a poetic embroidery over the objects; for art, and the inner urge that creates it, insist on this blending of external nature with the being of the artist; wherever this fusion is lacking, the great characteristic traits are rarely so sharply realized or thrown into relief, and in spite of the most laborious dissection and enumeration of the separate parts one can have no vivid idea of the whole. Anyone who knows how to judge what he reads will be able to see through the delicate veil of poetry, and know how to find the reality of nature in the ideal portrayal of it.[8]

It was just during Levaillant's lifetime that the confrontation began in France between the previously dominant school of Buffon and the Linnaean systematists. Buffon had died, full of years, in 1788, and immediately after his death the formerly suppressed opposition dared to emerge with the founding of a "Société Linnéenne" in Paris, whose members lectured to one another on "Latin descriptions drawn up according to the principles of philosophy and criticism." After the Revolution broke out, the group was better organized as the "Société d'Histoire Naturelle," and its members set up a bust of the famous Swede under Jussieu's well-known cedar in the Jardin des Plantes, on August 23, 1790, inscribed "To Charles Linnaeus, from the naturalists." It was intended as the signal for revolt, and from then on there were no limits to the French glorification of Linnaeus's work. It is not surprising, under the circumstances, that so accurate an observer as Levaillant became a fanatical defender of the procedure laid down by Buffon for true naturalists, because its opponents, like Bonnaterre, Lacépède, and Cuvier, with their passion for formalism, lacked precisely what Levaillant had effortlessly acquired in his early years — familiarity with living birds. Instead of representing natural relationships, their system was nothing but the result of pedantic efforts to arrange diversity according to a traditional method, which in use was constantly strained to the point of the ridiculous by new discoveries, without anyone's daring to give it up. His contempt for the "charlatanerie des méthodes" was concentrated chiefly on the current beak-and-foot systematics.

> The naturalists place the honeyguide — one cannot imagine why — under the cuckoos, though only in its foot structure does it resemble

> that group, and much as it differs in its other physical characteristics, it differs even more in its habits. At the risk of bringing down on myself the anathema of the keepers of scientific collections, I must continually repeat that fat books are nothing beside the great book of nature, and that an error does not cease to be an error though it be sanctified by a hundred eloquent pens.[9]
>
> All our usual classifications will remain inadequate so long as we do not possess a more complete knowledge of all species of animals, and particularly of the habits of each individual species. Exact observation of an animal's behavior, which is without doubt the most important way of discovering the true place that each species occupies in the natural order, has been completely neglected . . . I have noticed again and again in myself how decisive is the first general impression made by a bird as a whole, when one is accustomed to observe and study birds in their natural state; this first view is more certain and far less subject to error than the anxious searching out and verification of the generic characters that it has pleased our systematic nomenclators to establish.[10]

These attacks were clearly aimed at the large reference work, *Tableau encyclopédique et méthodique: Ornithologie* (1790–1792), and its author, the Abbé Bonnaterre (1747–1804), as well as at Desfontaines, who had become famous for his scientific expedition to "Barbary" (1789), and had been the first French ornithologist to abandon Buffon and adopt the Linnaean system. Bonnaterre had done so with certain variations that had been recommended to him by the perceptive anatomist, L. J. M. Daubenton; he employed 12 "classes" and divided them (closely relying on Latham, *Index ornithologicus*, London, 1790) into a total of 112 genera. Since these still included somewhat heterogeneous species, they provoked the hostility of the biologist Levaillant.

That he detested Linnaeus's binomial system with all his heart and soul was entirely logical. He followed Buffon's model closely and gave his birds individual names (like the "Broubrou" or the "Moqueur") which, by not indicating their place in the system, also were not misleading. He kept firmly to this procedure all his life. But then all he could do was look on while the systematists whom he ridiculed — J. R. Forster, Daudin, Dumont, Shaw, Stephens, Cuvier, Vieillot, Temminck, and all the other servants of the new order — fell on and appropriated his new species by naming them according to Linnaean precepts.

Levaillant showed himself an epigone of Buffon by taking no account of anatomical research — because he understood nothing about it. He was interested only in behavior. How differently, according to him, "modern" scientists conceived of their task: "Unfortunately, for many of our learned, the science of natural history is not the history of nature. A dry and arid nomenclature for the productions of nature, provided that it be arranged according to a tidy system — there lies the essence of the science."[11]

No wonder that such a lively mind was bored to death at the appearance of contemporary natural history collections. "In this accumulation of skins of exotic animals I saw only a strange warehouse, in which the different creatures, lined up without taste or selection, slumbered away quietly on behalf of science."[12] To interest and instruct visitors, he proposed nothing less than to show them the geographic variation of species. In a natural history collection

> it would be necessary to bring together examples of the same species from different areas. For instance, our kingfisher [*Alcedo atthis*] is found not only in the whole of Europe . . . but also in Egypt, China, Bengal, etc. If we were enabled to compare individuals collected from each of these different climatic zones, we should certainly acquire some solid basis for establishing the range of variation to which this species is subject, according to the different temperatures of its habitats, and the locally different kinds of food. But what a long way our collections are from providing such resources for serious study![13]

Such ideas, adopted from Buffon's *Histoire naturelle* and developed and followed up in many widely read books, could not remain ineffective for long, and Johann Wagler was not so far out when he noted in 1827, in his survey of the development of ornithology (*Systema avium*), "You can scarcely hesitate to say that with Levaillant's and Temminck's works on universal ornithology a new and happier era began." Levaillant's name would still be highly honored today if he had not taken matters too easily when it came to illustrating his program with numerous examples. He had intended his *Oiseaux d'Afrique* to be a collection of his biological observations and at the same time a model natural history of birds. Though there is much solid and reliable material in the six volumes, it becomes suspect because of the number of details invented by the author to impress his readers with the wealth and accuracy of his observations and the large number of his discoveries. Who can tell if the passion to attract attention at all costs and to become famous quickly existed in Levaillant from his earliest years, or if the great success of his first *Voyage* in 1790 led him astray? The swindle began in the *Second voyage* (1795), with its imaginary journey through the country of Great Namaqualand; and even though this may have been conceived by someone else, Levaillant adopted and continued it, without feeling any scruples about the mistaken impressions he was creating. Fate spared him the unraveling of his tissue of lies during his lifetime. But as early as 1832 Johann Wagler, who had had reason to make a detailed study of the *Histoire naturelle des perroquets*, wrote indignantly that "Levaillant has taken over plates of parrots from Buffon and Edwards without acknowledgment, and with some deliberate changes to disguise the fact."[14] Finally Sundevall, in his famous criticism of the *Oiseaux d'Afrique* published in 1857,[15] proved that Levaillant had very often given absolutely free rein to his imagination, and indeed that birds not belonging to the South African fauna, even "new species" that had been skillfully glued together from the feathers of several different species, were employed to increase his renown as a discoverer and particularly to make credible his journey across the Orange River and deep into Kaffir country. Perhaps it was only to reassure readers who might become skeptical that he complained about the many fake species of birds prepared by avaricious dealers, "samples of which are to be found in almost all collections. My vast experience of birds has given me such facility in detecting these fabrications at first glance that I have never let myself be deceived."[16] He found no difficulty in providing exact locations and attractive (but unfortunately fictitious) biological details for many of these imaginary species (about 79), as well as for others. And this was the behavior of a man who never tired of boasting of his own love of truth, and who enjoyed petty caviling and carping at the faults of others! Though in his youth, and even when he was in Africa, ornithology may have appeared to him as an august goddess, later — when he had been dazzled by his own too easily acquired and transitory fame — she became for him, in Schiller's words, "eine tüchtige Kuh, die ihn mit Butter versorgt" [an efficient cow, which provides him with butter].[17]

Even today only an expert can distinguish true from false in the cleverly concocted brew that Levaillant served up to his readers. But there is always enough useful material left to vindicate Levaillant's original reputation. He was really an excellent observer of birds and an understanding interpreter of their behavior, and in fact gifted as few others have been in communicating his ideas not only vividly but attractively. How gracefully in the *Second Voyage* he describes the movements of the touracos, how skillful and exciting is his account of a fight between a secretary bird and a snake! Hidden in his travel diaries are pearls that ornithologists have always overlooked, for who up to now has noticed that not Andersson (1872) but Levaillant first saw the lovebird, *Agapornis roseicollis*, nesting in the communal nests of the sociable weaver, *Philetairus socius?* It was on a night at the end of October 1783, at Löwenbach, a little beyond the lower Orange River. His account is worth quoting here, as an example of his skill in presentation:

> On the tree was one of those enormous nests that I have mentioned above, and that contained a republic of birds. Perhaps because the

smoke annoyed them, or perhaps because the light shed by our fire seemed to them like daylight, many of them fluttered among the branches, while others, twittering in a flock, made a confused though pleasant enough sound. It was a favorable opportunity for me to obtain some of them. I climbed a tree and slid my hand into one of the cells. But when, in spite of all my precautions, this movement shook the nest, they all tried to escape and from every hole at once there issued a prodigious quantity.

Nevertheless my hand kept reaching in farther. I had just touched something when I suddenly felt myself cruelly bitten, and this nip astonished me all the more because these nestbuilders, being of the same genus as the Cape sparrows, could not do that much harm. So there had to be a foreign species in the nest, which it would be curious to know about. It hung on; I did not let go my hold and in fact, with as much surprise as pleasure, I soon drew out of the nest two charming little lovebirds, male and female.

The presence of these intruders in a foreign republic seemed to me inexplicable. Only the Namaquas were not surprised. They knew from experience, and taught me, that when the republicans have finished their living quarters, sometimes birds of another species, stronger than they, arrive to chase them out and establish themselves, and as they multiply they live in the same way and in association with the others. So it is not only among human beings that the weak are oppressed, plundered, and exiled; among birds, too, the tyrants appropriate the fruits of the labors of others, and they do not even lack a logical excuse for proving that they had a right to do so.[18]

The direction of Levaillant's thought excludes him completely from the ranks of the systematists, who can be seen in this part of the book as handing on the torch to one another; he belongs firmly in the other camp, that of the biologists. Among these he was the last to try to seize command from morphology and to battle for the primacy of function over form. At that time a compromise between the two principles of classification seemed not yet possible; in fact, "morphologists" and "physiologists" were to draw even further apart.

6

Carl Illiger (1775–1813)

In Germany exotic ornithology became a subject for research much later than in England and France, where during the eighteenth century many a widely experienced ornithologist was hard at work who had acquired his knowledge by classifying and identifying the contents of large collections. But where in Germany at that time was there a collection of exotic birds that could have provided the basis for comparable undertakings? Not even the most lively interest and acute judgment could compensate for such a deficiency. Philipp Ludwig Statius Müller (1725–1776) and Johann Friedrich Gmelin (1748–1804), industrious professors of natural history at Erlangen and at Göttingen, did indeed name a good many exotic birds between 1773 and 1789, but circumstances forced both of them all their lives to remain compilers of inadequate information, without practical experience.

Only when the century was drawing to a close did it seem as if Germany would make up for lost time. About 1780 Duke Charles II of Pfalz-Zweibrücken equipped his castle of Karlsberg, which he had built at vast expense near Homburg in the Palatinate, with a natural history cabinet that was supposed to rival the richest of foreign collections, and entrusted it to Dr. François Holandre, who had been to Syria and Egypt. By purchasing the two large bird collections that had helped Levaillant in his apprenticeship, those of Jean-Baptiste Bécoeur (1718–1777, apothecary in Metz) and the encyclopedist Mauduyt de la Varenne (1730–1792), he obtained about 1200 species (including, in addition to those from Europe, "almost all those known from Cayenne, all those described by Catesby from Carolina,"[1] and many other exotics). But before any German scholar could profit from these treasures they were wiped out; the sansculottes who had irrupted into the Palatinate burned down the castle, together with the entire bird collection, on July 28, 1793.

In 1810 the much-hoped-for change finally occurred, with the founding of the Zoological Museum of the University of Berlin. It produced many important consequences, and its checkered early history was bound up with the names of Count von Hoffmannsegg and his friend Carl Illiger.

Johann Carl Wilhelm Illiger was born on November 19, 1775. His father was a merchant in Brunswick, who gave him a solid humanistic education and freely indulged him in his wish to observe and collect birds and mammals. In 1790 Carl entered the first class of the Catharineum, where the famous entomologist Professor J. C. L. Hellwig was teaching natural history. As soon as Hellwig noticed his new pupil's unusual knowledge of subjects in which he himself was interested, he encouraged him as much as he could, busied him with arranging the Hellwig insect collection, and finally took him completely into his own house. There Illiger, who had been prevented by a sudden lung infection from pursuing his intention to study medicine, spent several years quietly working on the collections. The compulsory rest helped him to recover from his illness, and when, on Hellwig's recommendation, Carl Wilhelm Ferdinand, Duke of Brunswick, made him a grant from the ducal purse for the larger part of his three years at the university, Illiger was able to enter Helmstedt in 1799. Here he completed the manuscript of a book that he had already been working on in Brunswick, called *Versuch einer systematischen vollständigen Terminologie für das Thierreich und Pflanzenreich* [An attempt to provide a complete and systematic terminology for the animal and vegetable kingdoms], Helmstedt, 1800.

In the spring of 1802, after he had finished his studies in Göttingen, he returned to Hellwig in Brunswick, still with a small grant from the duke, in order to help his fatherly friend in classifying and identifying insects. In the meantime an interesting man had joined the household, Count Johann Centurius von Hoffmannsegg (born at Dresden on August 26, 1766). He had come back from Portugal in August 1801, with a large entomological and botanical collection, and intended first of all to study the insects he had brought home, which he could do nowhere better than in Brunswick. None of the existing entomological collections was so well arranged as that of Court Councilor Hellwig, who, after cheerfully welcoming the count, left him to the company of Illiger because he was too old and too busy to keep up with his aristocratic guest's passion for work and for knowledge. Matters soon went beyond entomology; Hoffmannsegg was so delighted by the range of the young man's information that he sent to Dresden in 1802 for the rest of his zoological collection, and combined it with Hellwig's.

Included in it were the first bird and mammal collections that had been sent to him from Brazil.

At the very beginning of Hoffmannsegg's long stay in Portugal (1798–1801), his admiration and astonishment at the many examples of flora and fauna brought back by travelers from Brazil aroused a desire to study them.[2] Certainly not much was visible in the royal collection in Ajuda. "It will not, indeed, bear a comparison with that of Paris, or even that of Madrid; it is small, not a single class is well stocked, and it contains fewer specimens from Brasil than might be expected. There are, however, some curious specimens," was the verdict of Hoffmannsegg's traveling companion, H. F. Link.[3] For connoisseurs, the "curious specimens" were in fact sensational, because no other museum in the world possessed items worth mentioning from that sealed-off country; the only detailed knowledge of its fauna had found its way across the ocean in the writings of Georg Marcgraf (d. 1644).* Special privileges from the Portuguese government allowed the Count to break the taboo established 150 years before on everything that concerned Brazil. After nothing came of his original intention to go there himself, he entered into correspondence with a Portuguese named Beltrão, who lived in Rio de Janeiro, and asked him to send over birds. To his boundless joy Hoffmannsegg received a small shipment at Lisbon as early as 1800, which seemed to contain much new material. At the same time he became acquainted with a wealthy Brazilian, Francisco Agostinho Gomes, who was studying "the sciences" in Coimbra at the time and was planning to return to the town of Bahia. The count knew how to stimulate his interest in natural history; he taught him how to prepare birds and insects, introduced him to the most important books, and succeeded in extracting firm promises from him when he sailed back to Brazil from Lisbon in April 1800. Gomes was true to his word and from 1801 to 1803 sent the count quantities of birds, insects, mammals, and other creatures.

In this way Count Hoffmannsegg established links with Rio de Janeiro and Bahia, and in the winter of 1800–01 he sent his own servant, whom he had trained in collecting and preparation and whom he had taken on all his travels, to the province that promised the most interesting booty for the naturalist, Pará. Friedrich Wilhelm Sieber was the name of this courageous man, about whom little information has survived except that in his master's service he traveled in the lower

*See Chapter 2. When Napoleon's troops temporarily occupied the Portuguese capital in 1808, Étienne Geoffroy Saint-Hilaire and his preparator, P. A. Delalande, had the Royal Natural History Collection hastily shipped to Paris. There the Brazilian birds were immediately studied and described by Geoffroy, Levaillant, Cuvier, Temminck, and especially Vieillot. Pictures of some of them were also painted immediately.

Amazon region for a full 11 years, collecting animals and plants, and got as far as Santarem and Obidos.

Illiger, free to pursue his original inclinations again, now began to study the first Brazilian shipments of birds from Beltrão and Gomes with the same fervor that for years had distinguished his entomological work. A thorough study of the literature soon convinced him that the shipments contained many new species, which he promptly supplied with scientific names. The count saw to the provision of the necessary books. The exciting activity that reigned in the Brunswick house made an unforgettable impression on Hinrich Lichtenstein, son of the orientalist and naturalist in nearby Helmstedt, who in his own words (1802) "enjoyed the great privilege of being taken into this circle, directly after completing [my] academic studies and when I intended to make a journey to the Cape of Good Hope, in order to receive instruction and guidance in observing, collecting, and preparing African specimens."

In the meanwhile the count had met Link in Rostock in the summer of 1802, to settle the plan for a joint folio to appear under the title of *Flore portugaise*, and they had decided to order the drawings and copperplate engravings for it in Berlin. Because the execution of these plates, to be printed in color, did not in the least satisfy the count's high expectations of accuracy and beauty, he determined in the autumn of 1804 to move to Berlin, in order to found his own lithographic press.

At this time the name of Illiger, as editor of the periodical *Magazin für Insektenkunde* (1802–1807), had become well known among specialists and, at the recommendation of the famous Johann Christian Fabricius, the University of Kiel awarded the young entomologist an honorary doctorate in 1806. Everything seemed to be going well, when suddenly the unhappy events during the last part of that year deprived Illiger of his highly placed patron, the Duke of Brunswick, who succumbed to wounds received while he was commanding the Prussian army at Auerstädt. This also wiped out the small stipend that Illiger had previously enjoyed. He was in great straits, but Count Hoffmannsegg knew what to do, summoning his young friend to come and help him with the determinations of Sieber's first shipment from Pará, which had finally reached Berlin. The two were again rejoicing in their work together when, after seven months, Illiger's old lung trouble broke out afresh and forced him to return hurriedly to Hellwig's house in Brunswick. The zoological treasures from Pará lay untouched until in 1809 a large shipment of birds arrived from Sieber, just at the moment

when the founding of the University had been decided on and announced by the king. Count Hoffmannsegg recommended to State Councilor Wilhelm von Humboldt (then in charge of matters relating to the University) that a collection for the study of natural history

> should be established with Illiger as curator, and which he [Hoffmannsegg] would support by offering to it his recently arrived rarities from Brazil. The idea of helping a materially depleted state by establishing an academic center to enrich it intellectually, and equipping the new institution with every kind of necessary assistance, was a matter of such general concern that the recommendation, particularly in regard to the excellent qualities of the proposed director, could only be received by the king and his councilors with extraordinary favor and approval.[4]

All of Hoffmannsegg's proposals were accepted by the king in March 1810, and Illiger, whose health seemed permanently restored, moved to Berlin shortly thereafter. He wished to continue his former work with as little interruption as possible, because he had begun to develop a new system of classification for vertebrates, but Humboldt did not agree to this; he applied for a professorship in zoology for Illiger (in addition to the post in the Academy) which, in the course of the negotiations, he accepted. But because of either an innate distaste for the job or anxiety about his illness, he refused absolutely to teach in the university the subject of which he was such a complete master, and withdrew entirely to his research. Then Lichtenstein, who had been in Berlin for some time without a post, jumped into the breach and promptly declared himself ready to lecture on mammals, birds, and "amphibia."[5]

From then on Illiger fervently devoted all his energies and his remarkable gifts to the new foundation (which had been set up in the east wing of the second floor in the university buildings on Unter den Linden), and attempted, by evaluating the masses of collected material, to further the progress of zoology in various ways, always in close cooperation with the count. Inspector Rammelsberg, who had been called from Brunswick as a technical assistant in 1811, described the condition of the collections at the time:

> In 1811 the Zoological Museum consisted of one room in which the zoophytes donated by Dr. Gerresheim were exhibited. The remaining rooms were all still empty, and even the walls unpainted. The mammal and bird skins belonging to Count Hoffmannsegg were packaged in rows in Professor Illiger's house on shelves that he had had built for them. A stuffed hyena from the Royal Veterinary School was the final showpiece. In this year Superintendent Schaumburg arrived from Cassel; he had been engaged for six months to stuff the stored skins and to teach taxidermy at the same time. Quite a few recently received birds, presented by enthusiastic hunters, were stuffed by a guardsman called Mrotzek, who was good at handling fresh birds.[6]

In the spring of 1812 all Hoffmannsegg's Brazilian mammals and birds had been worked through, determined, and, wherever new, given names, a task interrupted in 1811 by Illiger's trip to Brunswick to marry Hellwig's eldest daughter, who had been a loyal nurse and helpmate during his earlier illnesses. In June 1812, a son was born to him on the day of Sieber's arrival at the museum after 11 years in Pará, to deliver everything that he had collected in the Amazon area since his last shipment, as well as Australian and North American birds that he had exchanged in London for birds from Pará. Illiger devoted himself to this new material with his usual passionate intensity, but his physical resistance stood up only a short time longer.

A sudden hemorrhage in March 1813 caused people to fear the worst. But the sick man remained at work as long as his strength allowed, and tried to cheer up his friends and make light of his condition. The end came suddenly. He died during the night of May 9, 1813, "just when the capital was closely threatened as a result of the battle of Grossgörschen. The news of his death was lost in the lamentations over the general misery; only three friends followed his hearse."[7]

Count Hoffmannsegg never recovered from the blow. A year earlier, on July 10, 1812, he had lost his close friend Willdenow the botanist in the prime of life. Only a less sensitive person could have found the courage to pursue his course without the two companions whose joint work and plans for the future bound them most intimately to him. The count could no longer bear the familiar places. The officious Lichtenstein, with whom he had never been able to make friends, took over—at first temporarily—the administration of the Zoological Museum, just as he had done in the Botanic Garden after Willdenow's death. From that time on Count Hoffmannsegg was rarely seen there, and after a while he handed over to the younger men who had been working for the museum in the last three years, "in an unforgettable hour, his work and the rooms arranged by him, never to enter them again."[8] At the end of 1816 he returned to Saxony, leaving the Hoffmannsegg-Hellwig Insect Collection on loan to the Zoological Museum, after he had made sure that it would be attached to a separate entomological department under the independent control of his proven friend Klug. After several years, on account of the high cost of producing his splendid *Flore portugaise*, which had been appearing in installments since 1809 and was not yet finished, he decided to sell the famous collection, on which the Hellwig-Hoffmannsegg-Illiger trinity had worked for so many years, to the Prussian government for 22,000 thalers. He spent most of his later years in the rural peace of his family estate, Rammenau, until he died at a great age on December 13, 1849, in the house where he had been born.

How much Illiger would have achieved if he had been fated to live as long as his friend! Two of his characteristic preoccupations — concentration on details and the search for the relation of the particular to the general pattern — are found together in all outstanding naturalists. This combination accounted for his success as a systematist.

Illiger's thoroughness completely dumbfounded Lichtenstein, who was 5 years younger. "He never left anything undetermined, or accepted it on mere conjecture, still less on someone else's authority; whatever could be established only by the most detailed, even microscopic, observation, and the most painstaking research in books, Illiger was certain to achieve."[9] The long and difficult training in entomology had fully developed these qualities to a maturity that greatly helped the young professor in his studies of the vertebrates.

This longing for precision had stimulated him as a student to write a *Terminologie* of German and Latin names, because he hated nothing so much as cloudy concepts. His aim was eventually realized, because German zoologists, and particularly ornithologists, agreed to his proposed terms. He was the first person to translate *uropygium* as *Bürzel* [rump], *crissum* as *Steiss* [under tail coverts], and so on; he also (following Kant's model in the *Critique of Pure Reason*, 1781) translated *Genus* as *Gattung* and *Species* as *Art*, to the annoyance of Blumenbach, who insisted on equating *Genus* with *Geschlecht* [race], and *Species* with *Gattung* [kind]. He also displeased Oken, who wanted to translate *Genus* as *Sippe* [clan].

The same feeling for law and order revealed itself in Illiger's principles of nomenclature. All his new generic names were honestly constructed and testify to a delicate sense of language, for example, *Tichodroma, Centropus, Pteroglossus, Pezoporus, Cathartes, Syrrhaptes, Pachyptila.* His aversion to the barbarisms that even then disfigured scientific nomenclature was so strong that, on the authority of Linnaeus's *Philosophia botanica* (1751), he recommended replacing them with better-chosen terms; this made him a founder of a school of nomenclatural purists that flourished for a long time, especially in Germany. What he thought about the use of the concept of genus, and the rules to be observed in choosing names, becomes clear from his long letter of September 12, 1812, to Temminck:

> In nomenclature, we seek to follow the plain and true path marked out by Linnaeus, which the moderns, especially the French, have neglected . . . I am more of an enemy to new names than you can possibly believe, and I have always sought to preserve the first ones given, but we must follow rules that have been approved in the classic works on our discipline. If we wish to have names remain unchangeable, we are obliged to choose them carefully. Names introduced into natural history seem to me to have the same dignity as a new word introduced into a language; it succeeds only when it is

well fashioned. I do not agree with those who believe generic names to be a matter of no consequence and absolutely immaterial.[10]

Toward the end of his short life Illiger was above all concerned to integrate intellectually his laboriously acquired knowledge. His mind, always conscious of the whole, must finally have wearied of the endless number of specialized studies in entomology. Therefore, under Hoffmannsegg's influence, he seized every opportunity to become familiar with higher classes of animals, particularly birds and mammals, whose wealth of forms seemed at that time still possible to take in at one view. On November 19, 1812, he presented to the Berlin Academy of Sciences a paper entitled "Tabellarische Übersicht über die Vertheilung der Vögel über die Erde" [A tabulated survey of the world distribution of birds],[11] with the comment that such undertakings as this might perhaps not only give clear hints to scientists writing new descriptions, on how properly to classify variant forms, but also "enable us to understand the specific variations that are dependent on climate and other influences, about which we are still completely in the dark." This treatise is the precursor of all general ornithogeographies.

Illiger was also interested in biology and anatomy. Nothing characterizes his attitude better than the statement: "Information about the habits and family life of captured insects is actually the most interesting of all. Natural history therefore preserves with particular gratitude the memory of those who make accurate and reliable contributions to this." In spite of frequent disappointments, he never wearied of bombarding Gomes, the Brazilian, with that sort of question in his letters. He asked him, "Do none of the local cuckoos lay eggs in other birds' nests?" In 1804 he inquired about the habits of sloths, the species of wild pigs and their anatomical characteristics, bloodsucking bats and their behavior, and desired precise descriptions of the anatomy of the manatee.

To Illiger's sorrow, his refusal to give lectures caused the Ministry to decide that all anatomical material, even the skulls and skeletons in the Royal Art Collection, should not belong to the Zoological Museum but should be handed over to Professor Rudolphi, the anatomist, as the foundation of a zootomic collection, because they should be principally an aid to education. This shortsighted directive remained in force despite his protest, strongly seconded by Pallas, because at that time only a few scholars understood that research into the systematics of vertebrates could not be limited simply to the study of physical externals.

These problems, and particularly a reform of bird classification, especially occupied Illiger's last years. He was convinced that the Parisian ornithologists had based their attempts at classification on false principles, but to confute them he needed a really representative collection. This was why he tried unceasingly to add to the number of species in the new museum, although this was extremely difficult during the years of the continental blockade. Pallas gave him great pleasure in May 1811 by personally handing over to him a present of a number of Siberian and northern Pacific species — the remnant of his own collection (1768–1773) and that of the Billings-Merck expedition (1790–1792). After Sieber's return home in June 1812, the Berlin Museum contained about 950 species of birds, including 350 South American (mostly Brazilian), 65 Australian, and 90 from Virginia and the Carolinas. Among them were no fewer than 150 species that Illiger identified as new and intended to describe.

Even before he had become familiar with the material, he ventured to publish *Prodromus systematis mammalium et avium* [Forerunner of a systematics of mammals and birds] (Berlin, 1811), the call to arms of the German systematists against the French. In it Illiger defended with great conviction the position of his intellectual precursors Blumenbach and Merrem. As far back as 1779, Blumenbach, inspired by Buffon, had attacked the Linnaeans in his handbook of natural history: "According to our conceptions of natural method, we must observe not single isolated characteristics, but all external distinguishing marks together, and the animal's entire *habitus.*" His disciple Merrem thought the same in 1788: "The similarities of all parts of the body taken together, not of individual parts, must determine genera, tribes, orders, and classes."[12]

Neither Blumenbach nor Merrem was able to follow these principles in a general overhaul of the Linnaean system, because the material available to them in the small collection of birds in Göttingen was much too scanty. The Frenchmen who, near the turn of the century, attempted new systems started with completely different theories; like Linnaeus, they wanted to make identification easier. The pendulum that Buffon had held back for so long now swung in the opposite direction. Georges Cuvier did not dare to alter much of the basic structure of the *Systema naturae* (1766) in his superficial outline of classification, published in 1797. Lacépède entered the field immediately after him. For his *Méthode*, published in 1799, he did indeed employ Linnaeus's binomial nomenclature, but without the slightest concern for systematic categories and principles of classification. Where Linnaeus in 1766 divided birds into 6 orders with 81 genera, Lacépède split them into 10 "divisions," 51 "orders," and 130 genera. In so doing he, like Linnaeus, used the most superficial "beak-foot systematics." *Upupa*, *Certhia*, and *Trochilus* were lumped together (in "Order XIV — with curved beak"), and *Serpentarius*, *Palamedea*, and *Glareola* (in "Order XXXI — with hooked beak"), while *Picus* and *Picoides* were separated from each other in different divisions, because one has four and the other three toes.

No more satisfactory was the system of Constant Duméril, whose *Traité élémentaire d'histoire naturelle* (1804) again grouped birds differently and, for example, divided the passerine orders, simply because of their beak shape, into seven "families" of "crenirostres, dentirostres, plenirostres, conirostres, tenuirostres, planirostres, [and] subulirostres."

Such principles of classification must have appalled a person like Illiger, who in 1799, at the age of 23, had written down "Einige Gedanken über die Begriffe: Art und Gattung in der Naturgeschichte [Some thoughts on the concepts of species and genus in natural history],"[13] and had taken the opportunity to identify a system as natural "when it accounts for all essential characteristics, or the whole *habitus* insofar as it is essential." An artificial system, on the other hand, is one that "is constructed only on the basis of a few parts." "*Genus* is the placing together of species that agree in their habitus. Genera distinguished only according to a few characteristics contradict our sense of fitness."

No wonder that Illiger's system (1811), because it was intended to establish these principles, took a completely different line from its precursors. He divided birds into 7 orders (Scansores, Ambulatores, Raptatores, Rasores, Cursores, Grallatores, and Natatores), 41 families, and 147 genera. The concept of "Family," with this meaning, had first been introduced into ornithology by Daudin (1800; Merrem in 1788 had called the same systematic category "Zunft" [tribus], Lácèpede in 1799 "Ordre," and Illiger in 1800 "Unterordnung" [sectio]); its general adoption by ornithologists can be attributed entirely and solely to Illiger's *Prodromus.* But the little book's lasting influence was not restricted to such externals; its historical significance rests on its final rejection of the dictatorship of the Linnaean classification, which German and English ornithologists had obeyed for decades, and on its clearing the way for an unrestricted study of relationships.

Guided by his fine feeling for systematics in the study of habitus (which he described in detail in his long and painstaking analyses), Illiger undertook many new divisions and groupings that today we know, or guess, correspond to historical relationships. He rarely recognized genera that his contemporaries had established, because he was reluctant to overload his system with their considerable number. In our day his almost complete disregard for such an unimportant characteristic as the presence or absence of a fourth toe, and his reunification of *Picoides* Lacépède with *Picus*, and *Ceyx* Lacépède with *Alcedo*, are counted very much in his favor. It is true that his "sense of fitness" sometimes betrayed him; it led him to equate *Anthus* with *Alauda*, *Saxicola* with *Sylvia* and *Motacilla*, and *Tyrannus* with *Muscicapa.*

His awkward monster genera, like *Falco* for all diurnal raptors, were broken up again by ornithologists soon after Illiger's death, first of all by Vieillot, who provided 273 genera in his *Analyse d'une nouvelle ornithologie* (1816); 40 years later Bonaparte distinguished ten times as many genera as Illiger, although in the same period the number of recognized forms had only doubled (from 3800 to about 8000).

Illiger did not live to see the rapid success of his tireless productivity. When, two years after his early death, the world had freed itself from the Napoleonic nightmare, the spirit of enterprise at once led naturalists overseas once more. The zoological museums even in Germany became popular establishments. The new Berlin Museum's bird collection began to develop rapidly, and institutions in other places also, stimulated by the Berlin example, began enthusiastically to cultivate exotic ornithology. By 1825 there were five new public centers in Germany: at Frankfurt (Senckenburg Museum) under P. J. Cretzschmar; at Darmstadt under J. J. Kaup; at Munich under J. B. Spix and Johann Wagler; at Dresden under Ludwig Reichenbach; and at Halle under Christian Ludwig Nitzsch. In addition there was the important private museum of Prince Maximilian zu Wied-Neuwied, who had quickly become famous for his journey to Brazil (1815–1817). The friends Hemprich and Ehrenberg were traveling on the shores of the Red Sea as Royal Prussian Naturalists; Friedrich Sellow was combing southeastern Brazil, also on behalf of Prussia; Ferdinand Deppe and the Count von Sack were collecting in Mexico, Ludwig Krebs on the Kaffir borders, and Eduard Rüppell in Kordofan; and the Freiherr von Kittlitz was preparing for a trip to Kamchatka. The lead in exotic ornithology gained by England and France over Germany dwindled with lightning speed, and in 1825 it looked as if Germany were ahead.

But the victor in this race was not London, Paris, or even enterprising Berlin; it was Coenraad Jacob Temminck, director of the Rijksmuseum van Natuurlijke Historie in Leiden.

7

Coenraad Jacob Temminck (1778–1858)

The pleasure of collecting natural objects was nowhere more widespread than in the Netherlands, ever since the period of the first "Indian voyages."[1] Clusius's thirst for knowledge was able to profit from it as early as 1600, and it spread from Holland in all directions, particularly toward England, Germany, and Scandinavia. Soon after 1600 John Tradescant (d. 1638), a gardener who had emigrated from Holland, founded at his own expense the first English museum of natural science, in South Lambeth, where it became famous during the seventeenth century as the Museum Tradescantium. In 1651 Duke Friedrich von Holstein-Gottorp bought most of the splendid cabinet of Bernardus Paludanus (1550–1633), a Dutch physician in Enkhuizen, and with it created the frequently mentioned "Gottorfian Art Collection," which Peter the Great later transported to St. Petersburg. When Peter returned to Holland for the second time in 1717, he considerably increased his collections by purchasing for 45,000 guilders the "Thesaurus animalium" of Frederick Ruysch, a Leiden professor, and that of the Amsterdam apothecary Albertus Seba. Encouraged by this honorable way of making money, the two lost no time before continuing their lucrative acquisition of natural history specimens, and were imitated by many of their countrymen. When Linnaeus came to Holland in 1735, he was repeatedly surprised at the quantity and excellence of collections of animals living and dead. Twenty-five years later, Pallas thought the same. In the meantime, Seba's second collection (1752) had been publicly auctioned by his heirs and in part acquired by an amateur, Aernout Vosmaer (1720–1799). When William IV's widow was acting as regent for her son, who was a minor, she also decided to establish a natural history collection (1756), and had the unfortunate idea not only of buying

Vosmaer's collection but of making him the director of the Stadtholder's Cabinet in The Hague. In this way an incompetent man reached an influential position and held it, until the invading French army in 1795 carried off the whole museum, which had meanwhile grown considerably, in triumph to Paris, in fact, for a 40-year stay. Though he should have directed his countrymen's acquisitive instinct into scientific channels, he was "extremely ignorant" (Pennant, 1765)[2] — and worse; he jealously persecuted anyone more knowledgeable than himself, or even, as with Pallas, stole his ideas.

For these reasons the eighteenth-century maturing of childish "natural diversions" into "natural science," which took place in other countries, did not affect the many Dutch amateurs; a mood, a caprice, that could be satisfied when the object was bought, owned, and contemplated, continued to inspire their collections of exotic birds; even the existence of these collections was made public only occasionally, in accounts of travels or in auction catalogues.[3]

The Amsterdam patrician house in which Coenraad Jacob Temminck was born on March 31, 1778, was also furnished with this sort of bird collection, not only because his father, Jacob Temminck, was keenly interested in collecting, but because, as treasurer of the East India Company, he possessed the income and the connections needed to obtain rarities.* Levaillant, who was a frequent guest at the Temminck house in 1780, never tired of admiring the splendid collection contained therein. And some ten years later, in his famous *Ansichten vom Niederrhein*, Georg Forster wrote:

> I should like to dwell on the beautiful collection of the treasurer of the East India Company, Mr. Temminck, and try to portray for you the inimitable perfection in the art of stuffing birds, attained nowhere else; I should like to expatiate to you on the number and the beauty of the new genera of birds with which the noble eccentric, le Vaillant, has enriched the collection of his first benefactor and protector.[4]

Though it was one of the most important, Temminck's cabinet, which he had started around 1770, was only one of many in the immediate neighborhood. His friend Joan Raye van Breukelerwaerd, who also lived on the Heerengracht (at number 29), owned a large bird

*Jacob Temminck was baptized in Amersfoort on November 15, 1748, and died in Lausanne on August 15, 1822. He married Alida van Stamhorst on April 17, 1774. His father, Coenraad Temminck, born on December 13, 1687, in Amersfoort, was councilor and assessor as well as burgomaster (1746–47) in his native city, where he died on February 9, 1762.

collection too.* "These two cabinets together would make the most beautiful collection in Europe,"[5] was Levaillant's opinion after he had visited them in 1780. Also in Amsterdam at that time was L. F. Holthuyzen's attractive cabinet of birds, rich in species;† in Hasserswoude near Leiden was the notable collection of the magistrate (*bailluw*) Boers (auctioned in 1794); and in Rotterdam (until 1787), "the agreeable collection of the burgomaster, Gevers Arnz. It is not very large, but contains only carefully selected and very rare items" (Levaillant).[6] But none of these amateurs could compare with the Prince of Orange. Georg Forster, after a visit in 1790, wrote, "The splendor, rarity, choice, lavish display, and careful maintenance of the cabinet belonging to the Stadtholder's heir are not only striking at first glance, but with longer and more precise study increase one's admiration."[7]

The beauty and wealth of the menageries competed with the choiceness of the collections, until the Revolution finished them. Jacob Temminck had an enormous flight cage near his house, where he kept exotic finches and also sometimes bred them. They included *Euplectes orix*, *Foudia madagascariensis*, and *Estrilda astrild*. The Prince of Orange, the magistrate Boers, and Mevrouw Backer in The Hague also possessed splendid menageries, but the most remarkable was the one set up by Arnoldus Ameshoff on his country estate, "Amstellust," described as follows by Levaillant:

*Joan Raye, Heer van Breukelerwaerd, born in Paramaribo (Surinam) in 1737, the son of the then governor, was an affluent plantation owner, who remained a lifelong bachelor. Rudolphi, who visited his collection in 1802, wrote: "Raye van Breukelerwaerd is a rich Dutchman who, like many others, leads a quiet life, though earlier he traveled a great deal and was even in Constantinople. He is only a dilettante and has collected almost exclusively very beautiful and rare items; he has obtained wonderfully much from Le Vaillant, who also prepared most of the specimens for him. Raye could not say enough in praise of Le Vaillant's skill in taxidermy; in a few minutes, he said, a small bird was most exquisitely prepared; he also gave me a letter to him in Paris. Unfortunately I could visit the precious collection only for a few hours, and there were proper labels on only a few of the specimens — most of them had only French names. He himself did not know anything more to say about them." (*Bemerkungen ans dem Gebiet der Naturgeschichte*, I, 1804, p. 109). About 1818 Heinrich Kuhl studied the parrots in the "Museum Rayanum." It was appraised by C. J. Temminck at 52,000 guilders when, shortly after its owner's death in Amsterdam on March 19, 1823, it was offered to the state. In 1827 the heirs asked 15,000 guilders for it. It was finally auctioned publicly on July 3, 1827, whereby Temminck acquired 7,000 guilders' worth of zoological specimens (A. Gijzen, *'s Rijks Museum van Natuurlijke Historie 1820–1915*, Rotterdam, 1938, p. 161).

†Holthuyzen's collection was sold at public auction in October 1793 at Hamburg, whither it had been shipped "together with the mahogany cabinets belonging to it." It was expected to fetch a higher price in Germany than in Holland, which was simply flooded with such things. The sale catalogue, *Catalogus rerum naturalium rarissimarum*, was made by A. A. H. Lichtenstein, father of Hinrich Lichtenstein, later the director of the Berlin Zoological Museum (see Chapter 6). Included were 500 species of exotic birds, of which 38 were described for the first time. The *Catalogus* has itself been "rarissimus" for many years. Holthuyzen lived until 1818 on the Heerengracht at the Amstel corner (communication from the Amsterdam City Archives.)

In a very large enclosure, surrounded with wire screening, and with a long pond in the middle, can be seen a prodigious number of aquatic birds from all over the world, among which it is astonishing to note splendid mandarin ducks [*Aix galericulata*], with fans of feathers on their backs (see Buffon, plate 805); the beautiful Louisiana duck [*Aix sponsa*], the pelican, etc. Most surprising to me was how well all these different species lived together; most of them multiplied just as they do in their native countries, and, more remarkable, even crossbred with other species. This pond alone offers a lifetime of observations for a naturalist. On another broad tract large open-air aviaries have been built in a row. Each one contains one or more birds of a particular species. In one of them I saw the chincou [lappet-faced vulture, *Torgos tracheliotus*]; in another helmeted curassows [*Crax pauxi*]; in yet another the common curassow [*Crax rubra?*]; in a fourth the Peruvian curassow [*Crax mitu*]. Not only has Citizen Ameshof succeeded in breeding young of all these species — he has interbred them and has raised crosses that were themselves fertile. In this same area I saw the king vulture [*Sarcoramphus papa*], the demoiselle crane [*Anthropoides virgo*], the American crane [whooping crane, *Grus americana*], and two species of Indian cranes, the flamingo [*Phoenicopterus ruber*], scarlet ibises [*Eudocimus ruber*], crowned pigeons [*Goura* sp.] from the East Indies, the secretary bird [*Sagittarius serpentarius*], male and female ostriches, which have bred here, a beautiful species of African bustard, the agami [chestnut-bellied heron, *Agamia agami*], the Chinese peacock pheasant [gray, or chinquis, peacock pheasant, *Polyplectron bicalcaratum*], etc. The spacious garden on the estate contains small aviaries, 10 feet square, always at a little distance from one another, and enclosed by screening. Each of them has a small water holder in the middle, and a partitioned-off section for the birds. Here male and female American jacana [*Jacana jacana*] can be seen; there a pair of purple gallinules [*Porphyrio porphyrio*]; in short, all the rarest and most beautiful birds. In a huge chicken yard are domestic fowl of all species and in countless varieties, which have arisen from crossbreeding within a single genus. The pheasantry is also very notable, and contains almost all known species of pheasant, together with the varieties that have occurred through the crossbreeding of the different species, including those from China as well as those from other countries. Here one notices the marail with the white head [white-crested guan, *Penelope pileata*], the hoatzin [*Opisthocomus hoazin*], etc. Among the pigeons, which are numerous, I saw eight Nicobar pigeons [*Caloenas nicobarica*], at least as many green ones from Ceylon [*Treron* sp.], and many other rare species from the East Indies. In separate cages are all the species of parrots and parakeets. Finally one arrives at the small bird aviary.

> This is built next to the main house and forms a part of it. There is a room with one large window looking into a hall, from which one can enjoy the spectacle, and which at the same time is connected with a spacious aviary outside. In the summer all the small birds are left in this aviary, which is planted with bushes, and in which many of them breed, even though they are in a climate very different from their own. During the winter the small birds are kept in the room, where there is a stove, and all the large birds return to specially constructed buildings which are suitably heated. The enormous sums that must be needed to satisfy this expensive taste are incalculable, particularly because Citizen Ameshof spares no effort in enlarging his menagerie and, quite apart from the purchase of animals, its mere upkeep must be excessively costly.[8]

"I can remember," wrote C. J. Temminck after many years (1813),

> in my childhood being a guest at a dinner given by Mr. Ameshoff who, in order to show off the magnificence of his menagerie, served up at his table, during a feast worthy of the times of Heliogabalus, not only various curassows and different species of exotic pheasant, but also mandarin ducks from China and wood ducks.[9]

In such surroundings the boy grew up, endowed by nature with all the gifts, and by his family upbringing with all the inclinations, for ornithology. Without completely different influences, however, he would have remained a dilettante all his life, like his father, who had not in the least intended him to become a naturalist but, following family tradition, a merchant. In 1795 the father's great influence secured for the 17-year-old the post of "Algemeen Vendumeester" (auctioneer) for the East India Company, described by Susanna in 1858: "The auctions of colonial produce took place in the presence of this official. At the close of business he was awarded a gold ducat for every stroke of the hammer."[10] This prosaic though lucrative job automatically came to an end when the East India Company was dissolved in 1800. From then on young Temminck lived only for his hobby, because he wanted to become an ornithologist like François Levaillant, the famous family friend who had published the first volume of his splendidly illustrated folio, *Histoire naturelle des oiseaux d' Afrique*, in 1796, and had dedicated it with gratitude to Monsieur J. A. Temminck.

For Coenraad Jacob Temminck, as for many others at that time, Levaillant's books kindled the spark of enthusiasm for ornithology. He said later (1825): "I owe to Le Vaillant and his writings my first thoughts about and my first sally into natural history." But another influence was needed to channel his youthful ideas into scientific work, and this came from Germany.

Dr. Bernhard Meyer, court councilor to the Prince of Isenburg, was born in Hanau on August 24, 1767, and worked there as a doctor from the time he qualified until 1796, when he moved to nearby Offenbach. There, after taking over an apothecary's business, he was freed from financial worry and could devote himself to his real interests, which he maintained to the end of his life (January 1, 1836). He had been a keen student of botany and zoology from his childhood, and, since there was no scarcity of fellow enthusiasts in the Lower Main region, he was soon able to form a circle of friends who developed research into local natural history further than almost anywhere else in Germany. With Dr. G. Gärtner and Dr. J. Scherbius, Meyer worked on the flora of the Wetterau, but at the same time he studied and collected the birds of woods and fields in his area, tireless as a hunter and observer.* An excellent ornithologist and physician from Hanau, Dr. J. P. A. Leisler (1771-1813), was one of his favorite companions and, like Meyer, acquired a noteworthy collection of birds and mammals and achieved important results in his detailed study of it.

Meyer had just agreed to work with a Nüremberg schoolteacher, Johann Wolf, in publishing an ambitious and handsome book, *Die Naturgeschichte der Vögel Deutschlands*, when he was visited by the young Temminck during the summer of 1804. Temminck had been married on April 3 to Dionysia Catharina Cau, and regarded the journey to Offenbach as a sort of honeymoon, because he brought his young wife with him and they both stayed for six months under Meyer's hospitable roof.† It was not Temminck's first trip abroad; he knew London and had been fascinated by the Museum Leverianum, but with Meyer he benefited for the first time from a systematic course of instruction in the theory and practice of ornithology. He was grateful to him ever after. He became a close friend both of Meyer and of Leisler (well known for his studies of molting) — a friendship commemorated today in Leisler's *Tringa temminckii.* In the spring of

*Meyer wrote to J. A. Naumann on February 28, 1804, "Like you, I have been a hunter ever since I was young, especially with a naturalist's intent. Neither rain, wind, nor snow, neither cold nor heat, could check my ardor. Whenever I look at my stuffed birds, I can see every bush and little piece of ground or water where I stalked or killed this bird or that. Only German birds are shown in my collection, and no foreign migrant is allowed to appear there." The "creator" of his collection was Schaumburg, an intendant in Hanau, who was "perhaps the best taxidermist in the whole of Germany. I don't know how to stuff birds myself."

†After 25 years of happy marriage, Temminck lost his first wife in 1829. His second wife, whom he married in 1830, died after a few years. His third wife, Anna Agneta Smissaert, born in 1806 and married in 1835, survived him, with two of their sons.

1805, Meyer set out for Holland to pay a return visit and to study sea birds and sandpipers indoors and out, under the guidance of his experienced friend. Numerous traces of this visit can be found in Meyer and Wolf's later *Taschenbuch der deutschen Vogelkunde* [Pocket book of German ornithology] (1810).

What Temminck had learned in Offenbach was not simply a refined sort of museum routine; above all it was the method of scientific investigation, comparison, and classification, and the use of specialist literature. When he reached home, he put these experiences to work in exotic ornithology, because, after he had taken over his father's large collection, it was the exotic birds that interested him most. In 1807 he published a catalogue of the almost 1100 species in the collection, which he had considerably enriched by his own acquisitions.* He correctly followed the Linnaean principles of nomenclature, but did not dare to name the species that he considered new. In spite of his still limited experience he had been working since 1804 on an ambitious plan; in order to commemorate the Dutch pheasantries of "the good old days," he wished to describe the gallinaceous birds of the world in a comprehensive monograph. It was to be a sumptuous book in two volumes, like the folios that had become fashionable among the newly rich in France, where they enjoyed the favor of the imperial court and particularly of the Empress Josephine, who had a passion for flowers and colorful birds. By 1806 Jean Gabriel Prêtre, in Paris, had prepared almost all of the 160 plates for it. During a new visit to Paris in 1807 Temminck met not only his father's much admired friend, François Levaillant, who was most helpful, but also a second well-traveled naturalist, Baudin's† companion Leschenault de la Tour,

*Temminck's father seems to have owned about the first 450 species listed in his son's "Catalogue" in 1807, and numbered according to their date of acquisition. Among the low numbers are many species that Levaillant had brought back in 1784. Shortly after No. 450 appear birds from Java (from Leschenault) and from Timor, collected by the Baudin expedition. P. L. A. Nemnich, who saw Temminck's collection in 1808, wrote (1809): "The preparation of the birds is either done by Mr. Temminck himself, or else he has it done in Paris. The very attractive enamel eyes come from Paris, but are very expensive, up to a louis the pair."

†Captain Nicolas Baudin, who died on September 16, 1803, on the Île-de-France (Mauritius), led two government-sponsored expeditions that proved very productive for ornithology. The first, to the West Indies, lasted from 1796 to 1798, when France was at war with Britain, which commanded the seas. Because he was not allowed to fly the tricolor at sea, he sailed from Leghorn under the Austrian flag, and in return for permission to do this he promised to send the material collected by his naturalist, René Maugé, to Emperor Francis II's natural history collection. Maugé was particularly successful during a long stay in Puerto Rico (July 17, 1797, to April 13, 1798), but the emperor got nothing out of it. Baudin delivered everything in Paris, and saw himself well rewarded by promotion to captain and the command of a new French expedition, which the First Consul, undiscouraged by the failure of his Egyptian adventure, ordered in the middle of the war. The expedition was supposed to explore the coasts of New Holland (Australia), not

who had just returned from Java in July 1807, and who with charming generosity made his notes and bird skins available to the young enthusiast. Nevertheless, this illustrated monograph on the Gallinaceae never saw the light, because something else suddenly interfered — pigeons!

Some years earlier Pauline de Courcelles (1781-1851), a gifted pupil of Barraband's, had begun painting birds and had collected 72 plates of exotics. These, with a commentary by a young man still unknown, A. G. Desmarest, were published in 1805 by Garnéry in Paris as *Histoire naturelle des tangaras, des manakins et des todiers* — much to Levaillant's annoyance. After that, she changed over to pigeons, and was looking for a specialist to write a text for her pictures. She was finally successful in securing Temminck as a collaborator. He started work at once, and the first installment of the expensive work, dedicated to Louis Napoléon, King of Holland, and entitled *His-*

merely as a geographic but also as a natural history survey. No fewer than nine zoologists and botanists, carefully selected by the Academy of Sciences, were allotted to the corvettes *Le Géographe* and *Le Naturaliste*, which were armed defensively. They sailed from Le Havre, under Baudin, on October 19, 1800. Four naturalists, including J. G. Bory de Saint-Vincent, who later became famous, had to be landed on Mauritius at the end of April 1801. The others were Maugé, Péron, Levillain, Lesueur, and Leschenault de la Tour (1773-1826). After Baudin had reached the southwest corner of Australia and his naturalists had collected extensively on the shores of Géographe Bay (so christened at the time) at the beginning of June 1801, he continued his researches along the west coast as far as the Shark Bay (Bernier Island) area, and from there sailed to the unexplored island of Timor, off which both ships lay at anchor during the autumn of 1801. Then they set course for the southern tip of Van Diemen's Land (Tasmania). After Timor, dysentery raged on board, and on December 29, 1801, Levillain the zoologist died of it. Soon afterward the expedition lost its experienced ornithologist, René Maugé; he died on February 20, 1802, on board the *Géographe*, off the east coast of Tasmania. After various landings on this island the ships put in to Port Jackson (Sydney), because news had arrived that England and France had finally concluded a peace. There the hospitable English provided the crews with the several months of convalescence that they badly needed. The *Naturaliste* was then sent home with the weakest members of the expedition, by the shortest route possible. The crew of the *Géographe*, on the other hand, with three live emus from Kangaroo Island on board, destined for the Jardin des Plantes, landed for three weeks on the shore of King George Sound, near what is now Albany. On April 30, 1803, they returned to Timor, and set out for Mauritius only on June 3, leaving behind the botanist Leschenault, supposedly because he was ill. Leschenault soon sailed to Java, where he stayed until 1806, assiduously collecting plants and animals. No comprehensive report of the ornithological results of the Baudin expedition appeared (it had returned home without its commander on March 25, 1804, shortly before the emperor was proclaimed). It was mainly Vieillot who, from 1805, gave some incidental attention to the material in the Paris Museum and described many new species; even Lesson (1831) and Swainson (1838) were able to find more new items (nothing had been labeled with data by the collectors!). A number of the Javanese birds discovered by Leschenault were described by Vieillot (1816-1819), Cuvier (1817), and Charles Dumont de Sainte-Croix (1817-1818). From 1817 to 1822 Leschenault was in the field again; he traveled from Pondicherry through Hindustan as far as Bengal and finished in Ceylon, bringing back once again rich zoological collections to Paris.

toire naturelle générale des pigeons, par C. J. Temminck, Directeur de l'Académie Royale des Sciences et des Arts de Harlem . . . avec figures en couleurs, peintes par Mademoiselle Pauline de Courcelles, appeared as early as 1808, with the plates printed in color. Certainly this was a most promising beginning for Temminck in scientific publishing, because his skillful treatment of the subject and the inclusion of many new species placed him among the leading ornithologists. But his pleasure was short-lived because, when he returned to Paris from Amsterdam in 1811, he discovered that the artist (who had married the Dutch flower painter, Joseph August Knip, in 1808), had swindled him by publishing the last installments (ninth through fifteenth) with her name as sole author on the title page. Probably she intended to make an impression on the new empress, Marie Louise, who appointed her "première peintre d'histoire naturelle de Sa Majesté." Temminck and Mme. Knip parted in anger, and the pigeon volume remained a fragment because, when the glories of the Napoleonic empire ended shortly thereafter, the period of luxurious books temporarily ended too — much to the advantage of scientists, who never had been able to buy such expensive folios. Unfortunately, Temminck's old project of a sumptuous monograph on the Gallinaceae vanished as well.

Only after a lapse of many years was Mme. Knip able to publish, from 1838 to 1843, a further 69 of her plates of pigeons, with text by Florent Prévost, assistant naturalist in the Paris Museum. She seems to have been reconciled with Temminck at the time, because he allowed her to portray some of the species of pigeons discovered by members of the Natural History Commission.

Temminck's original project was replaced by something much more modest. He had his entire text on the pigeons reprinted in octavo in Amsterdam, and added to it the text on the Gallinaceae. These three volumes of his *Histoire générale des pigeons et des gallinacées* (1813–1815), completed after long and painstaking preliminary work, including a visit to London (1812?), firmly established Temminck's reputation among ornithologists. Immediately thereafter appeared his *Manuel d'ornithologie, ou tableau systématique des oiseaux qui se trouvent en Europe* (1815), the first thorough handbook on European ornithology, patterned on Meyer and Wolf's *Taschenbuch der deutschen Vögelkunde* (1810), but covering a much larger geographic area. Its success was so great that Temminck promptly prepared a second and considerably enlarged edition, but without letting his other interests slide. "I am the factotum of my many collections, sometimes a general and sometimes a private," he wrote to Johann Friedrich Naumann on October 14, 1816.

> I am the one who mounts all the birds, fish, and quadrupeds; in search of one or the other I vary my trips, sometimes by the seashore,

> or beside our lakes, or near our swamps; all the many animal skins that I use for exchange or as presents for my friends are prepared by my own hands; in the winter I concentrate on mounting quadrupeds and exotic birds, which I receive dried and sometimes badly prepared; then I work at classifying and preserving my collections, and the evening is devoted either to my literary work or to my social duties . . . I still have not succeeded in getting a good assistant, because in our country there is absolutely no taste for the study of natural history. The only idol worshiped by my compatriots is gold; it is considered almost ridiculous for anyone to concern himself with science, except simply as a hobby and a relaxation from office work.[11]

To widen his experience he was not content merely to send written questions to the most knowledgeable ornithologists in many European countries; he preferred to take a long journey himself in his own carriage, specially equipped for transporting bird skins. At the end of September 1817 he set out with his wife for Switzerland, visiting Offenbach and Stuttgart on his way, to spend the winter in Lausanne, where his old father and one of his sisters lived. In Bern, at the end of November, he was shown the museum by Friedrich Meisner. The route then went over the Mont Cenis to Turin, capital of the kingdom of Sardinia and also headquarters of Professor F. A. Bonelli, the zoologist, who showed Temminck, in his well-stocked museum, several still undescribed species of southern European birds (mostly discovered in Sardinia by Alberto de la Marmora).* After visiting the Ligurian coast and the North Italian lakes they traveled to Trieste and the north Dalmatian coast. At the beginning of June 1818, they took their carriage over the Semmering to Vienna, where Johann Natterer's collection, full of new items from southern Spain, had just arrived. The Temmincks went down the Danube to Pressburg, in order to see a little of the life of Hungarian marsh birds, and returned to Vienna by way of the Neusiedler Lake. In early July, after Prague, they visited Berlin, where Temminck became familiar with Lichtenstein's museum and took the opportunity, when he was with K. A. Rudolphi, to leaf through the proofs of the *Zoographia Rosso-Asiatica.* In response to a long-standing wish of Johann Friedrich Naumann, his close correspondent, Temminck visited him in Ziebigk at the beginning of August 1818, and then traveled to Leipzig, where

*This museum, founded by Bonelli in 1811 on the lines of the Paris Museum, contributed greatly to the development of ornithological activity in Italy. Year after year its collection of birds was enriched, particularly by Alberto de la Marmora (1789–1863), who, because he had been involved as an officer in the constitutional struggles of 1821, was not permitted to leave Sardinia, and as a result acquired an outstanding knowledge of the island. Later he was able to make still further ornithological discoveries there (*Falco eleanorae* Géné, *Larus genei* Brème, and so on). Bonelli was followed as director of the Turin Museum by Giuseppe Géné (1780–1847) and then by Filippo de Filippi (1814–1867).

he was pleased to see a number of birds from Siberia and Kamchatka, including *Cinclus pallasii*, that had been sold to Professor C. F. Schwägrichen by the Leipzig engraver C. G. H. Geissler (who had been entrusted with them by Pallas as models for the plates to the *Zoographia Rosso-Asiatica*). On the journey back they were joined in Kassel by Bernhard Meyer and Heinrich Boie, with whom they visited at Neuwied Prince Maximilian zu Wied, who had come back from Brazil a year earlier.

In 1820 appeared the first two volumes of the new *Manuel*, eagerly awaited by fellow specialists. For a long time it remained the standard work on European ornithology. The last two volumes had to wait until 1835 and 1840, because by 1819 Temminck was already absorbed again in exotic ornithology, more rewarding to his enthusiasm than European. For a long time he had wanted to make his bird collection the most complete in the world. His only competitor among private collectors was William Bullock, in London, who had acquired the lion's share at the auction of the Museum Leverianum in 1806. A collector not only of birds but also of antiquities, paintings, curiosities, and every sort of stuffed animal, in 1812 he had built at his own cost an "Egyptian Hall," where he set up a "London Museum," and steadily added rare items to it. Among his many patrons was Sir Joseph Banks, who about 1812 presented him with his entire zoological cabinet, full of birds brought back from Cook's second and third voyages. Anyone wishing to see the South Sea birds described by Latham, or the newly discovered Australian species, went to Bullock, who owned more than 3000 birds. The British Museum, on the other hand, was still in a sad state, although William Leach had persuaded the trustees to acquire George Montagu's British birds in 1816. A German visitor wrote to Oken's *Isis* in 1821:

> In Room No. VIII one can see a collection of birds, incredible as belonging to the British Museum, because this Order, of which I had just received a remarkable impression [at Temminck's house in Amsterdam] is so absolutely miserably represented. In addition, the items are badly cared for, and do no honor to the institution.[12]

Then suddenly, in the spring of 1819, the whole "London Museum" came up for auction. This was Temminck's great chance. In the Egyptian Hall he found the élite among enthusiastic bird collectors, not only Englishmen like Lord Stanley* and William Leach (representing the British Museum), but also others who had come from a great dis-

*Lord Stanley, who had already appeared as a buyer for his private collection at the auction of the Museum Leverianum (1806), had built a museum and a large menagerie at Knowsley, the family seat. After he died in 1851, the city of Liverpool

tance. Fector was there for the Imperial Natural History Cabinet in Vienna; Hinrich Lichtenstein, the director, for the Berlin Zoological Museum; and the rich Baron Meiffren Laugier from Paris. Innumerable items were auctioned between April 29 and June 11,* and, in spite of sharp competition, Temminck was able to secure 536 birds belonging to 363 species, for which he paid the high price of £445. Time later proved how well he had chosen. His collection had now grown to over 4000 items and contained nearly all the genera then established; he even maintained that his cabinet was "incontestably the richest and most complete ornithologically of all those existing in Europe," though certainly the Paris Museum's bird collection must have been more important than Temminck's at that time (on January 1, 1809, it contained no fewer than 3411 specimens of 1903 species, and from then until 1819 it was enlarged still further). Temminck kept the collection to feast his own eyes on in his spacious Amsterdam house, to which the public was admitted twice a week. The birds stood on white pedestals, in brown cupboards painted white inside and equipped with glass doors revealing the whole of the interior.[13]

Now at last Temminck, assisted by his young friend Heinrich Kuhl, could start on a comprehensive and systematic book that was to be called *Index général d'ornithologie.* For this purpose, some months after returning from London, he went to Paris for a long stay, taking his wife; he had last been in Paris in the autumn of 1815, as a first lieutenant of volunteer cavalry in the occupying forces. In January 1820, he visited Baron Meiffren Laugier de Chartrouse who, like Temminck,

---

inherited his important zoological collections, including many items from Cook's voyages. One of the show pieces in his museum was the stuffed kiwi, on the basis of which George Shaw (in the year of his death, 1813) had created the species *Apteryx australis.* Lord Stanley, who had obtained this prize from Shaw's estate, could congratulate himself on owning the single known example (until 1837) of a species whose place in classification had given rise to most remarkable speculations (the elderly John Latham, misled by Shaw's plate, described it in 1824 as an "Apterous Penguin"!). A European first saw two live kiwis in 1836; Lord Stanley (who on his father's death in 1834 had become Earl of Derby) bought one of them and presented its body to the Zoological Society's museum, where Richard Owen made a detailed anatomical study of it in 1838.

Gleanings from the Managerie and Aviary at Knowsley Hall was privately printed (Knowsley, 1846), with text by John Edward Gray and illustrations from life by the remarkable animal artist Edward Lear, whose *Illustrations of the Family of Psittacidae, or Parrots* (London, 1832), also drawn from living specimens (in the aviary of the Zoological Society), had first attracted Lord Stanley's attention. Lear also contributed a series of striking plates (particularly of birds of prey) to the earlier works of John Gould, before weakening eyesight turned him to landscape and limericks. See Vivien Noakes, *Edward Lear: The Life of a Wanderer* (London, 1968).

*After this auction Bullock set out on his travels. As one result of his six months' stay in Mexico, where he speculated in gold mines, he brought a collection of birds back to London in 1823, from which Swainson was able to describe many new species.

owned a collection of exotic birds. Some time earlier Heinrich Kuhl had become his guest, and now the Baron steered the conversation toward a plan that he had been nursing for some time. This was nothing less than to continue the series of pictures in which the younger Daubenton, with his 973 plates of birds for the *Planches enluminées*, had sought to represent most of the species known in his day. Because the Baron did not feel capable of writing a scholarly commentary, the connection with Temminck, a specialist both learned and well-to-do, was very important to him. Temminck was ready to cooperate immediately, and he collected a number of birds to serve as models for the paintings and engravings that preceded the making of hand-colored plates. The artists were his old favorite, Prêtre, and Laugier's man, Huet. The work progressed rapidly, and on September 1, 1820, the first installment of a new and magnificent folio could be published, with the names of both authors and embellished with a foreword by Georges Cuvier. It was *Nouveau recueil de planches coloriées d'oiseaux.*

Further installments followed at short intervals.[14] At first none of them was accompanied by a text, but served only to illustrate and name species of birds that had been discovered only recently, because, as if Napoleon's fall had signaled to adventurous naturalists that their moment had come, new items soon began to arrive from all over the world. In 1815 Prince Maximilian zu Wied had started for Brazil and joined the scientific expedition of Friedrich Sellow and Bernhard Meyer's protégé, Freyreiss. Also in 1815 the Russian brig *Rurik* set out for the Pacific with Chamisso, Eschscholtz, and Choris on board, but, most important, the Dutch government began to sponsor systematic research into the natural history of the Dutch East Indies. This soon resulted in the foundation of a new repository for the material collected, the Rijksmuseum van Natuurlijke Historie.

Temminck originated the idea of this central museum. To realize his farsighted and ambitious plan, during the winter of 1818–19 he had started discussions with A. R. Falck, the minister for education, who willingly cooperated. So in January 1820, Temminck was able to start his journey to Paris full of confidence, and there, because he had been most generously entrusted by the directors of the Jardin du Roi with the labeling of the bird collection, he was able finally to defeat his opponent, Vieillot. "All of Illiger's names have been adopted, and all of Vieillot's new names canceled," he wrote on May 20, 1820, to Lichtenstein, glad of his visible triumph over his bitter opponent in nomenclature and systematics.

Louis Pierre Vieillot (born on May 10, 1748, in the Norman town of Yvetot), 30 years Temminck's senior, was, like Temminck, an amateur ornithologist who had originally been in business. Keeping caged birds soon stimulated an interest in ornithology that consoled him during a joyless life. About 1780 he seems to have traveled to Santo Domingo for the first time; after his return he offered Buffon his ornithological notes, but Buffon had already finished his book on birds. Shortly before the Revolution broke out he retired with his family to Santo Domingo to engage in business again. But it soon became clear that he had avoided Scylla only to fall into Charybdis. Bloody uprisings occurred even on that remote island, with blacks rebelling against whites, and in 1792 or thereabouts Vieillot had to flee from the appalling slaughter to North America. The rich tropical fauna of Santo Domingo had kindled his enthusiasm for the study of exotic birds; in the United States he continued his hobby, and when he returned destitute to Paris with his wife, in September 1798, (after his two daughters had died of yellow fever on the way), he had enough self-confidence to continue the *Histoire naturelle et générale des colibris, oiseaux-mouches, jacamars et promérops* that his friend, J. B. Audebert (who died in 1800) had already begun. Audebert was a famous painter of flowers and birds, and for his exquisite plates Vieillot wrote a text, starting with the "oiseaux-mouches" (hummingbirds). Charles Dumont de Sainte-Croix (1758–1830), a well-to-do amateur ornithologist, took him on and gave him a small job in his law office, so that he could continue educating himself in his chosen field through study and a lively correspondence. With the support of various collectors, chiefly Dr. Perrein of Bordeaux (who had returned about 1800 from a long stay in Malimba, near the mouth of the Congo, bringing with him many unknown species of birds), Vieillot was able to publish his own luxurious volume on tropical finches and weaverbirds (*Histoire naturelle des plus beaux oiseaux chanteurs de la zone torride*, Paris, 1805–1809). At about the same time (1807) he attracted attention with his *Histoire naturelle des oiseaux de l'Amérique septentrionale* (never finished), illustrated by Prêtre, in which he included many of his own personal observations. While he was writing it he finally gave up his opposition to binomial nomenclature. Saul became Paul.

Between 1812 and 1820 no one studied more eagerly than Vieillot the exotic birds exhibited in the galleries of the Paris Museum, to which were added in quick succession the collections of Macé, Maugé, Péron, Leschenault, Delalande, Auguste de Saint-Hilaire, and others. Though the scientific staff thought little of him, the number of bird species he described (especially in the second edition of the *Nouveau dictionnaire d'histoire naturelle*, 1816–1819) was very considerable. Not content with this, he attempted (under the influence of Illiger's revolutionary *Prodromus*, 1811) to reform the existing ornithological systematics. In his *Analyse d'une nouvelle ornithologie* (1816) he sought to remain true to the natural relationships by introducing numerous new generic and family names, which was at first almost universally resented. "In a state close to indigence" (Lesson's description of his situation), he lived out his days as an eccentric, "often depriving himself of necessities for the sake of his special interest." At the end of his life he was in Rouen, where he died, old and blind, at the beginning of 1831.[15]

Soon after Temminck returned to Amsterdam, a royal decree of August 9, 1820, named him director of an imperial museum to be organized in Leiden. The buildings assigned to it were in fact overflowing, because the natural history cabinet, which had already served the University of Leiden for decades, had recently been increased by the remains of the once imposing Stadtholder's Cabinet, brought back, by royal order, to Holland from Paris in October 1815, by Professor Brugmans (d. 1819). And now the Royal Museum of Natural History in Amsterdam, founded in 1815, and Temminck's large private holdings were to become part of the University Museum at Leiden, for Temminck had offered to present his collection to the state if, in view of its value, he were allowed a life annuity of 3000 guilders in addition to his director's salary of 2000 guilders. A new wing was thus urgently needed. Although started immediately, it was only in 1823 that the building was sufficiently advanced to allow Temminck to move his collection from Amsterdam to Leiden.[16]

If an outstandingly qualified collector was wanted for the direction of the new institution, containing 5800 mounted birds, there could have been no happier choice. Untiring in organizing expeditions, in arousing the interest of travelers, in corresponding with people all over the world, and in exchanging and selling specimens, Temminck ensured that for decades his museum, in which the love of order and beauty were ruling passions, should take precedence over all others, and his knowledge of the world, his wit, and his lively and spontaneous charm won him friends everywhere, not least at court.

In future these qualities, rather than his scientific performance, distinguished Temminck. He had received very little formal education, and though he had had a French-Swiss tutor as a boy, he knew neither Latin nor Greek (which was noticeable when he named new species), and he had never shared in the blessings of a properly organized course of university study. So his understanding of the tasks of the zoologist progressed very little beyond the simple collection and description of specimens according to their external characteristics, a limitation which, however, combined with his excellent memory, enabled him soon to excel all his contemporaries as an expert on birds and mammals, and to maintain for many years the reputation of being the greatest authority in this field.

In the meantime, scientific tasks connected with the directorship of the steadily growing museum arose, such as only an extremely versatile and thoroughly trained zoologist could have mastered. Recognizing in time the limits of his own capability, in the summer of 1821 Temminck engaged as curator the young Dr. Heinrich Boie, from Heidelberg, with whom we shall deal in the next chapter. At Boie's urgent suggestion, the osteologist Dr. Heinrich Christian Macklot, one of his Heidelberg friends, was added to the staff in 1822, and the admirable Salomon Müller joined them as preparator in 1823. It was with mixed feelings that Temminck saw these three valuable assistants depart for Java in 1825. But before matters reached this point he had looked about for a substitute, and when an odd character called J. J. Kaup, a student from Darmstadt, did not work out well, he turned, in 1825, to the young Hermann Schlegel, who remained with him during his lifetime and became his successor.

The rapid development of the Rijksmuseum is intimately connected with the zoological exploration of the Dutch possessions in the East Indies, and although the first inspiration for it did not come from Temminck, his museum reaped the harvest right from the beginning and, from 1820, he himself provided the impetus for all further scientific expeditions.

8

The Natural History Commission of the Dutch Indies

By the London Treaty of August 13, 1814, England restored to Holland her former possessions in the Malay Archipelago.[1] Soon afterward King William I, who was interested in science and concerned about the economic development of the colonies, was persuaded to send Professor Carl Reinwardt to accompany the first governor general of Java, Baron van der Capellen. Reinwardt, born at Lüttringhausen near Barmen on July 3, 1773, was a learned and versatile university teacher and the director of the Royal Cabinet of Natural History in Amsterdam, which had just been founded. As "director of matters pertaining to agriculture, arts, and sciences on Java and the neighboring islands," and lavishly supplied with funds, he was supposed to spend several years on research into natural resources. He started on his journey on October 29, 1815, and, with the best of support from his friend the governor general, from 1816 to 1820 pursued a wide variety of studies in Java; in 1818 he founded the Botanic Garden at Buitenzorg (Bogor). In 1821 he voyaged on the 200-ton brig *Experiment* to the eastern part of the Malay Archipelago (Sumbawa, Timor, Banda, Amboina, Ternate, and northern Celebes) and, after another period in Java, returned to Holland at the end of October 1822. There he handed over his material to the new Rijksmuseum and assumed the chair of natural history at Leiden left vacant by Brugmans' death. Because hardly any zoologists had been to the islands before him, his collections consisted almost entirely of new species. From 1823 on Temminck published

Reinwardt's most startling ornithological discoveries in his *Planches coloriées*, but without making proper use of his notes.

In the meantime a new enterprise had begun that promised still greater results for zoology. Two young scientists, Dr. Heinrich Kuhl and Dr. Johan Coenraad van Hasselt, had sailed to Java in 1820, and the hope of sensational surprises kept Temminck in a fever of expectation.

Although this expedition had begun before Temminck's nomination as director of the Rijksmuseum in Leiden, it was made possible only by his vigorous help as a private scholar. Kuhl was the first in his long series of assistants, almost all from Germany. Their academic training had roused the first stirrings of a lofty ambition which, once kindled and nourished, could no longer be confined within their native land.

Even at 23, when he undertook this important journey, Heinrich Kuhl was a famous scientist, an honorary doctor of philosophy and natural history, a member of the Imperial Leopoldine-Caroline Academy of Natural Scientists, and author of notable treatises on zoology and anatomy. A son of a provincial law court director, he was born at Hanau (the city of the immortal Rumphius) on September 17, 1797. He came very early under the influence of the Wetterau naturalists, so that his boyish enthusiasm for the study of animals and plants was guided into orderly channels by such famous men as Leisler and Bernhard Meyer. Kuhl was particularly attracted to Leisler, after whose death on December 8, 1813, he regarded himself as the guardian of his beloved mentor's intellectual inheritance. When Goethe visited Leisler's house on July 28, 1814, "to see the deceased's stuffed animals," he noted in his diary the presence of "Kuhl, a well-informed pupil of Leisler's."[2] Further, Goethe reported in 1816 that

> in his last years Leisler devoted himself to the study of bats but, trusting to his excellent memory, wrote nothing down. All his results would have been lost to us if a young man, the last of his pupils, had not mastered so much himself that he was able to write a monograph on these unusual creatures, which is soon to appear.[3]

This monograph, Kuhl's famous first fruits, was published in Hanau in 1817 as *Die deutschen Fledermäuse.* In the meantime he had left the Gymnasium of his native town in 1816, intending to study medicine in Heidelberg. But a meeting with Theodor van Swinderen, professor of natural history, traveling in Germany for further study, who had followed Bernhard Meyer's advice in going to Hanau especially to visit the "extraordinary youth," convinced Kuhl that he should follow van Swinderen to Groningen and there devote himself to zoology and anatomy. Arriving at the end of September 1816, he struck up a close friendship with a contemporary born in Doesburg in 1797, Johan Coenraad van Hasselt. He was hard at work on anatomical research, and was able to offer copious new stimuli to Kuhl's omnivorous hunger for knowledge. They went together on a long walking tour of Germany (Bremen, Brunswick, Berlin, Halle, Ziebigk, Jena, Heidelberg), during which they visited ornithologists of note. When, after their return, the two friends wished to use their museum experience to study nature in the tropics, and to follow the example of Rumphius and Reinwardt by traveling to the East Indies, they found influential backers in van Swinderen and Temminck.* First of all Kuhl received a grant from the Dutch government that enabled him to go to London with Temminck for the spring and summer of 1819, and the next winter to go to Paris with van Hasselt. Everywhere he enthusiastically studied exotic animals, not only because he wanted to prepare himself for the Dutch colonies but also because he had decided to treat monographically the birds of prey and the parrots.† He worked with such zeal that the first proofs of his *Conspectus psittacorum* reached him from Bonn before he had left Paris, at the end of February 1820. Those who met him personally were strongly impressed by his uncommon gifts;‡ Prince von Wied

*As early as March 5, 1817, Temminck had written to J. F. Naumann, "Young Kuhl from Hanau . . . will accompany me and will also help me by working for me this year; he is a young man of whom I think a great deal and whom I wish to train, in order some day to send him to travel in the little-known regions of our colonies in the Indies." (Archives of the Naumann Museum, Köthen.)

†Kuhl, eagerly looking for new species of parrot, found some surprises awaiting him in London at the Linnean Society, which in 1817 had bought an important collection of vertebrates from Australia for its small museum, at a cost of £220. It had been assembled by George Caley (1770–1829) who, on behalf of Sir Joseph Banks, had collected chiefly plants but also animals from 1800 to 1810 in the area around Port Jackson and in Tasmania, using Parramatta as his base. After Friedrich Wilhelm Sieber had bought a number of his bird skins for the Berlin museum in 1811, Caley had been unable to find another buyer. Several years elapsed before all his ornithological discoveries were studied by Vigors and Horsfield and were published, with pertinent biological details, in an excellent treatise in the *Transactions of the Linnean Society*, 15 (1827): 170–331. In 1817 the Linnean Society had been presented with further ornithological novelties (68 birds) from Australia by the famous botanist Robert Brown, who had accompanied Captain Flinders on his voyages of exploration from 1801 to 1803, on board the *Investigator*, along the south and east coasts of Australia, and through the Torres Straits into Arnhem Bay.

‡On March 6, 1819, Heinrich Boie wrote from Heidelberg to Johann Friedrich Naumann: "At the moment Kuhl is receiving money from the Dutch Government, to last until a year from Michaelmas, so that he can go to Paris and London and then leave for the East Indies — lucky man! If Heaven will only preserve his cheerfulness and his health his name will be eternally celebrated! In fact, I could wish that the whole of Europe could show a young man of his age with the same knowledge of all branches of zoology! Tiedemann is not entirely satisfied with his comparative anatomy, but he is working on that now more than anything else, and will soon rank with the best. At present, and for the last month, he has been helping Temminck with the working out of new systematics." (Peter Thomsen and Erwin Stresemann, eds., "Briefe, gewechselt in den Jahren 1818 bis 1820 zwischen Heinrich Boie und Johann Friedrich Naumann," *Journal für Ornithologie*, 94 [1953]: 15.)

(whom Kuhl had visited in Neuwied in August 1817, immediately after the prince's return from Brazil), Lichtenstein, Oken, Robert Brown, Cuvier, and Alexander von Humboldt received him warmly, and no one was surprised when a royal decree of May 2, 1820, officially commissioned him and van Hasselt to "travel in the East Indies at the direction of our governor general of the Dutch possessions, for the particular purpose of increasing scientific knowledge of the natural resources of these countries," for a term of four to six years.

Fate had been kind to them: now there was no time to lose. Even during his farewell visit to his parents in Hanau, where he had gone from Paris, he quickly put in order some finished and half-finished manuscripts. Besides many zootomic studies, they included a revision of the shearwaters and petrels (Procellariidae and Pelecanoididae). He had them printed in Frankfurt under the collective title of *Beyträge zur Zoologie und vergleichenden Anatomie*, for who knew if he were destined to return from his long journey?

On July 11, 1820, the ship that was to take Kuhl and his friend van Hasselt, a brand-new doctor of medicine, to Java weighed anchor in Texel. Accompanying them were the "dissector" Gerrit van Raalten as preparator and J. Keultjes as draftsman. The high hopes of the natural scientists followed them. The Heidelberg anatomist Tiedemann wrote that "no one ever undertook a scientific journey so well prepared," and the accuracy of his judgment was proved by Kuhl's first reports, sent off during the voyage from Madeira, from the Cape, and from the Cocos (or Keeling) Islands. In December 1820 they arrived. Baron van der Capellen, a sensitive and perceptive patron of all scientists, provided the young men with every conceivable assistance and found them a house in Buitenzorg, where the endless virgin forests on the nearby gigantic volcanoes promised many secrets to explore. The Javanese mountain fauna was almost entirely unknown. Most of Reinwardt's Javanese collections had disappeared beneath the waves when four of the ships to which he had consigned them were wrecked on the voyage to Holland; the remainder, which had arrived in 1820, were still waiting to be worked through. Scientists knew just as little about the years of research devoted to Javanese natural history by Dr. Thomas Horsfield (a physician born in Pennsylvania), because the magnificent results described in his *Zoological Researches in Java and the Neighbouring Islands*, achieved principally in the period from 1811 to 1818, began to appear in London only in 1821. So the joys of discovery almost overwhelmed the two friends, and urged them to the utmost effort. After only a few months Kuhl wrote to Temminck that, from now on, the birds of western Java would be almost as well known as those in any part of Europe. As he wrote in a letter from the foot of Pangrango on July 18, 1821:

> We are continuing in the same way as we began. Hardly a day goes by without our establishing a new genus or determining a new species, and we are still collecting a great deal to which we pay no special attention at the moment, but intend to study only in Europe. The plants provide almost as much material for research as the animals, or even more, and each area, each mountain, brings us new forms almost all unknown in Europe. Also I hope that, once I get back to Europe, I shall not be considered one-sided because I am attempting a detailed study not only of the fields that I specially enjoyed working on in Europe, but am trying to pay attention to everything, and have taken up several very difficult branches of study of which in Europe I had scarcely an idea.[4]

On August 8 Kuhl wrote home: "We have just returned from an extremely fatiguing mountain trip. In fact, in spite of all the attempts that failed, we finally succeeded in reaching the peak of Pangrango, which is 8500 feet above Buitenzorg and therefore 9400 feet above sea level." It must have been his last communication, because his happy and sun-drenched life came to a sudden and gloomy end. Shortly after he had returned to Buitenzorg, laden with new plants and animals from one of his strenuous journeys, he took to his bed with an inflammation of the liver, which killed him a few days later, on September 14, 1821.

Kuhl's early death was an indescribable loss to the natural sciences. Surveying all that he accomplished in his short existence, one understands his teacher, Professor van Swinderen, who compared him to Pallas and declared that he had expected still greater things from the young man.[5]

For a treatise on the Javanese fauna that Kuhl and van Hasselt had planned to write jointly, they had prepared a large quantity of specimens, drawings, anatomical descriptions, and anatomical sketches, in addition to their collections. When these arrived in Leiden, Temminck wrote to Lichtenstein, in a letter dated July 30, 1824:

> On my return about two weeks ago I found 32 cases arrived from Java, the Moluccas, and Japan, and we expect another 35 from the same countries. What I have seen of this brilliant collection, put together by our dead friends Kuhl and van Hasselt, shows the indefatigable zeal of the two naturalists; as a result of their efforts we shall have a fauna of the island of Java, and the natural products of of this region will be better known than all our European ones. We have already counted twenty new species in the genera of *Chiroptera* alone. The fish collection is enormous, and every species is represented by skeletons, by prepared specimens as in spirits of wine, and

> by drawings made in lifelike colors at the places where they were caught. We have received more than two hundred species in skeletons of mammals and birds, from the largest to the smallest; everything was prepared with a care and completeness that leave nothing to be desired.[6]

According to a later count, Kuhl and van Hasselt sent from Java 200 skeletons, 200 mammal skins representing 65 species, 2000 bird skins, 1400 fish representing 420 species, 300 reptiles and amphibia representing 90 species, many insects and molluscs, as well as manuscripts and 1200 drawings!

But Temminck did not have much idea of what to do with this treasure. A small number of the bird species he published serially, accompanied by meager descriptions, as plates in his *Planches coloriées;* he also described some of the new mammals — and that would have been all, if Heinrich Boie had not taken over the reptiles, and Cuvier and Valenciennes the fish. Nothing more was heard of the manuscripts and illustrations.*

Van Hasselt remained behind in the deepest depression. Death had robbed him not only of his beloved friend, but also, two days later, of another companion, the competent artist Keultjes, whose job was taken over by van Raalten to the best of his ability. On his deathbed Kuhl wrote a farewell letter to Baron van der Capellen in which he recommended Dr. Heinrich Boie as his only suitable successor — a last wish that the Baron promptly forwarded to the minister for the colonies. But to van Hasselt's extreme disappointment Temminck rejected the proposal out of hand.†

*Providentially, Baron van der Capellen requested that Kuhl's and van Hasselt's manuscripts and drawings in Buitenzorg be copied, which was done by van Raalten and the Baron's physician, Dr. Kollman. The originals came into the possession of Dr. Blume (who had been the director of the Buitenzorg Botanic Garden since 1822). He took them with him to Holland without giving them any more than casual attention thereafter. Even the copies had been made in vain because, after Boie had used them, they disappeared until 1838, when they were accidentally discovered in a chest that came to light during the demolition of a coachhouse in Buitenzorg, used until then by Diard. According to Junghuhn they were then taken over by the "hortulani" of the Botanic Garden, Teysmann and Hasskarl. One of the two versions must finally have reached the Leiden museum, because Finsch in 1906 ("Zur Erforschungsgeschichte der Ornis Javas," *Journal für Ornithologie*, 54: 317) asserted that Kuhl had written on 211 Javanese birds, and had given the first descriptions and names to 68 of them.

Dr. Kollmann selected a number of duplicates from the collection of the two dead friends and sent them to Dr. Johann Wagler in Munich, enabling him to describe new Javanese birds (for example, *Columba unchall* and *Leptopteryx cruentus*) in his *Systema avium* (1827). Others went to Jacob Sturm in Nuremberg.

†On September 28, 1822, Boie wrote from Leiden to Lichtenstein: "But what shall I tell you, dear Professor, of the deep grief into which we have been thrown by the death of the irreplaceable Kuhl? We thought of him in all the work we were doing, and we looked forward to his return with pride and joy, and now he has

In a letter to Boie on December 25, 1822, from Tjikandje, van Hasselt complained:

> I am now entirely on my own and practically without help, in the midst of a natural world so indescribable and rich that when Kuhl was alive he and I used often to say to each other: there is no end in sight, it is impossible to grasp everything. Therefore do everything you can to put an end to this — to me unintelligible — governmental procedure.

Temminck had to be prodded again and again before he finally decided to yield; on December 5, 1823, the king signed a decree commissioning Dr. Boie, Dr. Macklot, and Salomon Müller to proceed to the Dutch East Indies for a period of six years, to continue the researches begun by Kuhl. This was too late for the unfortunate van Hasselt, for shortly thereafter it was learned that he had died in Buitenzorg on September 8, 1823. He had ranged over the Residency of Bantam for more than a year, had climbed the volcanoes of Karang and Pulosari, and had got as far as the uninhabited western tip of Java. Baron van der Capellen buried him next to his friend Kuhl in the Buitenzorg Botanic Garden.

In fact, two whole years elapsed between the royal decree and the actual embarkation of the three appointees. During that time they continued as Temminck's assistants in the Rijksmuseum and helped many scholars who had traveled long distances to study the now world-famous collections. Among them were Valenciennes, who hurried over from Paris to fetch the Javanese fish; Jardine and his friend Selby who, after their visit to the Leiden museum in August 1825, themselves undertook to publish something like Temminck's *Planches coloriées*, which soon afterward began to appear in parts as *Illustrations of Ornithology*, 1826–1843; and Dr. Johann Wagler, a German systematist who later became famous.* He never forgot his experience; he wrote in 1826:

already finished a career that had hardly begun. When he was dying he believed that I should be able to continue at least part of his work, and asked the Governor van Capelle [*sic*], who was favorable disposed, to recommend me to the present Ministry as his successor, without knowing that I had already received an appointment in Holland and apparently may still be able to speed the fulfillment of his wishes. But these very circumstances together with the lack of staff in our museum which seems to make my presence necessary for the moment, are the justification for leaving his last wish unsatisfied, and perhaps it can only be satisfied after years have elapsed, and I must console myself with the conviction that later I shall at least be better trained to follow in the footsteps of my unhappy friend." (Archives of the Berlin Zoological Museum.)

*Johann Wagler, born on March 28, 1800, the son of an eminent lawyer in Nuremberg, studied natural history in Erlangen from 1818 to 1820. In the latter

> I am seized by nostalgia whenever I think back to that happy time when Boie and the worthy Macklot would return with me from collecting on the beach, and we would begin intimate discussions, working through the system of Amphibia devised by the good Scheuchzer and Merrem, so passing the hours in the most pleasant manner imaginable.

Heinrich Boie, the oldest and most experienced of Temminck's assistants at that time, came from an extremely cultivated background. He was born on May 4, 1794, at Meldorf in Holstein, the younger of two gifted sons of a well-known father, who was none other than the "Hainbund" poet, Heinrich Christian Boie (1744–1806), editor of the *Musen Almanach* and friend of Lessing, Gleim, Bürger, J. H. Voss, and many other philosophers and poets. His brother Friedrich (1789–1870), a lawyer by profession, is known in the annals of ornithology for his long series of articles and his friendship with J. F. Naumann. But Heinrich Boie, who died young, was more gifted intellectually and more widely learned. Having to choose a paying profession after he finished the Gymnasium in Hanover, he began to study law much against his will, first at Kiel and then, after 1814, at Göttingen. But the lectures by the famous Blumenbach proved more fascinating than the lectures on law. Birds had been Boie's chief interest ever since childhood, and when he went on a walking tour along the Rhine during the autumn holidays in 1814, he did not fail to stay in Offenbach, to make the acquaintance of Court Councilor Meyer and his unique bird collection. Here he also met Heinrich Kuhl for the first time. In the autumn of 1815 he moved to Heidelberg for further legal study, where he was looked after by his uncle, Johann Heinrich Voss, the translator of

year Professor Johann Baptist von Spix, who had just returned with Martius from Brazil, invited him to transfer to Munich and become an assistant in the Museum of the Academy of Sciences. In Munich, and soon elsewhere, his remarkable gifts and his work on the snakes that Spix had brought back attracted so much attention that, in 1825, King Maximilian Joseph gave him a stipend for travel and study in foreign museums. He went first to Étienne Geoffroy Saint-Hilaire and Lesson in Paris, then for a longer time to Temminck in Leiden, and finally to London, where he became a friend of Robert Brown, Vigors, and Gray. When Spix died on May 15, 1826, Wagler became professor extraordinary of zoology in the newly founded University of Munich, and director of the attached Zoological Museum. One of the products of his journey was the *Systema avium*, published in 1827, the first attempt since Latham's *General Synopsis of Birds* (1781–1785) to list all known bird species, together with their exact descriptions. The enterprising scheme bogged down, perhaps because in 1828 Lesson published a practical little handbook (*Manuel d'ornithologie*) of a similar kind; it was continued only in the reworking of some genera (after Wagler had visited the Berlin museum in the autumn of 1828) and in the excellent "Monographia psittacorum" (1832). Shortly after this appeared, Wagler succumbed, on August 23, 1832, to injuries received a few days earlier when he had been hit by a stray shot while hunting. In him Germany lost an experienced and gifted systematist of vertebrates who, if he had lived longer, would probably have become outstanding.

Homer. However, after he had followed his natural inclinations by attending zoological lectures by the anatomist Tiedemann, who had begun teaching at Heidelberg in 1816, he could no longer resist the compelling urge to become a zoologist by profession. Tiedemann did everything he could to strengthen his decision and in the autumn of 1817 even got Boie a post as curator in his newly founded zoological collection, so that he was now free from financial pressure and could devote himself entirely to his favorite occupation. In the summer of 1818 Kuhl appeared from Groningen with his friend van Hasselt, attracted to Heidelberg by Tiedemann's famous lectures and his zoological collections, and was overjoyed to be able to talk to such an expert authority as Boie was, even at that time. In the meantime Kuhl had probably also heard of him from Temminck, who, hard upon the publication of his *Manuel d'ornithologie* in 1815, had received from Boie a criticism as thorough as it was outspoken. Temminck, far from resenting adverse comment by an unknown, immediately started to correspond with Boie and in August 1818 got to know him personally. As a result of all these connections, and after a disappointment in the summer of 1821, when he was found unsuitable as a traveling companion for Eduard Rüppell, of the same age and son of a Frankfurt banker, who was going to northeast Africa, he was appointed a curator in the Leiden Rijksmuseum, with the promise that he, like Kuhl, would be able to go to the East Indies at government expense when the opportunity arose. At his departure Heidelberg graciously bestowed on him an honorary Ph.D.

In his new post Boie soon so distinguished himself that, according to a Dutch commentator, "no small part of the fame enjoyed by the Leiden museum is attributable to his energy and his gifts."[7] He now attacked the study of reptiles and amphibia, and during the following years prepared the text and plates for a major work, the *Erpétologie de Java*, in which all the discoveries of Reinwardt, Kuhl, and van Hasselt were described and illustrated. Before his departure for Java he entrusted this precious legacy to Temminck, who proved a poor custodian; he had a few pages printed in Brussels in 1830, and that was that.

Heinrich Christian Macklot, Boie's future traveling companion, was born in Frankfurt am Main in 1799 and entered Heidelberg in 1818 to study medicine and natural history. There he soon became a friend of Boie's, at whose prompting Temminck appointed Macklot (after he had passed his doctoral examinations in the summer of 1822) to head the osteology department of the Leiden Museum. He was a good anatomist and very skilled as a mineralogist and physicist. Schlegel admired "his almost foolhardy brilliance and his Herculean strength."

The youngest of the group, Salomon Müller, came from a humbler background. The son of a master saddler in Heidelberg, he was born on April 7, 1804. He met his later companions by a remarkable accident, when Boie and Macklot, as students, were discussing the results of a collecting trip at a Heidelberg inn, and noticed that the boy waiting on them was listening very closely to their conversation. Soon they discovered that he not only shot birds in his spare time but was also a remarkably good hand at stuffing them. Thus an alliance was formed. And when a preparator was needed, in the summer of 1823, for the journey to Java, Temminck took Boie's pressing advice and sent for Salomon Müller to come to Leiden.

So the Heidelberg trinity was assembled again. But what awaited them was not simply the glowing colors of adventure; there were also dark threats of danger. No one felt this so strongly as Boie. At a time when the journey was supposed to begin in the coming summer, he wrote from Paris to Lichtenstein on April 13, 1824:*

> I am seizing this opportunity, dear Professor, to say what may turn out to be a warm farewell to you, because in a few months I shall be traveling to Batavia to take Kuhl's place and perhaps there is little hope that I shall see Europe again, to say nothing of you. Van Hasselt's unfortunate death, which the learned world must lament the more because his last days gave promise that he would far exceed all expectations, has prompted all my acquaintances to urge me to withdraw from such a dangerous career, but that lies beyond my powers, and I approach my fate with truly cheerful calm, content if for only a while I may serve in good health the cause of science, which gives me more pleasure every day.[8]

The three companions were finally able to set sail on December 21, 1825. A fourth member had been added to the group, the draftsman Pieter van Oort. During a fifteen-day calm off the Cape of Good Hope they were busy collecting. Many naturalists had done the same since the days of Sparrman and Levaillant, though none so continuously and methodically as Dr. Andrew Smith (1797–1872), who some years before had taken over the administration of the museum that the South African Institution had set up in Cape Town.† Boie made the acquaintance of this "most devoted amphibiologist" during his stay, and also that of a compatriot from Württemberg, Carl F. H. Ludwig (1784–1847), who showed him a large collection of birds destined for the Stuttgart Cabinet of Natural History. Temminck's emissaries arrived

*In the spring of 1824 Boie accompanied Temminck to Paris for four months.

†Smith led an expedition into the interior of South Africa in 1834–1836, publishing the results as *Illustrations of the Zoology of South Africa*, 5 vols. (London, 1838–1849). He was later under fire for his administration of the medical department in the Crimea, but was knighted in 1859.

in Java only on June 6, 1826 — the most unfortunate possible moment, because Baron van den Capellen, who had always welcomed young naturalists like a benevolent father, had been recalled after 10 years' service, and had been replaced by a Walloon named Leonard Vicomte Du Bus de Gisignies, a dry calculator who as general commissioner was supposed to set the finances in order. He had not the slightest comprehension of the needs and problems of the newly arrived German scientists. They could look out for themselves, or better still, start home again without delay. The little group, which had been joined by van Raalten, the last of Kuhl's companions, was nearly forced to do so because payment of the money allotted to them by the minister for the colonies was stubbornly refused, and only after months of pressure could they gain possession of Kuhl and van Hasselt's house at Buitenzorg, which had been serving other purposes. The unfortunate men would have been condemned to months of inactivity had they, and especially Boie, not used the time to learn more and more about the habits of the Buitenzorg fauna, particularly the birds. Temminck's energetic intervention finally resulted in the paying over of their money, on February 27, 1827. Now free to move about, they decided to undertake a joint expedition to Sumatra, encouraged by Major A. H. Henrici, a compatriot who was stationed at Padang and who had started to collect birds there. A young botanist from the Buitenzorg Botanic Garden, a Würzburger called Zippelius, wanted to join them. In this way a whole group of scientists emerged which, regardless of its component parts, was generally referred to as the "Natural History Commission" ("Natuurkundige Commissie van Nederlandsch Indië").

While the enterprise was temporarily delayed by Macklot's falling ill with fever, Boie, Müller, and van Oort went to Tapoes, at the foot of Pangrango, on July 4, 1827. Boie wrote a wonderful diary in the form of letters to his friend Hermann Schlegel.[9] It reveals a devoted scientist with the sensibility of a poet, a man who searched through the "divinely, for a zoologist often unbearably, beautiful world of plants," who followed the blue whistling thrush over "heaped-up masses of rock, with crystal-clear water foaming down among them," and whom the "indescribable splendor" of this solitude reminded of Schiller's "wo die Wälder am dunkelsten nachten" [where the woods turn to night at its deepest]. It provided him with "a truly splendid sight" as he watched a flock of fiery-red *Pericrocotus miniatus* (Sunda minivets) on the mountain slope and saw how "they fluttered shimmering in the morning sun from one peak to another." When he had shot a few members of the "cheerful armies" of his "favorite dicaeids" (flowerpeckers), he was astonished "to find that almost all the specimens had in their stomachs unripe fruit the size of peas, and very rarely insects."

Only someone who has shared a similar experience can completely relive the fervent enthusiasm that drove the band of young men unceasingly on through the virgin forest of the giant volcano, with its many voices. But again death reached out and snatched away the best of the group — its leader, Heinrich Boie. On August 23 he had to attend a celebration of the king's birthday in Buitenzorg, where he suddenly fell so ill of a "gall-fever" that he died only a few days later, on September 4, 1827.[10] However, he left to his grieving companions not merely the sad memory of a brilliant and lovable friend, but also many detailed descriptions of the morphology and behavior of most of the West Javanese birds, for a projected complete ornithology of the area. The material for several monographs, and descriptions of several new genera, were also found among his papers. Macklot took over the whole stock, so that he could later follow his friend's wishes for its use. How these hopes were wrecked we shall soon learn.

Although Boie's death had shocked them all profoundly, the members of the group were still hoping to set off soon for Sumatra, when Macklot was suddenly ordered to proceed to Batavia, along with Müller, Zippelius, van Raalten, and van Oort, and board a merchant ship for an expedition to the west coast of New Guinea.

This undertaking, begun by secret orders of December 31, 1827, had been conceived in response to the strong English threat to the territorial pretensions made by the Dutch, who nevertheless had not yet set foot in western New Guinea. A fortified settlement was to be established as soon as possible in a suitable place, to check English ambitions. Temminck may have got wind of this just in time, and urged that the members of the commission be allowed to take the opportunity of seeing New Guinea. His attention had just been drawn to this country, full of zoological surprises, because in 1826 Lesson, the naturalist on board the *Coquille*, had begun to describe rare birds and mammals from Dore.

The French corvette *Coquille*, commanded by L. J. Duperrey, sailed from Toulon on August 11, 1821, with two naturalists on board — the medical officer, P. Garnot, and the naval apothecary, R. P. Lesson (1794–1849). After rounding Cape Horn the ship cruised in the Pacific, visited the southern Moluccas in the autumn of 1823, and in January 1824 put into Port Jackson for repairs. There Garnot was forced by illness to leave the ship and travel home as fast as possible on a merchantman sailing to Europe. Almost everything that the *Coquille* had collected until then was given to him, and was lost when his ship was wrecked in July 1824, Garnot barely escaping with his life. The *Coquille*, on the other hand, with Lesson on board, proceeded to New Zealand and the Caroline Islands (Kusaie), and gave him a chance to make important discoveries near Dore in New Guinea (from July 20 to

August 9, 1824). She returned to Toulon on March 31, 1825, and in 1826 Lesson published the first new descriptions. Before that he married Clémence, the daughter of the ornithologist Charles Dumont de Sainte-Croix, and a pupil of Huet, the painter of birds. Rendered an enthusiastic ornithologist by the impressions and successes of the journey, Lesson devoted himself henceforth to the study of birds until his death, and lost no opportunity to describe new species. For some time hummingbirds were his particular favorites, as they were beginning to be for several French private collectors (such as Prévost, Longuemare, and Bourcier). Deprived of his pension by the July Revolution, Lesson was forced to give up his independent life in Paris in 1830 and earn his living in his native town of Rochefort, first as a naval apothecary and then as professor of chemistry in the naval school.

Soon after the *Coquille* came home, her first mate on the voyage, J. Dumont d'Urville, was commissioned by the French government to go round the world again on the same ship, rechristened *Astrolabe.* She left Toulon on April 25, 1826. The two naturalists were J. P. C. Quoy and J. P. Gaimard, who had distinguished themselves when they accompanied Freycinet on the *Uranie* (1817–1820). This time the voyage proceeded by way of Cape Town to Port Jackson, and from there by various detours to Dore Bay (September 6 to 27, 1827). After long cruises in Melanesia and Polynesia the *Astrolabe* returned home on April 2, 1829, and the zoological results began to be published in 1830.

Meanwhile the Dutch commission, in accordance with its instructions, left Batavia on February 2, 1828, for Makasar, where Müller collected as many birds as he could before the company assembled on board the *Triton*, a naval corvette whose captain was charged with carrying out the secret orders. On May 9 the ship reached the estuary of the "River Dourga" [Digoel, at the northern entrance to the Princess Marianne Strait], on the southern coast of New Guinea, and then set a northerly course. On June 13 the group landed at the estuary of the River Utanata [Oeta?], and on June 29 reached Aidoema Island, sailing into the deep bay opening behind it (henceforth known as Triton Bay) which leads into the Lobo region. Here the captain decided to found the necessary settlement, which was to be called Fort Du Bus, but before construction could be finished the *Triton* had to sail on August 31, leaving a garrison, because there were 64 cases of malaria on board that had to reach medical help in Ambon as soon as possible. Twenty-one sailors had already died. Although the stay in New Guinea had been so short, it had allowed Macklot and Müller to collect 17 mammals of 6 species, 81 reptiles of 27 species, and even 341 birds of 119 species, which, though worked on long after the zoological collections made by the *Coquille* and *Astrolabe*, nonetheless contained many previously unknown species.

The number of sick men forced the *Triton* to remain anchored in Ambon Bay for four weeks. She then set course for Timor, where the members of the commission were left on shore in the harbor of Kupang to explore the island thoroughly. They were all suffering from fever; Zippelius the botanist died of it on the last day of 1828. On April 27, 1829, the sixth victim of the murderous tropical climate died, Gerrit van Raalten, the last of those who had responded to the lure of distant lands in 1820. "Oh send no more men here who are not brutes and can be useful to the world; they will be lost beyond all hope of rescue," wrote Macklot to Temminck when he had to tell him of Boie's death. "The climate is not suited to the active European; all precautions and the observance of the best rules are vain, and only a higher power or chance can decide one's fate." He, too, was soon to die.

While Macklot continued to study the geology of western Timor, Müller worked on ethnology, and above all on zoology. In August and September 1829 he undertook a long march into the interior and reached the Mutis range, a feat equaled by no other naturalist until the Haniel expedition of 1911.

One can imagine the surprise and joyful welcome given to the three survivors when they returned to Java in November 1829, for nothing had been heard of them for a long time and a rumor that Macklot was dead had been believed. Diard, a Frenchman who was an old tropical hand and an experienced collector, had been named to the commission during their absence, to fill the place left vacant by Boie's death. It was now agreed that he and Macklot should have equal privileges as joint heads of the expedition.

The three companions, exhausted by the great trials of their long enterprise, spent a considerable time in Buitenzorg, and sent off no fewer than 46 chests of specimens to Leiden, mostly from New Guinea and Timor. Not until March 18, 1831, did Macklot recommend government sanction of a journey to the eastern end of Java, along the north coast, to be made by Müller, van Oort, and himself. After the plan had been approved, they spent a little more time in Buitenzorg and then set off north.

Accompanying them were three new arrivals who had sailed from Holland in November 1830 to fill the commission's ranks: P. W. Korthals, a botanist; van Gelder, a draftsman; and Overdijk, a preparator. Either shortage of money or frequent illness held them up for a long time in the regencies of Preanger and Krawang. This proved disastrous for Macklot, for during the night of May 8, 1832, the Chinese workmen in Krawang rose in revolt and marched on Purwakarta, where they burned down all the houses belonging to Europeans, including the one in which Macklot was staying. He barely escaped, and had to watch while not only his instruments and books but also all his written descriptions, including the diaries of his New Guinea and Timor journeys, were devoured by the flames. And as if that were not enough, even his friend Boie's manuscripts, which he had preserved faithfully for five years, perished miserably. He was beside himself with rage and despair, so that three days later, when a handful of Europeans banded together to attack the Chinese, the gigantic German was among them. Krawang was invaded in reprisal and the rebels were driven back in hand-to-hand fighting as Macklot, run through by many spears, sank to the ground dead.

As a result of this catastrophe Diard became head of the commission. To rejoin him, the rest of the members sadly returned from Preanger Regency to Buitenzorg, and from that time on nothing more was said about reconnoitering the farthest eastern regions of Java. The original feverish impulse that had swept them onward had long since been spent, and someone other than Diard would have been needed to kindle a new flame. More than a year passed after Macklot's death before a new enterprise finally got under way in June 1833, and the plan to explore Sumatra was finally realized. Müller, Korthals, and van Oort sailed to Padang,* where they worked up and down the coast for a while until Müller returned to Java in October, to send off the collected material that was still stored there. He came back again, with Overdijk and van Gelder, in January 1834. It was nine years since he had been permitted to accompany Macklot and Boie as a humble preparator, and in that time the boy had become a mature man, the apprentice an independent researcher and an expert more knowledgeable than anyone else about the East Indies, for nobody could dispute the fact that he was intellectually the head of the commission. It was entirely due to this outstanding man that further plans for the study of Sumatra were made and carried out, lasting another full two years, until the beginning of 1836. The expedition began with the northern part of the uplands around Padang, where Müller climbed the volcano of Marapi.

When they returned to Padang they had there to mourn the loss of another of their members, Pieter van Oort, the last of the group that set out with Müller in 1825, who succumbed to malaria on September 24, 1834. Even though Salomon Müller had been toughened by many trials, he was disheartened by the death of his great friend; as he wrote to Temminck, the length of his stay and his life in these unlucky

*During the first months they were accompanied by Heinrich Bürger, an apothecary and the loyal companion of Philipp Franz von Siebold, who had invited him to Japan in 1824. When Siebold had to leave the country in 1829, Bürger continued to collect Japanese animals for another three years. Many of the bird species described and portrayed in the *Fauna Japonica* were sent to Temminck not by Siebold but by Bürger, after 1830.

countries had now begun to grow almost unbearable. Therefore he asked leave to return home, if he could be guaranteed a livelihood. Useless! Temminck, who realized all too well that Müller had become the driving force behind the commission, paid no attention to his wishes, although in the contract he himself had limited the time to six years, which were long over.

The year 1835 was mainly devoted to journeys north of Padang and to the residency of Tapanuli. When Müller finally took his group back to Buitenzorg at the beginning of 1836, a new member of the commission, who had arrived from Holland some months earlier, was introduced to him, Dr. Ludwig Horner, a physician and geologist, born in Zürich in 1811. Shortly after the meeting Korthals, Müller, and Horner were ordered to travel together to Bandjarmasin on the south coast of Borneo, where they explored the country from August to December 1836, as the first naturalists to arrive there. Their chief task was supposed to be the preparation of cartographic data, but Müller made remarkable zoological collections along the lower reaches of the Barito, up which he traveled with Korthals as far as the entry of the South Tewe, and along the coast. Soon after returning, Dr. Horner set out with Overdijk for Sumatra, where he worked for almost two years, especially in the Padang uplands, before he too fell victim to the tropical climate in Padang on December 7, 1838. But Müller and Korthals were at last allowed to return to Holland, where they arrived in good health on August 22, 1837, the first (and almost the only) members of the commission who were destined to see home again. During his 11 years in the Indies Müller, first with Boie and Macklot and then alone, had collected vast quantities of material for the Leiden museum, including 1,100 skins and 300 skeletons of mammals, 6,500 skins, 700 skeletons, 150 nests, and 400 eggs of birds, about 2,300 reptiles and amphibians in alcohol, 3,000 fish, 35,000 insects, and so on.

After Horner's death, the only professional scientist of the Commission left in the Indies was Diard. Let us take a closer look at this man, who for years had remained in the background while the events described were taking place.

Incomparably efficient as a collector, unexcelled as a preparator, but lacking the ambition to publish his own findings, Diard was exactly the sort of man Temminck liked. Over a period of 20 years he had visited many unexplored countries and islands in southeast Asia and had delivered a large number of new bird and mammal species to the Paris and Leiden museums. Although he became the object of jealous competition among French, Dutch, and English institutions and turned this to his own advantage, not without causing much ill feeling at first, in the end he managed to gain considerable influence and respect in Java, thanks to his intelligence and ready tongue.

Pierre Médard Diard, born on March 19, 1795, at Château-la-Bresse (Indre-et-Loire), had traveled east in 1817, after serving his apprenticeship in the Jardin du Roi, to collect fauna and flora in the fabulous land of India. In Calcutta he met a young compatriot, Alfred Duvaucel (born in 1793), who had gone to the tropics for the same purpose as "Naturaliste du Roi" at the suggestion of his stepfather, Georges Cuvier. They quickly became friends, and traveled together to nearby Chandernagore. There they made the acquaintance of the "conqueror of Java" and then governor of Benkulen (Bangkahulu), Sir Stamford Raffles. The district of Benkulen, on the west coast of Sumatra, had been since 1685 in the possession of the East India Company, which had built Fort Marlborough there. The London Agreement of August 13, 1814, had settled that Benkulen should be exchanged with the Dutch for Singapore, but the transfer was delayed until 1825. Raffles, with admirable energy, used this interval to make a scientific study of the country.

This extraordinary man, the son of a ship's captain, had been born at sea on July 5, 1781. At 14 he entered the service of the East India Company, which sent him to Penang as assistant secretary in 1805. During a period of convalescence in Malacca he became acquainted with the famous polyhistor William Marsden, the author of a *History of Sumatra* published in 1783, when he was secretary to the Benkulen government. He was exactly the right man to arouse his young compatriot's curiosity about scientific matters. As a reward for his great services in the capture of Java, Raffles was named lieutenant governor of Java and its dependent islands in 1811, with his residence in Buitenzorg, where he lived until the beginning of 1816. A most generous patron of scientific investigation while he was in this high office, he had become interested in botany and zoology through Dr. Thomas Horsfield, who had gone to Java as a doctor in 1802 and had undertaken, as his chief task, to study the natural history of the island. The two men became close friends, and when Raffles, after a year's stay in London, where he was knighted in recognition of his political services, landed in Sumatra in November 1817, as governor of Benkulen, he resolved to do for the natural history of Sumatra what Horsfield had done for Java. Before Horsfield finally left the Malayan archipelago to settle in England (where he presented his collections to the museum of the Honourable East India Company), he went from Bangka, in July 1818, to visit his friend Raffles in Benkulen, and accompanied him on a short trip across the country. In the meanwhile the naturalist Dr. Joseph Arnold, who had been traveling in Sumatra at the governor's request, and had discovered the famous plant *Rafflesia arnoldi*, died at Padang. Soon thereafter Raffles went to Calcutta and, with the agreement of the East India Company, brought back Diard and Duvaucel to replace Arnold. He needed such well-trained collectors just at that moment, because he had planned a study of the natural history of Benkulen. His choice was soon proved admirable, because Diard and Duvaucel amassed large zoological collections, which Raffles intended to work through. But he was disagreeably surprised when he found out that some of the newly discovered mammals had been described in Paris by Étienne Geoffroy Saint-Hilaire and Frédéric Cuvier, because the two collectors had secretly sent home a large part of their material.* The contract was thus broken, and this well-known or even notorious part of their careers ended in a major scandal. In 1820 both were expelled from Benkulen.

Nevertheless, their collections enabled Raffles to present to the Linnean Society, as early as March 1821, a *Descriptive Catalogue of a Zoological Collection Made in the Island of Sumatra and its Vicinity.* He kept the collections themselves, in large part, and steadily added to them. On February 2, 1824, he and his family (three of his four children had died in Sumatra) finally started for home. All the scientific material that he had been industriously amassing for many years accompanied him on Board the *Fame.* Soon after she sailed fire broke out on board and spread to the whole ship. A boat rescued him and his family and took them to land nearby, but all his possessions went up in smoke. "All, all has perished; but, thank God, our lives have been spared, and we do not repine," he concluded with unfaltering hand in his report of the horrible experience. What treasures they were: all his notes (enough to provide a detailed description of Sumatra and many other islands); grammars and vocabularies of the local languages; a large map of Sumatra which he had continually added to; and, not least, his natural history collections, dating from 1820 or earlier, in addition to upwards of 2000 drawings of zoological and botanical items and all the notes of his dead companion, Arnold. "And, to conclude, I will merely notice, that there was scarce an unknown animal, bird, beast, or fish, or an interesting plant, which we had not on board: a living tapir, a new species of tiger, splendid pheasants, etc., domesticated for the voyage; we were, in short, in this respect, a perfect Noah's ark."[11]

After he had finally reached home with the next ship, Sir Stamford allowed himself no rest. One of the plans that he worked on undeterred was the founding of a zoo in London, to be even more magnificent than the Jardin des Plantes in Paris. "The increase of zoological knowledge by the study of the living beings" was particularly dear to

*Valenciennes (1826), Georges Cuvier (1829), and Lesson (1830 ff.) described some Sumatran bird species from the Diard and Duvaucel collection.

him. So the Zoological Society of London came into existence, but its spiritual founder and first president did not live to see more than the highly promising beginning of a magnificent development, for on July 5, 1826, he died of a stroke, on his forty-fifth birthday. His trusted colleague Nicholas Vigors, who in the same year presented his own private collection of birds to the Society as the foundation of a museum, furthered, while he was secretary of the Society (until 1833), the rapid realization of Sir Stamford's ideas.[12]

In 1828 Vigors engaged the young John Gould (1804–1881), an assistant gardener who was very skillful in stuffing birds, as keeper and taxidermist ("preserver") of the Zoological Society's museum. Instructed by his patron in exotic ornithology, he became superintendent of the Ornithological Collection in 1833 and found an opportunity to display his outstanding ability. In 1831 (two years after marrying Miss Elizabeth Coxen) he had already surprised the specialists with the first installments of a splendid folio, called *A Century of Birds from the Himalaya Mountains* (London, 1831–1832), whose illustrations in color had been made by Mrs. Gould, after sketches by her husband, and then lithographed, while Vigors wrote the scientific text. It was the first of many successful folios published by the productive author. Very soon Gould added scientific knowledge to his artistic skill. To acquire quickly a knowledge of the variety of species, he traveled to the Continent in 1833 and visited the zoological museums in Paris, Munich, Vienna, Berlin, and Leiden. *A Monograph of the Ramphastidae* (1833–1835) and *A Monograph of the Trogonidae* (1836–1838), both splendidly illustrated, prove how much he had learned. He was therefore able to work on the many new discoveries in the bird collection brought back from South America and the Galapagos Islands by Charles Darwin in 1836, following the voyage of the *Beagle*, and presented by him to the Zoological Society's museum. In order to prepare a book on Australian birds, Gould resigned his position in favor of G. R. Waterhouse in 1836. Waterhouse in turn passed it on in 1843 to Louis Fraser, who had done so much for ornithology during his African travels. On Gould's retirement the Society's bird collection already outnumbered that of the British Museum (whose Zoological Department was headed from 1822 to 1840 by the incompetent J. G. Children and remained in the antiquated Montague House until 1845), but according to H. E. Strickland's expert opinion, it was greatly inferior in size and splendor to that of the Paris Muséum d'Histoire Naturelle.

To return to Duvaucel: he succeeded in getting back undisturbed from Padang to Calcutta, and for several years most successfully continued his work for the Paris museum, as witnessed by the *Histoire naturelle des mammifères*, by E. Geoffroy Saint-Hilaire and F. Cuvier

(Paris, 1818–1825). He visited Sylhet and the Khasi Hills, and had reached the foot of the Himalayas* on his way to Tibet when an attack of dysentery hurried him back to Calcutta. From there he took ship to Madras where he soon died, in 1824, aged not yet 32.

Diard seemed at first to have had worse luck than his friend after being cast off by Raffles, because he was arrested in Padang on a charge of spying and sent to Batavia. He was released only on Reinwardt's intervention, but soon afterward received a grant for a journey to Bantam (Banten). In 1821, after exasperating Temminck by sending off a large Javanese collection to Cuvier,† he traveled with the French consul in Cochin China from Batavia to Hué, but in 1824 reappeared in Buitenzorg after an antiforeigner uprising had driven him from Cochin China. From then on he was entrusted with commercial jobs, going in April 1825 to Bangka, which had just been given to the English in exchange for Cochin China. In August 1826 General Commissioner Du Bus (the same who had treated Boie and Macklot so unpleasantly) assigned him to accompany the resident, L. G. Hartmann, on a trip to Pontianak, from which he returned with the first collection of animals ever made in Borneo. Temminck was alarmed when he heard of it, since he feared that not he but his rivals in the Paris Museum would enjoy the new material. In this situation Diard quickly yielded, canceled the contract with Paris that had brought him in 6000 francs a year, and sent everything that he had collected in Pontianak to Leiden. He was well rewarded, for Temminck, wishing to purchase further cooperation, or at least neutrality, entered Diard as a member of the Natural History Commission, which he headed after Macklot's death in 1832. From that moment he felt no more urge to risk life and limb on difficult and dangerous journeys; while his colleagues were running these risks, he stayed in Batavia and Buitenzorg.

Dr. Horner's death in 1838 would therefore have meant the end of the commission's active work if in the meantime a younger scientist, Dr. Forsten, had not introduced himself to Temminck. He was irresistibly attracted toward the Indies by the chance of achieving fame as a discoverer, even at the cost of an early death.

Eltio Alegondus Forsten (born at Middleburg in 1811) had barely received his doctorate in medicine from Leiden, in January 1837, when he asked to be sent to the tropics as a member of the commis-

*Just at this time zoological and botanical exploration of the Himalayas had begun, primarily through the activities of the Danish botanist Nathaniel Wallich, who as director of the East India Company's Botanic Garden in Calcutta journeyed to Nepal, and gave some of the animals he brought back not only to General Thomas Hardwicke but also to Alfred Duvaucel.

†Drapiez (1823–1828) and Lesson (1826 ff.) described a number of new bird species from this Javanese collection of Diard's.

sion. Temminck made the appointment conditional on a year's trial apprenticeship in the museum, so Forsten actually sailed only in September 1838, under orders to explore Celebes. But immediately on arriving in Java his patience was sorely tried. Horner had died in Sumatra, as he now learned, Diard could hardly be bothered with him, and the commission's funds were supposed to be fully committed to two botanists who had been taken on for a year, Franz Junghuhn and J. K. Hasskarl. Not until December 1838 did he receive orders from the East Indies government clearing the way for the Celebes journey. Diard wanted to assign Overdijk to him as preparator, but Forsten disliked him and took a German instead, a "cannoneer in the artillery" whom he had recently come to know and appreciate in Batavia. This was Heinrich van Gaffron. Forsten embarked with him and seventeen "inlanders," as hunters and collectors of insects and plants, on January 30, 1840. They arrived in Manado on March 22 and shortly thereafter took up residence in an idyllically placed government building in Tondano. The dysentery with which Forsten had fallen ill in Java here became so painful that he could no longer walk, and had to be carried about in a chair. Nevertheless, his tremendous energy enabled him, in Gaffron's company, to spend the first year going over almost the whole of Minahassa, thereupon deciding to make Gorontalo his base for further research. But his ship was driven by contrary winds into Ternate, where he landed on June 19, 1841. Because he was housebound by his illness, he left it to Gaffron to make several voyages to unexplored Halmahera; then, in mid-September, in passable health again, he set out for Gorontalo and Tomini Bay. On April 14, 1842, he sailed on to Ambon and the Banda Islands, and then added a trip to the south coast of unexplored West Seram. This was his last expedition, because after a short time there his illness rapidly became so much worse that he had to be hastily transported back to Ambon, where he died on January 3, 1843.

The news of the loss of this exceptional explorer, who had enriched the Leiden museum with previously undreamed-of treasures, was a fresh and bitter blow to Temminck. The days of the Natural History Commission seemed at last to be numbered. Even Diard could no longer be counted on, because in the summer of 1841 he had taken several years' leave to recuperate in Europe, and in December 1842 Korthals had asked to be pensioned off. It was therefore very opportune for Temminck when a new candidate for the commission appeared in the autumn of 1841, Dr. C. A. L. M. Schwaner. Born at Mannheim in 1817, he had been trained chiefly as a geologist and mineralogist, but he was also well up in zoology. His wish was quickly granted, and he arrived in Batavia promptly in August 1842, where he was assigned to prospect for coal deposits. He chose to do this in Borneo and, accompanied by

Heinrich von Gaffron, who had come back to Java from Ambon with Forsten's collections, he reached Bandjarmasin at the end of 1843. During his four-year stay Schwaner became the "scientific discoverer of Borneo," not only because he was the first to succeed in crossing the huge island from Bandjarmasin to Pontianak, but also because he made a discovery important for the future of the Dutch colonial empire, of coal seams worth mining in southwest Borneo, on the Riam Kiwa river. On returning to Java in 1847 Schwaner worked through his scientific material. After three years he was commissioned to make a new journey to Borneo, but before he could start, and before his important book, *Borneo*, could be printed, he died of fever in Batavia on March 30, 1851.

A year earlier, on April 17, 1850, the Natural History Commission had been disbanded by royal decree. There were at the last four extant members: Salomon Müller (still on leave in Leiden), Schwaner, Diard, and (since May 1845) the geologist and botanist Franz Junghuhn (born in 1809 at Mansfeld). Temminck could not see that any of them could be of further use. He had no understanding at all of Junghuhn's magnificent work in botanical geography and geology in Java and the Batak lands of Sumatra (1840–1848), because all that mattered to him was zoological eyecatchers for his museum. He was just as unsatisfied with the versatile Schwaner. And also, because in old age he had let himself become resigned, he put up only weak resistance when the axe was laid to the tree he had planted and protected in happier times.

The Natural History Commission had existed for 30 years; many highly gifted scientists had given their lives in its service, and huge sums of money had been spent (around 600,000 guilders in the years between 1820 and 1825 alone) — but what had been achieved?

During Temminck's lifetime, zoology profited only meagerly from all this expense, even though enormous zoological collections and an inexhaustible wealth of descriptions had come into his possession. An exceptional personality with solidly based and far-ranging knowledge was needed to administer such treasures. Temminck did not have that sort of intellect. He remained what he had been in 1820, when he took the post that he himself had created — a well-informed zoological amateur. Like the predecessors who had evoked his youthful enthusiasm, he was above all concerned with acquiring and describing new mammals and birds, a collector who, through his stubborn persistence and clever calculation, had been fortune's darling and had been showered with gifts, though his insatiability grew with his means.

By 1850 Temminck had for a long time not been living up to his reputation as the most significant of living ornithologists. In many countries rivals had appeared to diminish his glory in his own special field of acquiring new and rare species. The English ornithologists, who allowed irreplaceable treasures to leave the country when the famous collections of Sir Ashton Lever (1806) and William Bullock (1819) were broken up, were very active again, so that the museums of the Zoological Society and the East India Company, as well as the British Museum, were supplied from a source that remained almost blocked for Temminck: the British Empire, and especially India and Australia. Brian Hodgson (1800–1894) earned immortality by his study of nature in the Himalayas. He was British resident in Nepal for 10 years (1833–1843), and on his return presented the British Museum with more than 2000 skins of local birds as well as a large collection of excellent watercolors of the local fauna. Competing with the museums in England were private collectors like Sir William Jardine, whose collection rapidly came to be measured in the thousands, or John Gould, who, in order to write a thorough account of the birds of the fifth continent, visited Australia from 1838 to 1841, accompanied by his wife. Matters were little different in France. To the Paris Museum,[13] directed after 1841 by Isidore Geoffroy Saint-Hilaire (in succession to his father Étienne), were allotted hundreds of rare species from the government expeditions of the *Coquille* (1821–1825), the *Astrolabe* (1826–1829), and the *Astrolabe* and *Zélée* together (1837–1841), which amassed a wealth of new birds from the coasts and islands of the world's oceans. Added to these were the astonishing novelties collected by the daring "traveling naturalist" Goudot and the ship's doctor Bernier, during the 1830s, on the still mysterious island of Madagascar. Beginning with Alcide d'Orbigny (1826–1834), Justin Goudot (1827–1843)*, and Claude Gay (1830–1842), explorers of the Andean countries were mostly French. In the meantime thousands of birds were gathered into the

*Justin Goudot, who as "zoologist at the Bogotá Museum" was corresponding with Lesson as early as 1828, traveled about New Granada (Colombia) for at least 15 years to collect natural history specimens, especially birds, and returned to Paris in 1843 with a rich haul. In all probability, the many new bird species from Colombia that from 1839 on were sold by Parzudaki, Boissonneau, and other dealers and described by French ornithologists (mostly with the locality given as "Nouvelle Grenade" or "Bogotá") were discovered by him. He was certainly the person who taught Indians how to prepare birds and thereby opened a new market, in feathers from Bogotá skins, which soon assumed gigantic proportions. According to a list published by Sclater in 1855 (*Proceedings of the Zoological Society of London*, 23: 131–164), 435 species of birds had been found in such Bogotá collections; Lafresnaye described over 70 of them. In the meantime Justin Goudot had gone back to Colombia, around 1848, where he died.

His brother distinguished himself as a bird collector in Madagascar, and was one of the first foreigners to penetrate as far as the capital, Tananarive. Soon after 1838 he died on the island.

The wholesale market in exotic bird skins, in the interests not of science but of fashion, had started in the eighteenth century, much earlier than the arrival of the Bogotá skins. Levaillant (*Histoire des perroquets*, p. 68) saw about 1779 at a feather dealer's in Paris more than 12,000 skins brought back by a traveler from North America (probably Louisiana). Many species were represented by hundreds private museums of rich collectors, like the Comte de Riocour in Aulnois (Meurthe), where the aged Vieillot had made studies; or Victor Masséna, Duc de Rivoli (and son of Marshal André Masséna), who for a time employed the ornithologist Louis Kiener as a private curator for his galleries; or Baron de Lafresnaye, an outstanding expert on South American birds. In Germany the Berlin Museum profited significantly from the major expeditions (financed by the Prussian government) of Friedrich Sellow (1816–1831) and Hemprich and Ehrenberg (1820–1825); the Senckenberg Museum in Frankfurt became notable through the northeast African collections of Eduard Rüppell; while Vienna preserved as its greatest treasure the 12,000 bird skins that Johann Natterer had collected, on orders from the emperor, for 17 years (1817–1835) all over Brazil.

It was impossible to keep pace with the rapid increase in knowledge of species, and the intention to reproduce even the most striking examples among the new discoveries in the *Planches coloriées* became more and more clearly utopian, because the numbers rose annually by the dozens. Temminck kept to the old plan until 1838 and then gave up.*

The good old days were long gone when scarcely a wish of his was not gratified, when the museums of Paris, Berlin, and Vienna, then still keen on new species, exchanged most generously with him, or when Eduard Poeppig from Cuba and Franz Sieber from New South Wales handed over their new material in return for hard cash. The Dutch government, concerned over the large outlay for the Natural History Commission, became more and more reluctant to make special grants, and Temminck's attempts to utilize Dutch consuls or doctors in government service throughout the world as collectors for him all fell through.

Only the Dutch overseas colonies, especially in the East Indies,†

---

of specimens. For example, he had over 300 of the Carolina parakeet, *Conuropsis carolinensis*, now completely extinct, and over 200 of the cedar waxwing, *Bombycilla cedrorum*. Another Parisian feather dealer of the time bought more than 6,000 heads of the Carolina parakeet, to be used for the trimming of dresses.

*Lesson estimated the number of bird species described by 1844 as 6266 (and of "real" genera as 1075). Only 1681 species had been illustrated in the books of color plates by Daubenton and Temminck. Oeillet des Murs's *Iconographie ornithologique*, which began appearing in Paris in 1845, was intended as a contination of Temminck's undertaking, but in 1849, after 72 plates (by Alphonse Prévost and Paul Louis Oudart) had been issued, the project collapsed because the limited sales did not cover the expenses. *Esquisses ornithologiques*, by Bernard de Bus de Gisignies, published in Brussels from 1845 to 1851, reached only 37 plates, chiefly of Central and South American birds.

†For a long time the Dutch had owned some trading stations on the Gold Coast (whose fauna had first been studied in the 1830s). They served as depots for merchants sailing to the East Indies, and remained Dutch until 1870. A young official who had worked for Temminck in the Leiden Museum for a while before going to Africa, H. S. Pel, was very active there from 1840 to 1854, and sent back valuable

remained his special province, where he was able to retain what amounted to a monopoly on their yield in specimens.

Not only did the shrinking of his influence steadily dim the brilliance of his reputation; in the course of time ornithological aims had become more complex, and the language of ornithology more finely articulated, without his having been able to keep up with the changes in the formulation of questions and in methodology. His favorite work, the *Planches coloriées*, struck contemporaries as old-fashioned, because ornithologists, particularly in England, had grown much more demanding. In 1836 Swainson commented:

> The figures are stiff and formal, and they are all put into nearly the same attitudes. The descriptions of the birds are meagre, and for the most part related to the mere colour of the plumage . . . The total absence of synonyms, specific characters, and scientific descriptions of the form, structure, or habits of the birds themselves, renders this work far inferior to what it might have been, and what the scientific world expected, from the reputation of its authors.[14]

At one time Temminck was considered right when he strongly attacked Vieillot for introducing a number of new family and generic names in his *Analyse d'une nouvelle ornithologie élémentaire* (1816), because the existing general groups seemed to him to have been awkwardly and unmethodically arranged. But later Temminck was forced to watch the "delirium of genera" spread contagiously, and to see his conservative nomenclature, which he himself hardly dared to defend, shoved disrespectfully aside. The new generation, full of ideas foreign to him, sprang up even in his immediate neighborhood. It must have been almost an echo of his own sentiments when his contemporary Lichtenstein wrote to him from Berlin on July 27, 1846:

> My dear old friend . . . Although I have no personal complaints to make, feeling myself still quite healthy and lively and also enjoying much pleasure in my family, nevertheless the real joy of living and being active is gone with my youth. Work is not so quick to get through as it once was, and yet there are more things to do. One demands as much of oneself as before, but one does not satisfy these expectations as well as formerly. Many outlines for articles are lying about unfinished and will die off quietly. The new era is getting beyond me, and however much one reads one feels that it is impossible to make progress, and one seems to oneself quite

zoological collections, made partly on his trip to Kumasi, the Ashanti capital. Temminck devoted his last publication (1853) to the mammals from this collection, and in 1854 invited Gustav Hartlaub to Leiden so that the birds could be worked on by an expert.

> stupid and ignorant in the new jargon of systematics and nomenclature, or at least too weak to swim against the stream, and through a sort of disgust frightened away from even attempting any sort of opposition. There is something revolutionary in every contemporary activity; we shall not live to see the settling of this ferment, but I cannot say that I cherish any great expectations of the final result, and I rejoice therefore that I have lived in a time when there were still scientific authorities and when everything proceeded more tidily.[15]

While Temminck was still full of "the real joy of living and being active" he had jealously kept his assistants away from the birds and had left amphibia and fish, to which he was indifferent, to such as Boie, Kaup, and Schlegel. He even locked all Siebold's Japanese animals away from Schlegel, doling them out only piece by piece and incompletely for the *Fauna Japonica.* "He tried to prevent me from using the bird collection by sealing the cupboards, on the pretext of keeping out harmful insects, although, as I warned him beforehand and as was proved later, this actually helped the little pests to carry out their work of destruction in the utmost peace and quiet."[16] On one occasion Temminck wrote to his wife: "The package with the eagles is probably from Herr Brandt in St. Petersburg. Just open it and tell me what it contains, but don't let Schlegel or Susanna see anything." And another time: "The parcel from the colonies and everything coming from there you must quietly hide, without a word to anyone."

Yet in the end the new species from Java, Japan, New Guinea, Timor, Sumatra, Borneo, Celebes, Halmahera, Seram, and Guinea that flowed steadily into the museum smothered Temminck's enthusiasm instead of fanning it. Schlegel and Müller might then work on ornithology if they liked — he withdrew more and more definitely to the mammals, and by 1839 had almost completely ceased to work on birds. Certainly this was not generally known at first, because Temminck named himself as well as Schlegel as author of the five-volume *Fauna Japonica*, including the section on birds (published 1844–1850), although he had worked only on the terrestrial mammals and had done nothing at all with any of the other groups. The three heavy folios of *Verhandelingen over de natuurlijke Geschiedenis der Nederlandsche overzeesche Bezittingen door de Leden der Natuurkundige Commissie in Indie en andere Schrijvers, uitgegeven op last den Konig door C. J. Temminck* [Treatises on the natural history of the Netherlands' overseas possessions, by the members of the Natural History Commission and other authors, published at royal expense by C. J. Temminck], printed between 1839 and 1844, and designed to sum up the results of the long-standing Indies project, carried on the title page only the names of Temminck and his long-time secretary, J. H. Susanna, although all the scientific work had been done not by them but by others: zoology chiefly by Müller and Schlegel, botany by Korthals, and geography and ethnology by Müller. No wonder that such behavior created enemies in Temminck's own camp, and as important a man as Junghuhn refused with thanks the honor of being allowed to contribute the fruits of his labors to Temminck's omnium gatherum. As he ironically put it, he was not presumptuous enough to appear as the servant of a coediting naturalist.

All these circumstances — Temminck's greedy wish to work over all the material by himself (despite his many distractions) without bothering about quick publication, then his dwindling interest, and finally his inability to stimulate others to willing cooperation — ensured that only a very small part of the scientific results of the exploration of the Dutch East Indies, for which he was responsible, became generally available. A comparison of what the young scientists achieved on the spot with Temminck's published yield tempts one to ask whether Kuhl and van Hasselt, Boie and Macklot, Horner and Forsten did not die in vain for their idol, science.

Schlegel and Salomon Müller (after his return from Java in August 1837) indeed did what they could for years to salvage the enormous harvest. However, when the Natural History Commission was disbanded in 1850 and Müller was pensioned off, he did not stay with Temminck but went to Freiburg im Breisgau, where he died, in modest and retired circumstances, in the spring (?) of 1864; only Schlegel stuck it out.

By now the halls full of stuffed animals had grown silent, and weariness had succeeded surfeit, when unexpectedly a storm wind swept through the ghostly rooms — the tremendous figure of Prince Charles Bonaparte, who wanted to catch up in a matter of months with everything that the "tardigrade Dutchman" (as a disrespectful contemporary across the Channel called him) had been neglecting for years. Bonaparte's famous name opened all the cupboards that Temminck's autocratic and arbitrary selfishness had hermetically sealed against other scientists.

9

Charles Lucien Bonaparte (1803–1857)

During the autumn of 1803, the Parisian coffeehouses rang with a new scandal, this time in Bonaparte's own family. Lucien, the third son of Carlo and Letizia, had enraged his powerful brother, who the year before had been elected consul for life, by persisting in a love affair. Instead of obeying the command to marry the king of Etruria's widow, Lucien had secretly wedded the charming Alexandrine de Bleschamps in Plessis on October 23, 1803. She had been his mistress since the spring of 1802, while she was still the wife of the bill broker Jouberthon. On May 24, 1803, she had given birth to a child, christened Charles-Lucien-Jules-Laurent, who was now legitimated by the marriage.

The revolutionary tyrant, already greedily contemplating the crown and therefore endeavoring to increase his power in every possible way, never forgave his more republican brother for this self-willed marriage. When Napoleon became emperor he excluded his brother from the succession and denied him the title of Imperial Highness. But in April 1804, after Lucien had escaped the dangerous proximity of the emperor by putting himself and his family under papal protection in Rome, Napoleon attempted to get his way by fair means, offering Lucien a throne if he would get a divorce. It was risky to oppose such pressure, but Lucien remained firm and felt secure in Rome — wrongly, as it soon turned out, because the quarrel between emperor and pope ended with Pius VII's imprisonment on July 6, 1809, and his deportation. It was necessary to act at once, so Lucien decided to escape to America. But he got only as far as the Mediterranean, where on August 1, 1810, the whole family was seized by English warships and taken to England, where they were held until Napoleon's downfall in 1814.

Only then was Lucien Bonaparte permitted to return to his possessions in the Papal States, which he had acquired years earlier with the millions resulting from skillful exploitation of Napoleon's power after the invasion of Italy. On August 18, 1814, he was specially honored by Pius VII, who felt bound to Lucien through hate and suffering, with the title of Prince of Canino. From then on he lived in princely splendor, partly on his country estates, partly in England, and partly in his palace in Rome, where he died on June 30, 1840. His wife and their nine children survived him.

His eldest son Charles, who had been granted the papal title of Prince of Musignano, early felt attracted to natural history. Plants and insects satisfied him at first, but soon birds fascinated him more and more, and before he was 20 he decided to prepare an *Ornitologia Romana.* He had an invaluable beginner's guide in the first two published volumes of Temminck's *Manuel d'ornithologie*, but there was one bird, which he had taken in the Roman Campagna, that he did not succeed in identifying, so he sent it straight off to the famous author in Leiden. This species, the moustached warbler, was indeed new to science and, as *Sylvia melanopogon*, appeared, in December 1823, on Plate 244 of Temminck's *Planches coloriées.*

But an *American Ornithology* replaced the *Ornitologia Romana.* In June 1822, when the prince was barely 19, he went to Brussels to marry his cousin Zenaide. She was the firstborn daughter of Napoleon's oldest brother Joseph, who had been transferred from the kingship of the Two Sicilies to the throne of Spain and had finally been brought down in Napoleon's collapse. While the *Bellerophon* was carrying the fallen giant to St. Helena, Joseph quietly took ship for America. From that time on he lived on his large estate near Trenton on the Delaware River, 25 miles from Philadelphia, as the Comte de Survilliers. He was far away from his wife and daughters, who had found asylum first in Frankfurt and then in Brussels. Whether it was the family or Charles's own adventurous spirit that inspired him to visit his father-in-law, in the spring of 1823 he left Italy and embarked with his "Zenaida amabilis" for Philadelphia, where a son was born to them in 1824.

The prince indulged in ornithological activity on the voyage by carefully observing the storm petrels that appeared near the ship, and even succeeded in capturing a few for his collection. His first ornithological publication, "An Account of Four Species of Stormy Petrels,"[1] deals with them; it is a remarkable scientific achievement for a 20-year-old beginner. In Philadelphia he began at once to study American birds, which had only just become a subject for research. It was not 10 years since Alexander Wilson, the "father of American ornithology," had died at the height of his career, and it is not surprising that *Ameri-*

can Ornithology, the great work to which the former weaver's apprentice and poet had devoted his energies, with its good descriptions and colored plates, and above all with its admirable portrayal of birds' habits, soon became Bonaparte's favorite book. It would have been difficult for his interest in zoology to develop in the intellectual narrowness of the Papal States, but here in Philadelphia, the cultural center of a rapidly developing republic, it was encouraged in many ways. An Academy of Natural Sciences, with a good library and its own *Journal*, had existed here since 1812, and its members met regularly to discuss zoological subjects, including ornithology. A letter by an American, Dr. Edmund Porter, who was passing through the city and attended a meeting on October 11, 1825, mentions — among other scientists present, such as Richard Harlan, Edward Harris, Titian Peale, Thomas Say, Charles Alexandre Lesueur, and Constantin Samuel Rafinesque-Schmaltz* — young Bonaparte:

> He is a little set, blackeyed fellow, quite talkative, and withal an interesting and companionable fellow. He devotes his attention to ornithology and has published a continuation of Wilson's work on the above subject . . . [He] read a memoir on the "golden Plover." To a novice it seems curious, that men of the first intelligence should pay so much attention to web-footed gentry with wings.[2]

The existence in Philadelphia of an ornithological collection, and quite a respectable one at that, was very serviceable to the inclinations of the "interesting fellow." It formed part of the Philadelphia Museum, until recently still known as the Peale Museum because it was founded by the portrait painter Charles Willson Peale (1741–1827). He had opened it in 1784 to exhibit his own pictures, but had gradually increased its attractions by adding "various natural curiosities," until by 1825 it contained more than 1000 mammals and birds. This was where Alexander Wilson had studied. After his death it was enlarged by the scientific collections made during the expedition of Major Long to "the Rocky Mountains" (or, more exactly, to the Arkansas River), in 1819–20. He had been accompanied by Thomas Say as zoologist and Titian Peale, the 19-year-old son of Charles Peale, as "assistant naturalist," meaning preparator and draftsman. As a gifted systematist, Bonaparte soon recognized among Long's birds several species not described by Wilson, and by April 1824 he had definitely decided to

*The zoologist Charles Alexandre Lesueur, who had accompanied Baudin's expedition to Australia (1800–1804) as a draftsman, had been in North America since 1815 and was living in Philadelphia at the time. Rafinesque-Schmaltz, who when young had been a friend of Swainson's, had emigrated from Sicily to the land of liberty in July 1815, because he was afraid that after the English withdrew he would be persecuted for his liberal opinions.

complete Wilson's much-admired work with supplements. In that month he published his first article for the purpose in the *Journal* of the Philadelphia Academy; at the same time he established contact with artists and engravers, because he envisioned nothing less than a series of volumes illustrated as well as Wilson's. Titian Peale was engaged as the painter and Alexander Lawson, who had already worked for Wilson, as engraver; Audubon, whose drawings seemed to the prince not bad, might have made his debut here, but Lawson refused energetically to lend his needle to the reproduction of such depraved art.[3]

To add material for the supplementary volumes, Bonaparte sent Titian Peale, as draftsman and collector, to eastern Florida in the winter of 1824–25, together with George Ord, an able ornithologist who had brought Wilson's work up to date with an eighth and a ninth volume in 1814, and was now busy with a new edition. In St. Augustine both of them were guests of the prince's cousin, Achille Murat, who was the elder son of the unfortunate king of Naples.

The first volume, entitled by Bonaparte *American Ornithology; or, The Natural History of Birds Inhabiting the United States Not Given by Wilson*, appeared in the summer of 1825. In spite of its external similarity this work is completely different from its model. It is not the product of a field ornithologist, fascinated by living birds (for Bonaparte only now and again took his fowling piece for a walk in the neighborhood of Philadelphia, or at his Uncle Joseph's in New Jersey), but of the systematist and animal geographer, who felt at his best among museum cupboards and bookshelves, and satisfied his urge for work in the comparison of bird skins and citations from ornithological literature. Even this early work anticipates the *Conspectus generum avium*, which was published 25 years later.*

The more he plunged into these studies, the more clearly he realized that in Philadelphia he could not solve the problem of providing a complete generic survey, so, equipped with notes for the later volumes of his work, he went to England, landing at the beginning of December 1826. From there he first visited his mother-in-law, who was living in Florence as the Contessa di Survilliers, and then his parents in Rome. Because he had to travel through Switzerland (France under the monarchy being closed to him) he took the opportunity of visiting the Senckenberg Museum in Frankfurt, which had been opened in 1820, "an institution which has risen up with such wonderful rapidity," and whose director, Dr. Cretzschmar, gave him for his collection two black-winged kites that had just been sent from North Africa by Rüppell.

*A first draft, in Bonaparte's hand, of volumes 2–4 of *American Ornithology* was found many years later in the California homestead of James G. Cooper, presumably given to his father, William Cooper, by the author (W. O. Emerson, "A Manuscript of Charles Lucian Bonaparte," *Condor*, 7 (1905): 44–47).

Shortly afterward, on January 12, 1827, Cretzschmar wrote to Lichtenstein:

> An agreeable visitor was Prince Charles Lucien Bonaparte who came here from New York eight days ago and whose closer acquaintance I made. He is unquestionably a very scientific man, as he has demonstrated in his works. This patron of ornithology has joined us in our project and will announce his entrance into the Congress* from North America, and he offers to work on all species of United States birds. This circumstance seems very favorable to our matter, and I believe that the prince, with his fiery enthusiasm, would give you pleasure. What a pity that you have no opportunity to know him better![4]

On his return to Rome, Bonaparte greeted the Italian specialists with a *Specchio comparativo delle ornitologie di Roma e di Filadelfia*, published at Pisa in 1827. Then he traveled back to England and enjoyed himself there in libraries and collections. In the British Museum he made the acquaintance of John Edward Gray; he studied in the Museum of the Zoological Society, which shortly before had been enriched by the Raffles collection from Sumatra, and delighted in the "endless treasures of Mr. Leadbeater," London's most respected dealer in natural objects. Here he again met John James Audubon, who had crossed the ocean to find a publisher for his eccentric and mammoth work and to collect subscriptions to it in England and France. "His fine head was not altered, his mustachios, his bearded chin, his keen eye, all was the same," wrote Audubon from London on June 18, 1827. But four days later, when he met his old acquaintance in the company of Vigors, the secretary of the Zoological Society, and Children, Leach's successor at the British Museum, he had to admit with astonishment that the prince had partly relapsed into being a European. "I cannot tell you how surprised I was when at Charles's lodgings to hear his manservant call him 'your Royal Highness.' I thought this ridiculous in the extreme, and cannot conceive how good Charles can bear it, though possibly he *does* bear it because he *is* good Charles."[5]

In London the most important thing for the prince was to complete the manuscript of his *American Ornithology*. After that was accomplished, he took it to Philadelphia with him in the fall of 1827, where two further volumes (II and III), with plates by Alexander Rider, were published in 1828 under the supervision of his friend William Cooper, of New York. The prince himself stayed only a short time in America

*Cretzschmar planned nothing less than the founding of a society of German ornithologists and the publication of a "Synopsis" of world birds, with the participation also of foreign experts such as Temminck and Joseph Natterer.

and returned to Europe for good in 1828, with his wife and young son, choosing Rome for his permanent residence. There he actively pursued his beloved taxonomic studies and research, which extended more and more to birds of the whole world and, by 1831, had made him so famous among ornithologists that Swainson could write, "To Charles Lucian Bonaparte, Prince of Musignano, not only the eyes of America, but of Europe may be turned, as to one who seems destined by Nature to confer unperishable benefits on this noble science."[6]

But soon ornithology alone was not enough; he finished his American book when he sent off the manuscript for a fourth volume (on aquatic birds, published in 1833). Now he intended to increase the knowledge of all branches of the vertebrates by his researches and theoretical work; with his *Iconografia della fauna italica*, begun in 1832 and completed in 1841, he added to the understanding of Italian mammals, reptiles, amphibia, and fishes, as well as birds. During this period the prince tried to learn more by travels to other countries, which all became important to ornithology because he rarely visited a museum without noting down diagnoses of new species. From October to November 1837, he was in London and took part in several meetings of the Zoological Society, where he met Yarrell, Blyth, and George Robert Gray, and became a friend of John Gould, whose recent studies were leading up to his magnificent work, *The Birds of Australia.* Bonaparte was now working on a new plan; in 1838 he invited his old friend William Swainson to move to Rome and join him in compiling a catalogue of all known species of birds (estimated at 7000 to 8000). The *Conspectus* might have taken shape then and there if Swainson had not been so inconsiderate as to price his collaboration rather high; he demanded that the prince buy his collection for £500.

After the death of Lucien Bonaparte on July 29, 1840, his eldest son inherited the title of Prince of Canino, and from then on styled himself "Charles Lucien Bonaparte, Prince of Canino and Musignano." He continued to work hard on further zoological treatises, including a *Systema ornithologiae* (Bologna, 1840). When he returned from a meeting of German doctors and natural scientists in Freiburg im Breisgau in 1838, he obtained permission from the Grand Duke of Tuscany to establish regular congresses of Italian scholars, on the German pattern, and he acquired far-reaching influence as the president of their branch. In 1840 a fellow ornithologist, the Marchese Carlo Durazzo, described him as "the greatest master of natural science," and in 1843 he was elected an honorary member of the Berlin Academy of Sciences. Even in Louis Philippe's France this Bonaparte became a corresponding member of the Institut de France in 1844, after he had already (shortly after the July Revolution) received unrestricted permission from the government to visit French zoologists (like Isidore Geoffroy Saint-Hilaire) from time to time.

Evidently the general public had become accustomed to viewing the nephew of the world shaker as only a temperamental scholar. But now it realized that slumbering within him was a fighter for the ideals of the Great Revolution, who was awakened by the first powerful stirrings of liberalism. Brought up by enlightened parents, acquainted in Pennsylvania with the blessings of a democratic system, "good Charles" had grown to cherish republican institutions. When he saw the caprice of despots beginning to hem him in again, he waited with concealed impatience for the dawn of liberty. Encouraged by the liberal reforms introduced into the Papal States by Pius IX, and carried away by his sympathy for the nationalist movement in northern Italy, he thought the moment had come in 1847. He allowed political allusions to creep into his address to the congress of scholars in Venice, which provoked his prompt official expulsion from all Austrian territories. When he came back to Rome, the agitation for freedom had spread and had turned against the pope, who now attempted vainly to exorcise the spirits he had himself called up. The sole effect was that the radical party, among whose leaders the Prince of Canino was now numbered, attracted more and more support. Finally, to avoid something worse, Pius IX had to leave the field to the republicans and flee to Gaeta on November 15, 1848. Named as vice president of the legislative council on February 16, 1849, the Prince of Canino five weeks later signed the proclamation that called the Romans to arms on behalf of a united Italy. He must have been as unprepared as were his friends for another Bonaparte, his cousin Louis Napoleon (who three months earlier had emerged victorious from the French presidential election), to stab the Italian freedom party in the back in order to win over the clergy — but the unexpected happened. On April 24, 1849, a French expeditionary force landed at Civitavecchia and almost immediately attacked the Eternal City. Although courageously defended by its inhabitants and the Garibaldi legion, it had to capitulate to superior force on July 3. The prince had believed in a republican victory until the last minute; only when everything was lost did he embark secretly, like his father 40 years earlier, from Civitavecchia, intending to live in France. But scarcely had he landed in Marseilles when his cousin's government handed him an order of expulsion. Though he continued his journey unconcerned toward Paris, he was arrested in Orléans and taken under guard to Le Havre, where he was forced to take ship for England.

Many another man would have been crippled by so much ill luck, but not Bonaparte. After passing through London, where he was welcomed by his friends, he went directly to a meeting of the British

Association in Birmingham, and then to Scotland to visit Sir William Jardine, who had just received from Blyth a selection of rarities from the Himalayas. From there the refugee went to Holland, where he was hospitably received. After a short while he planned firmly and finally to write in exile the book he had been thinking about for over 20 years — a synopsis of vertebrates, and primarily of birds. All the notes that he had deliberately gathered during this long period, his collection (which had been growing steadily ever since 1823), and his beloved books, he had had to leave behind in Rome, but the 12,000 birds of the Leiden museum, most of them not yet studied, offered him the most complete substitute that he could wish. Nothing attracted him so powerfully as chaos waiting for a mighty organizing hand. He began work at once. His *Conspectus generum avium*, written in Latin to emphasize its international significance and the world citizenship of its author, was intended to include all known species of birds and to facilitate a survey by its arrangement of small, closely related groups. This forced him to create many genera and subgenera, even when they were connected by transitional species. As early as 1825 he had written: "If we proceed to the abolition of all artificial distinction between genera united by almost imperceptible gradations, *Sylvia* would be joined to *Turdus*, *Myiothera* to *Troglodytes*, *Lanius* to *Muscicapa*, the whole of these would be confused together, and, in fact, orders and classes would be considered as genera."[7]

But the task that he had set himself included also the distinguishing of species in themselves, either by reference to the literature or, if they were not yet described, by a diagnosis, even if very brief. He was extremely happy to find a congenial spirit in his contemporary Schlegel, who advised him on all difficult questions, and who had even provided him with the decisive impulse to write the *Conspectus*.

Bonaparte's political optimism had meanwhile been considerably dampened by experience. In 1850 he remarked, referring to the bird of paradise that he had named *Lophorina respublica* (which has a red, instead of a yellow, collar, like its nearest relative):

> There are those who are much inclined to name their most beautiful species for princes; since I am not in the least enamored of the authority of any princes, I have adorned this extraordinarily beautiful bird of paradise with the name of Republic: of that Republic which would be a Paradise, if it had not been made into a Tartarus by the evil intrigues and the ambition of Republicans who are unworthy of the name they have taken.* But since there is not to be

*This thrust was directed at his cousin, Louis Napoleon Bonaparte. Charles Lucien despised, as a betrayer of the republican idea, the man who several years before had begged the Prince of Canino for a loan and then had behaved so shamefully to him. He saw in him the son not of Louis Bonaparte but of Mijnheer Verhuell, and later in Paris never went near the imperial court.

a paradisiacal Republic, there shall at least be a republican bird of paradise [sed cum non existat Respublica paradisea . . . existat saltem Paradisea respublica!].[8]

Ornithologists came to visit and console the famous scholar and unfortunate exile in Leiden. Professor Carl Reinwardt, already 76, did not come empty-handed but presented Bonaparte with his earlier writings on the birds of Java. The Vicomte Bernard Du Bus de Gisignies, director of the Brussels Museum, came accompanied by Baron de Selys-Longchamps, who 12 years earlier had been a guest in Bonaparte's Roman palace. From London came his old friend John Gould, happy to be able to show such a sympathetic expert some new and attractive items from the Andes and Australia.

As if he could not wait for the zoologists to learn of his activity in Leiden, Bonaparte gave his acquaintances portions of his *Conspectus.* Already at the beginning of February they received the first sheets, and the first gathering (up to sheet 34) was sent off on July 24, 1850.

In the meantime, since he was interested in making his synopsis as complete as possible, he decided in May to travel with Schlegel to Berlin, because Lichtenstein had collected treasures there that were not much inferior to Leiden's, and the experienced scholar was needed to clarify the arbitrary nomenclature of the Berlin dictator. Certainly the two friends did not spend much time over it. The prince's restlessness limited them to looking through a few families of Passeres, with Cabanis, before hurrying on to the house of Schlegel's parents in Altenburg. On the way they saw Poeppig in Leipzig. Then they visited Johann Friedrich Naumann in Ziebigk on June 16, broke the remainder of the journey in Frankfurt, Darmstadt, and Mainz, and spent several days as guests of Prince Maximilian at his castle of Neuwied. When they had returned to Leiden by way of Brussels, Bonaparte wrote a few hundred manuscript pages (sheets 35 to 44) in nervous haste and also, in collaboration with Schlegel, completed an illustrated *Monographie des Loxiens,* evaluating the Berlin and Frankfurt experiences. It appeared as early as November 1850.

In the summer of that year, during this intense activity, which often kept him at the museum until late at night, Bonaparte received an important message: the French government would allow him into France. So it had not taken long for his cousin the president to change his mind. He must need a large and dependable entourage, and who should seem more reliable, in the inevitable political battles, than his own family?

Without taking much time to think it over, Bonaparte packed and left for Paris, in order to set up a new home. But even this fresh convulsion could not stem the flow of his productivity. Scarcely had he arrived in the city of his birth when he visited Isidore Geoffroy Saint-

Hilaire and Pucheran in the Muséum d'Histoire Naturelle, where he worked through the material for the last sheets (45 to 68) of the first volume, "Insessores," so that it appeared before the end of 1850.

Hartlaub, the "praeceptor mundi," greeted the *Conspectus* in 1851 as

> an exceptional work, which, for the first time since Latham's *General Synopsis of Birds* (1781–1785), succeeds more or less completely in organizing and critically working over the ornithological material, monstrous alike in the enormously proliferated mass of species and the chaos of synonyms . . . The number of genera prescribed . . . is vast. The 543 pages of the first part of the work . . . contain no fewer than 1075! The 350 new species, most of which belong to the archipelago of the East Indies and to West Africa, are described briefly from originals in the Leiden collection and in other museums.[9]

As a center for ornithological activity and the residence of private collectors and respected dealers (like the Verreaux brothers* and Émile

*Maison Verreaux, Place Royale 9, was at that time the most important shop in the world dealing in natural history specimens and had as clients all the significant zoological museums and private collectors. During its greatest prosperity in the 1850s it was supposed to have 3,000 mammals and 40,000 birds stored in its warehouses. The brothers Jules and Édouard Verreaux directed it. Their father, Pierre Jacques Verreaux, who had carried on the business since 1800, had married the sister of Étienne Geoffroy Saint-Hilaire's preparator, P. A. Delalande. When Delalande was sent to South Africa for three years in 1818 by the Paris Museum, young Jules Verreaux (1807–1873), who was only 11, was allowed to accompany him. The experience decided his career. After Delalande's death he went back to the Cape in 1825 and stayed there for 13 years, proving himself a skillful preparator, to Andrew Smith's delight. At first he bought up large animal collections in South Africa, and in 1830 brought over his younger brother, Édouard (1810–1868), to escort them home. Édouard was encouraged by his mercantile success to go out to the Cape again in 1832, and to sail in a French ship as far as China. But in 1833 he returned to Paris and took over his father's business. Jules came home only in 1838, having earned his keep until then in Cape Town as a preparator and exhibitor of zoological curiosities. He had also acquired the birds collected by Victor Sganzin from 1831 to 1832 on the east coast of Madagascar. But he was unlucky enough to lose all his collections, because the ship to which he had entrusted them was wrecked on the cliffs of La Rochelle. In 1842 Jules undertook his third and last voyage to distant countries and collected for five years in Australia and Tasmania, on behalf of the Paris Museum. Afterward he devoted himself to the scientific direction of the Maison Verreaux, and acquired a great reputation as an expert on exotic birds; he himself worked on the collections of successful travelers (du Chaillu in Gabon and Captain Saisset in New Caledonia), which the firm was expert in obtaining, and he was supposed to have helped Prince Bonaparte considerably with the second volume of his *Conspectus*. Jules Verreaux succeeded Florent Prévost as assistant naturalist to the Paris Museum in 1864, and worked with Henry Milne-Edwards, concerning himself chiefly with the rich bird collections sent back from west and central China after 1862 by Father Armand David, which were full of novelties. When German troops approached the French capital in 1870, he fled to London, where he found refuge with the young Bowdler Sharpe. Sharpe said of him that "he possessed an immense knowledge of birds, probably greater than any man of his generation."

Parzudaki, among others), Paris provided the restless prince with plenty of stimulation. Above all, Isidore Geoffroy Saint-Hilaire supplied his eagerness with new tasks and recommended to the prince that he arrange the families within their orders and genera within their families not in continuous lines, but in parallel columns, according to an evolutionary principle. Under the influence of this important zoologist Bonaparte foreshadowed the theory of evolution.

In Paris there seemed to be no limits to Bonaparte's productivity. A vast quantity of small articles, which appeared in the minutes of the meetings of the Paris Academy of Sciences and elsewhere, helped him to reform avian systematics, to clear up synonyms, and to publish descriptions of new species. Soon he also began to rearrange the entire bird collection of the Paris Museum according to the order of succession in his own classification, because Isidore Geoffroy had given him a free hand. Even before his cousin had been proclaimed emperor of the French on December 2, 1852, the plain Charles Lucien Bonaparte, as he had styled himself after leaving Rome, had become "S. A. le Prince Bonaparte,"* and after a while (in 1854) he received the title of Monseigneur. J. W. von Müller reported in 1852: "In the French metropolis he is now the leader, the supporter, and the protector of all foreign naturalists, whom he receives in his hospitable house."[10] Famous visitors like Sundevall, Sclater, Hartlaub (1856), and John Gould (1856) were his guests.

Baron von Müller, who had met Bonaparte in Paris in 1851, and who has provided us with much valuable information about contemporary museums and ornithologists, was an adventurer whose chief merit was having led Alfred Brehm and Theodor Heuglin on to the path of fame.

His grandfather, Johannes Müller, after emigrating to South Africa as a young man, had grown very rich through his banking activities in Cape Town, and was thus able to buy Schloss Kochersteinsfeld in Neckarsulm, Württemberg, in 1824. Johann Wilhelm was born there on March 4 of the same year. He finished his schooling at the scientific school run by the Paulus brothers in the Ludwigsburg Salon, which from 1839 to 1845 was directed by Christian Ludwig Landbeck (1807–1890), well known among German ornithologists. It was certainly Landbeck who succeeded in kindling an enthusiasm for ornithology not only in young Müller, but in one of his classmates, Theodor Heuglin (born March 20, 1824). While they were still at school the two friends concocted romantic plans for an expedition of discovery to Africa.

Verreaux died in Paris in 1873. His post in the museum was taken by the 30-year-old Émile Oustalet (died 1905), who was associated with Father David in the publication of a pioneer work on the birds of China (*Les oiseaux de la Chine*, Paris, 1877).

*Prince Bonaparte was the title conferred by Napoleon I on his brother Lucien, when they had become reconciled during the Hundred Days in 1815. The son revived it because he no longer wished to carry a papal title.

After Müller had finished at the university he put these plans into action and in 1845 traveled through France to Morocco and Algiers. For his future enterprises he needed a skillful taxidermist, so he approached C. L. Brehm in Renthendorf. The result was that the "bird parson's" second son Alfred (1829–1884), who was only 18, was attached to the "second African expedition" as Müller's "secretary." On July 6, 1847, they left Trieste and crossed Greece to Alexandria. They reached Khartoum on January 7, 1848, passed through Kordofan in April and May, and then embarked on the Nile for Wadi Halfa. They returned to Alexandria by way of Cairo. In Alexandria, Müller said goodbye to Alfred Brehm in mid-February 1849, promising to come back soon. Accompanied by his noteworthy collections, he passed through Vienna, where he had himself fêted as a great explorer and appointed "Imperial Consul for Central Africa."

He knew how to attract attention when he got home, too. In his school friend Heuglin, who was allowed to help him look through and work on the ornithological collection, he stimulated a burning desire for emulation, so that he joined the "Third Scientific Expedition of Freiherr Dr. John Wilhelm von Müller to Central Africa." The other members had been waiting in Egypt for several months for the "Chief"; Alfred Brehm had been joined in November 1840 by his brother Oskar and a doctor and ornithologist from Köthen called Richard Vierthaler, recommended by Baldamus. But they still had a long wait ahead of them. At the beginning of 1850 Müller got into financial difficulties, which he kept secret, and instead of arriving in person he sent Alfred Brehm a small sum of money and some instructions. These directed the expedition to proceed under Alfred's leadership to Khartoum, where the Baron would meet them in July to discover the sources of the Nile and then go on to the Gold Coast (!). The further development of this project, which cost Oskar Brehm and Vierthaler their lives and kept Alfred Brehm in northeast Africa until 1852, can be omitted here. Under these circumstances Heuglin could not count on Müller's help, so at his own expense (but on behalf of the Imperial Academy of Sciences in Vienna) he set out at the end of 1850 on the journey to northeast Africa, which he was determined on.

In the meanwhile Müller was again presenting an imposing figure at home. After purchasing the Royal Printing House, Im-Guttenberg, in Stuttgart, he began publishing there in the autumn of 1849 a new ornithological journal, *Naumannia*, edited by Eduard Baldamus. He visited famous ornithologists in France and Germany, and worked hard at an ambitious illustrated book, *Beiträge zur Ornithologie Afrikas*, which never got beyond five parts (Royal Printing House, Stuttgart, 1853–1870). The lamentable end of his adventurous "Central Africa Expedition" led to a break with the *Naumannia* group in 1852, but in the same year Müller enriched his museum by buying the private collection that Landbeck wished to sell before emigrating to Chile. It included 2000 birds, mostly mounted. In the autumn of 1852 it seemed as if Müller finally intended to settle down, because he became director of the Brussels Zoological Garden, but he resigned in 1854. Two years later he was on another long trip, visiting the most important North American museums and making a journey to Mexico, which resulted in his *Reisen in den Vereinigten Staaten, Canada, und Mexiko* (3 vols., Leipzig, 1864–1865). He died on October 24, 1866, at Schloss Kochersteinsfeld, when he was only 42.

As if he wanted to fill the jar of the Danaids, Prince Bonaparte now struggled with the completion of the unpublished second volume of his *Conspectus.* In September 1852, he revisited the museums of Mainz and Frankfurt and lectured to the twenty-ninth meeting of German natural scientists and doctors in Wiesbaden. Everywhere he made notes for the second volume, which he began to send off to various ornithologists sheet by sheet in January 1855.

In 1853 Bonaparte had continued his researches in London, and in the summer of 1855 was there again (after visiting Spain and Portugal) to take part in a British Association meeting at Glasgow. In 1856 he visited the museums of Germany, Belgium, and Holland, and made a vivid report on them to the Paris Academy.

His first goal was apparently Köthen, where the tenth meeting of the Deutsche Ornithologen-Gesellschaft took place, June 2–5, 1856. The leading specialists in Germany, including J. F. Naumann, C. L. Brehm, J. H. Blasius, Bernard Altum, C. L. Gloger, and Eduard Baldamus, and the Frenchman Léon Olphe-Galliard, had come because the topic for discussion was, "What is a species, particularly in ornithology?" Although many German participants defended their views incisively and persuasively, Bonaparte alone could deal with the question in a fundamental way, because only he, as a disciple of Isidore Geoffroy, understood existing species as the result of a long evolution, the product of a struggle between the conservatism of heredity and the modifications of environment. His address to the meeting included the following:

> Will one oppose to our obvious theory of limited mutation the theory of successive creations and the theory of the so-called transmission of acquired characteristics? And are the creatures that died out a long time ago supposed to have left no progeny at all, though we see creatures like them everywhere? No! We will state with unanimous conviction that the antediluvian crocodiles, elephants, and rhinoceroses were the ancestors of those living in our day; and

> these animals would not have been able to continue to exist without the manifold mutations that their systems produced to adapt themselves to the environment, and that became second nature to their descendants. I say manifold mutations, because clearly the number of species seems more likely to increase than to decrease . . . If the environment remains the same, so do the species. The stabilizing influence is then by itself all-powerful. The mutating influence can succeed in opposing it only when the whole world surrounding the species changes . . . But races, however different in characteristics they may be, vanish entirely or at least do not long survive as soon as the environment that produced them ceases to be the same . . . The transitions between the different races and their type are the best evidence that we can supply to set aside putative species, which are to be relegated to races, with which the painstaking zoologist must nevertheless occupy himself just as earnestly.[11]

From Köthen the prince traveled to Berlin. Here, from June 9 to 13, he is supposed to have been extremely busy with an exact comparison of all real or possible new species in the Zoological Museum, "grâce à la cordiale réception de son érudit directeur, le célèbre Lichtenstein, et à la puissante coopération de son ornithologiste par excellence, M. Cabanis."[12] On June 14 he proceeded to Brunswick, where the museum, headed by Professor J. H. Blasius, "contient une des meilleurs collections d'oiseaux d'Europe." In Leipzig he was taken about by Poeppig, in Dresden by Reichenbach, and in Strassburg Schimper showed him "la richesse de ses magazins." His fourth visit to Frankfurt found "M. Rüppell absorbé par ses études numismatiques." In Neuwied he dropped in on his brilliant friend, Prince Maximilian. In Bremen he was astounded by the large number and the excellence of the individual specimens that Hartlaub had assembled. He left no natural history museum unvisited in Wiesbaden, Antwerp, Brussels, and Ghent. But especially he was enchanted all over again by Leiden, where he was received with the usual cordiality by his old host Temminck, "le Papa (et non le Pape) de tous les oiseaux," and his dearest friend Hermann Schlegel, "le premier zoologiste de notre époque."

Immediately on his return Bonaparte completed further sections of his manuscript, for which the experiences of this important journey had been useful. They were published as sheets 20 to 29 of Volume II only in the autumn of 1857, with the following publisher's note:

"The illness of the distinguished author delayed publication, although he was still working [at the book] on his deathbed; since this premature death interrupted it finally, I am publishing everything left to print, as far as the manuscript goes."

Charles Lucien Bonaparte had lost his race with death, in whose shadow he had been living for a long time. This had urged him to the fantastic speed that had characterized his scientific activity after he had begun living in Paris.

> The illness allowed its victim to take no more rest, either mental or physical, during his last years; with dreadful haste, driving him on from one hardly started task to another, it tortured and hounded him, in spite of his tremendous resistance, into the grave. "The more I have to put up with, the more I work," said Bonaparte, when I found him, during one of my last visits, writing in his bath (Hartlaub, 1858).

But now what he dreaded had happened; when he breathed his last, on April 29, 1857, the *Conspectus* still lacked auks, loons and grebes, ducks and geese, rails, grouse and pheasants, tinamous, and other groups, but nobody could be found to complete the torso. And even so, according to the author's magnificent plan, this *Conspectus* was supposed to be only the precursor of an "Ornithologie générale et particulière," in many volumes, for which the zoologists had for years had great hopes. Other work, too, that Bonaparte had begun shortly before his death had to remain unfinished: an *Iconographie des pigeons* (begun in 1857 and continued until 1858 by Alfred Moquin-Tandon), and an *Iconographie des perroquets* (begun in 1857 with Charles de Souancé, nephew of Victor Masséna, Duc de Rivoli, and left unfinished in 1858).

None of those who encountered this great ornithologist could resist the spell of his personality, which combined a fiery Roman temperament with intelligence and sound judgment, and a cheerful disposition. Only his looks betrayed his ancestry; small and stocky like Napoleon I, he resembled him also in feature more than any other member of the family. This likeness was not only attested by J. W. von Müller,* but also still startles anyone who sees the portrait drawn of him by J. H. Maguire† in England in 1849. But what a contrast between the politically minded scholar who, full of goodness himself, wished to believe unfalteringly in the goodness of mankind, and his uncle, the successful politician, the expert in and exploiter of human weaknesses!

An attractive memorial to Charles Lucien Bonaparte was left in the reminiscences of one of his closest friends, Hermann Schlegel:

*Müller tells the story of the Napoleonic veteran who mistook the prince for his uncle, shouting "vive l'empereur!" *Naumannia,* 2, 1 (1852): 92–93.

†A large lithograph based on this pencil drawing was on sale at the time, and a copy exists in the archives of the Berlin Zoological Museum.

> With absolute certainty, this remarkable man found the right way into the profoundest depths of science, because of his all-penetrating perception and his almost excessively intense activity. To the highest degree pleasant in company, fonder of children than most men,* and in his family circle the most amiable and unreserved imaginable; in politics he was by contrast what the French call a "farouche républicain."† But he was wild only in speech and not in action . . . In his leisure hours he was contented and relaxed, forgot his destiny, and at most complained of the many enemies he had, without realizing that he was his own worst enemy. During our walks together conversation usually revolved around science, but for a change he spoke from time to time about something else and became very amusing. Then he could sing a Parisian street song in his own particularly apt and lively manner, and again suddenly return to his beloved dream of a universal republic.[13]

Bonaparte's achievements in ornithological systematics ensure him an outstanding place in history. His work left deep and indelible marks; but there has never been any lack of ornithologists who disliked them. For "while Reichenbach and Cabanis are endeavoring to keep nomenclature as strictly as possible in conformity with grammar and the classics, Bonaparte treats it with the most extravagant license" (Hartlaub, 1854). He was the mischievous parent of such monsters as *Lichtensteinipicus*, *Reinwardtoena*, *Kaupifalco*, and *Schiffornis*, which can never be expunged from our nomenclature, and he gave free rein to his love for new names. However, Elliott Coues seems to me to go much too far when he says:

> In my view Bonaparte's services to the science of Ornithology ceased in 1850. The sum total of his subsequent contributions to the subject, until death cut short his schemes, is not only a worthless but a pernicious aggregate. In his later years, Bonaparte simply played chess with birds, with himself as king: *le roi s'amuse!* Scheme followed scheme, tableau tableau, conspectus conspectus, with perpetual changes, incessant coining of new names, often in mere sport — it was nothing but turning a kaleidoscope. It may have been fun for him, but it was death to the subject.[14]

*By Princess Zenaide he had eight children, three sons and five daughters, all of whom survived him. She died on August 8, 1854, in Naples. The line is now extinct.

†To his sorrow, his own sons became decided opponents of his political convictions and allied themselves with the "tyrants" and "hypocrites." During the Roman republic the eldest belonged to the papal party; the second became a priest in 1853 and shortly thereafter private chamberlain to the same pope whom his father had helped to banish.

What Coues as a horrified censor of nomenclature here caricatures, I find of secondary importance. There is no denying to Bonaparte the fame of having outstripped all his contemporaries in knowledge of species, especially since he studied in more museums in all sorts of countries than any other scholar; and because his *Conspectus* is essentially not a product of literary diligence but the precipitate of experience, it is often consulted even today by systematists. His arrangement of genera and their relationship to families and orders differs from our "modern" system less than other contemporary efforts, because Bonaparte, already inspired by the theory of evolution and uninfluenced by the dogmas of natural philosophers and anatomists, could rely for essentials on his eye, which he had been sharpening for years during his careful comparative studies. Certainly this self-confidence not infrequently led him into error, but in such cases only the comparative anatomy of a later period was in a position to confute his views.

Bonaparte wisely circumnavigated the dangerous rocks on which many systematists of that time foundered, although even he was once in great danger of harkening to the sirens' song of the natural philosophers. In 1827, when he had come under the influence of Vigors and Swainson in London, he wrote:

> An extensive reform is evidently needed in the department of classification that relates to genera; and we propose, with this in view, to undertake at some future period a general work, when, erecting our system on a more philosophical basis, though we may restrict some genera, and enlarge others, we shall . . . at least place them all on an equal footing.[15]

But not only did he never realize the proposal, he later (in 1850) energetically attacked the natural philosophers who faithfully respected the "revelations" of such men as Oken and Kaup:

> As far as concerns this author, a declared enemy of this bastard system which men have vainly attempted to ennoble with the title of "golden mean," he protests more than ever against the matching of two and two or three and three, with the sole purpose of diminishing the number of groups of species, which in themselves are sufficiently natural, but which become artificial when one forces them to comply with preconceived ideas.[16]

These "preconceived ideas" will form the subject of the next chapter.

10

The Effect of Natural Philosophy

Two different trends in systematics began to diverge as early as the eighteenth century.[1] The adherents of one believed in Aristotle's inductive method and proceeded according to the basic Linnaean precepts, though without completely accepting the criteria that he had prescribed for the distinction between orders and genera. These had all been established from the externally visible parts of a bird's body, particularly the shape of the beak and the foot. In contrast to Linnaeus, Merrem had emphasized in 1788 that the systematist must take all parts of the organism into account. His view gained ground slowly among ornithological systematists, and especially after Georges Cuvier made a similar pronouncement in 1806. However, not enough students were able to use the theory, because most systematists lacked anatomical training, and most anatomists the total view of the wide diversity of forms. Many authors, like Lacépède (1799), Illiger (1811), Vieillot (1816), and their large band of followers, decided to retain the old systematics based on the preserved skin, either as a temporary stopgap or as an "expedient" procedure, but in so doing they reached widely varying results. Others tried to form systems based on "anatomy," and plunged into the study of certain obscure morphological details, like the shape of the sternum (Blainville, 1815; L'Herminier, 1827), the tongue (Wagler, 1830), or the nasal glands (Nitzsch, 1820), in which they succeeded as little in agreeing as the nonanatomists, even in main outlines. The eclectics (Merrem after 1781; Nitzsch, 1827) tried to improve this poor state of affairs by seeking their points of reference in the entire avian morphology, but as the diagnoses of orders, families, and genera became steadily more complicated, and the proliferation of the different groups steadily greater, they departed further and further from the state of general agreement that prevailed among the orthodox disciples of Linnaeus from 1758 to 1811.

The supporters of the second tendency were not content to describe facts and organize species according to the Aristotelian principle, that is, according to their degree of "resemblance." They were not satisfied to have established that species, genera, and orders were separated by gaps, and that these gaps were sometimes narrow and sometimes wide; they asked questions about the basic reasons for such an arrangement. In the beginning there were still theoreticians who believed with Leibniz that these gaps would gradually be closed by further discoveries, but soon the extent of their error became clear. Only rarely were new species described that seemed to fit into the gaps between the higher taxa already known. Although owls, parrots, pigeons, swallows, and many other groups could be arranged with some others in series according to resemblance, any connecting links between these series remained undiscoverable.

Because this conclusion contradicted the teachings of philosophy, the theoreticians were dissatisfied with it, believing it to be based on an error, and looked to comparative methodology to explain the disparity between the result and their expectations.

Aristotle had once undertaken to systematize nature as a ladder of entelechies.[2] This idea had been exalted into a dogma by his interpreters and was finally incorporated into the foundations of philosophical systems. In 1635 the Jesuit Eusebius Nieremberg maintained, in volume III, chapter 3, of his *Historia naturae*, that "Nature proceeds as in a continuous weft, without gaps. There is no hiatus, no break, no dispersion of forms; instead, they are connected as one link is with another. This golden chain embraces the universe." After the great Ray, in his *Historia plantarum* (1686), had reiterated the thesis in the words "Natura non facit saltus," Leibniz took it up and established the law of continuity as the logical revelation of the divine ordering of the world. In the harmonious creation, he taught in 1687,

> link must be fastened to link without a break, creatures, like events, must shade off in infinitely small differences and thus form a continuous series of steps, in which the individual steps are only quantitatively and not qualitatively distinguished. Because if the differences were not infinitely small, breaks would occur, a "vacuum of forms" in the realm of creation, "a possible living creature would be wanting." But that would be a contradiction of the complete being of the Creator — which is impossible.[3]

According to Leibniz the universe is a ladder whose rungs stand above and below one another. The rungs themselves, that is, the beings, and the distances between them are eternal and immutable. This immutability is an integral part of Leibniz's concept of the monad.[4]

The theory demonstrates in a fascinating way how natural phenomena are to be systematized. It was left to the natural scientists to prove its accuracy. They exercised their scholastic abilities on the problem for a hundred years, and tried to arrange creatures according to "resemblance" in continuous series, networks, or osculating circles, without considering the objection made by Kant in the *Critik der reinen Vernuft* (1781):

> What may seem to us small differences are usually in nature itself such wide gaps, that from any such observations we can come to no decision in regard to nature's ultimate design — especially if we bear in mind that in so great a multiplicity of things there can never be much difficulty in finding similarities and approximations.[5]

The Genevan entomologist Charles Bonnet (1720–1793) was one of the first zoologists to apply Leibniz's ideas to the observation of nature. First in his *Traité d'insectologie* (1745), and then in *Contemplation de la nature* (1764), he illustrated the *lex continui* according to a ladder of objects, but did not "pretend in the least to establish the real gradations of Nature . . . this is merely a way of envisioning beings and moving from one to the other."[6]

> Without presuming to establish the progressive order of Nature, nevertheless we shall place birds between fish and quadrupeds . . . In this order, aquatic birds will be ranked immediately above the flying fish. The amphibious birds, those that live equally on land and in the water, will occupy the next rung, and will thus become the link between the regions of water and the regions of earth and air . . . The bat and the flying squirrel are those odd animals that are so suitable for demonstrating the gradations among the works of Nature. The ostrich, with goat's feet, which runs more than it flies, seems to be another link between birds and mammals. The carnivorous quadrupeds correspond to the birds of prey. Quadrupeds that live on grass or grain correspond to the birds with similar diet.[7]

Bonnet's scheme was thus that of a single series of steps. Though it might be possible to arrange the principal groups of the "natural objects" in this way, nevertheless most of the subordinate groups could not be fitted into such a simple pattern. Donati (1750) therefore conceived the notion that every being is a knot in the web of nature, and its resemblance to other forms may be compared to the threads running between the knots. This may have given Linnaeus the idea of demonstrating "analogies" between six of his seven orders of mammals and his six orders of birds; specifically, in the twelfth edition of his *Systema naturae* (1766), he compares, with disastrous effect, the Accipitres with the Ferae, the Picae with the Primates, the Anseres with the Belluae, the Grallae with the Bruta, the Gallinae with the Pecora, and the Passeres with the Glires.* In his *Tabulae affinitatum animalium* (1783), Johannes Herrmann, professor of natural history in Strassburg, based his design for a network of relationships among animals on Donati's idea by indicating also the "affinities" of the species and genera of birds known to him. For example, *Trochilus-Certhia-Picus-Corvus* are connected in a vertical row, and *Certhia-Merops-Alcedo* in a horizontal row. The logical consequence of such ideas had to be a three-dimensional representation of the resemblances and connections; it appeared first in a work by a Frenchman, Jean Baptiste Robinet, called *De la nature* and published in five volumes in 1763. Robinet's scheme is a three-dimensional lattice, composed of connecting ladders with a net at the base. The wise man's more penetrating glance

> will one day see in the diversity of creation an enchanter's net composed of innumerable threads and intermingled knots, with one depending on all and all on one, where our limited vision now believes that it sees only the juggling of an untiring imagination, which has haphazardly emptied out its cornucopia,

prophesied Georg Forster in *Vom Brodbaum* [On the breadfruit tree], with poetic pathos.[8] He was to be proved only too right because, when French zoologists at the end of the eighteenth century decisively rejected the concept of a network of affinities, since Vicq d'Azyr and Cuvier had pointed out other possibilities, German and English scientists became all the keener to develop it. Vicq d'Azyr observed:

> Anyone who considers a quadruped after having acquired exact knowledge of human anatomy finds such great similarities between them that he can turn from examining one to the other without surprise. But the chain is broken between quadruped and bird; even the ostrich is no use as a link because, except for its pubis, it possesses none of the characteristics proper to a quadruped. Skeleton, lungs, stomach — everything excludes it from this class of animal. Therefore the anatomist who has learned from experience and is

*Linnaeus grouped together under "Ferae" the predators, *Phoca*, the insectivores, and the marsupial rats; under "Bruta" the edentata, rhinoceros, elephant, and walrus; under "Belluae" *Equus*, *Hippopatamus*, and *Sus*.

> rigorous in his comparisons will reject rough equivalents, and will take good care not to reunite what nature has separated.[9]

In Germany, soon after 1800, the suggestions for systematics provided by the *lex continui*, together with ideas from Platonism and Spinoza, were amalgamated into "Natural Philosophy." At that time Schelling had begun to propagate his theory of a metaphysical order in the universe, and of prototypes represented in nature. Its remarkable language and inspired audacity could not fail to impress budding natural scientists, for it showed how, with imagination and deduction, one could with little effort reach the high goal toward which the prophets of the study of causality, so much esteemed during the Enlightenment, had made barely visible progress: the understanding of the divine ideas that order the world of phenomena.

The natural philosophers saw the eternal prototypes manifested not only in the species but also in the higher categories; they were all members of a harmonious world fabric whose structure should now be investigated. Genera, orders, and classes were thus logically conceived as realities, although, unlike individuals, they were not substances but only ideas, because for Schelling, who assumed himself to be able to derive the natural world from the intellectual, thinking and being were indivisibly united.

The dogma provoked real mass hysteria. K. A. Rudolphi complained in 1812: "The mysticism now accepted by so many requires ignorant people, and finds them in plenty. O young men, who read these words: through faith you will never arrive directly at truth; only doubt will lead you to it!"[10] Vain warning! Even the great chemist Justus von Liebig, who studied in Erlangen in 1821–22 and attended Schelling's lectures, succumbed to the force of his eloquence. He wrote:

> I spent part of my time as a student in a university where the greatest philosopher and metaphysician of the century compelled the students to admiration and imitation. Who at that time could protect himself against the contagion? I too lived through this period, so rich in words and ideas, so poor in real knowledge and sound study; it robbed me of two years of my life, and I cannot describe the alarm and horror with which I awoke to consciousness from this intoxication.[11]

Therefore it was not surprising that many zoologists remained captives of natural philosophy for their entire lives, especially after Schelling had found his scientific partner in Oken, yet another, according to Liebig, of those "swindlers who, without any conscience whatsoever, break the first law of natural science and philosophy, that of accepting as true only what is demonstrable or has been demonstrated."

One of the first of Schelling's students to take part in extending natural philosophy, with particular relation to animal systematics, was Johann Baptist von Spix (1781-1826), who later traveled with the botanist Martius. In his *Geschichte und Beurtheilung aller Systeme in der Zoologie nach ihrer Entwicklungsfolge von Aristoteles bis auf die gegenwärtige Zeit* [History and evaluation of all zoological systems in their course of development from Aristotle to the present day], he distinguishes between the "natural" system (whose advantage is its "inner worth") and the "artificial" systems (pp. 9-12). For him

> form, in the animal world, of necessity implies the rest of what is known about an animal, so that, just as one can indicate a figure by means of dots, so one can deduce from form enough to establish the nature of a particular animal [pp. 147-148]. Anyone who investigates natural phenomena must be powerfully struck by the observation that, from the lower and simpler forms up to the higher, though they steadily become more beautiful, nature repeats herself. Indeed, a test for any system could be the correspondence between classes, orders, families, genera, and the way in which they succeed one another [pp. 273-274].

Emboldened by the authority of Linnaeus (1766), Spix explains that

> birds of prey correspond to predators among mammals, sparrows correspond to rodents, ostriches to pachyderms, gallinaceous birds to ruminants, waders to horses, geese (especially auks and penguins) to cetaceans [p. 274]. The animal kingdom is distinguished from the other kingdoms of nature by being a more completely developed circle, whose area is filled by smaller and smaller circles, and these in turn by figures with various angles. To the degree that these circles grow smaller, in an exactly corresponding degree animal nature withdraws into narrowness and darkness. Starting from the outside, these circular divisions of the kingdom are as follows: class (*classis*), order (*ordo*), family (*familia*), genus (*genus*), species (*species*), variety (*varietas*).* Nevertheless, the organization of each variety is just as full and complete as in the kingdom as a whole; only just as either an angle or a circle can enclose exactly the same space, though its appearance and manifestation (form) are different, so there the organic form is more ignoble, but here more noble and complete. Therefore the kingdom simultaneously has its roots in the variety and its flowers in the class [pp. 6-7].

*This is the hierarchy of concepts established by Linnaeus in 1751, with the later addition of "Familia."

At this point Spix attacks Blumenbach, Illiger (1800), and especially Lamarck, who in his *Philosophie zoologique* (1809) had stated that all systematic categories above the concept of species are artificial constructs and instruments invented by us, that we cannot do without.

> In contradiction to this theory [Spix's own], we learn in our day from the writings of the foremost scholars in every country the general thesis that this architectonic interlinking of creatures is not founded in nature but is the product of a mere caprice and poetic fancy of the natural scientist; that there are neither kingdoms nor classes, neither families nor genera in nature, but only in the minds of scholars. What happiness for natural history, where a mere chance is recognized instead of order in nature, instead of an immutable obedience to law only a fantastic caprice! Just as animal bodies are subdivided into organs, the human state or a military group into ranks, so are animals organized in nature; and just as notes in music combine to form a concert, so in nature this harmonious structure is expressed in concrete forms [pp. 7–8].

The old argument between realists and nominalists, which had inflamed men's minds as early as the fourteenth century, had flared up again.

Related ideas had already appeared in the work of Gotthelf Fischer von Waldheim (*Tableaux synoptiques de zoognosie*, 1806), who placed man at the center of his system and allotted the central circle, itself subdivided into a series of osculating circles, to mammals. These figures of circles, answering the need for harmony and, when allowed to osculate, giving pleasure by satisfying the demands of the law of continuity, were an expressive attempt to trace out the divine order among living creatures according to a geometric principle. The idea was soon carried further, creating an English trend and a German trend in natural philosophy. In England, where the idea of the "ladder of natural objects" had long been accepted among theoreticians, and William Smellie in his *Philosophy of Natural History* (Edinburgh, 1790) had devoted an entire chapter to the "Progressive Scale or Chain of Beings in the Universe," the zoologist W. S. MacLeay (in his *Horae entomologicae*, 1819), believed that during his research on two genera of beetles he had discovered that all really natural groups can be arranged in circles; in fact, there are nine interlocking circles, beginning with the kingdom and ending with the subgenus. Each of these circles can be subdivided into ten (a number conforming to natural law) osculating circles, five large and five small. Thus MacLeay unknowingly returned to the ideas of the Pythagoreans. Those fathers of idealism had already taught that number is the form underlying all forms and the absolute cause of things. The harmony of the universe, in fact, is produced by its mathematical relations. N. A. Vigors, a learned ornithologist and a friend of MacLeay's, seized on the idea and published a treatise, "Observations on the Natural Affinities That Connect the Orders and Families of Birds,"[12] in which he maintained that all really natural groups of birds take on the form of a circle when one arranges their representatives according to the characteristics of their true relationship, their "affinity" (in contrast to their "analogy," resemblance arising from external causes, functional resemblance). The circles of equal rank touch one another. There are five orders of birds: Raptores, Insessores, Rasores, Grallatores, and Natatores, each of which is subdivided into five tribes of five families each.

The new dogma immediately found an ardent proponent in William Swainson, at that time one of the most prolific and widely read ornithological writers in England, who until 1838 tried again and again to hammer the new theory into the public consciousness and to explain it with examples and diagrams. He was not content with Vigors's simple ideas, but subdivided the original circle into a sacred trinity. According to Swainson, these three circles are for types, subtypes, and aberrations, and the last of these encloses three smaller ones. These in turn are counted as three, alongside the first two, so that the space inside the original circle is filled by 2 + 3 = 5 circles (Fig. 1). The five constituents of each circle are analogous to the corresponding constituents of all the other circles, so that one is enabled to assign each constituent to its natural position within the circle.

Possessed by the fanaticism of a prophet and far more extreme than his predecessors, from whom he parted company in 1827, Swainson maintained, "we are upon safe and solid ground, although these principles are of very recent discovery"; in fact, "we become as much inclined to question the circular progress of the planets round the sun, as the circular development of the variation of forms in the animal and vegetable creation."[13] According to him, the task of the systematist is to establish the correct position of species, genera, familes, and so on, on the circumference of the circle by means of their affinities and analogies. To this end Swainson, aided by his remarkable talent for drawing, studied the external characteristics of birds (plumage, beak, foot) in as much detail as possible; it was necessary to consider only these and the birds' habits, because "the study of the inward organization is not so essential to the zoologist as that of the outward organization." In so doing he found an analogy between penguins and tortoises, whereas affinity links birds with pterodactyls, or *Struthio* with *Ornithorhynchus*. The classes of birds are subdivided into five

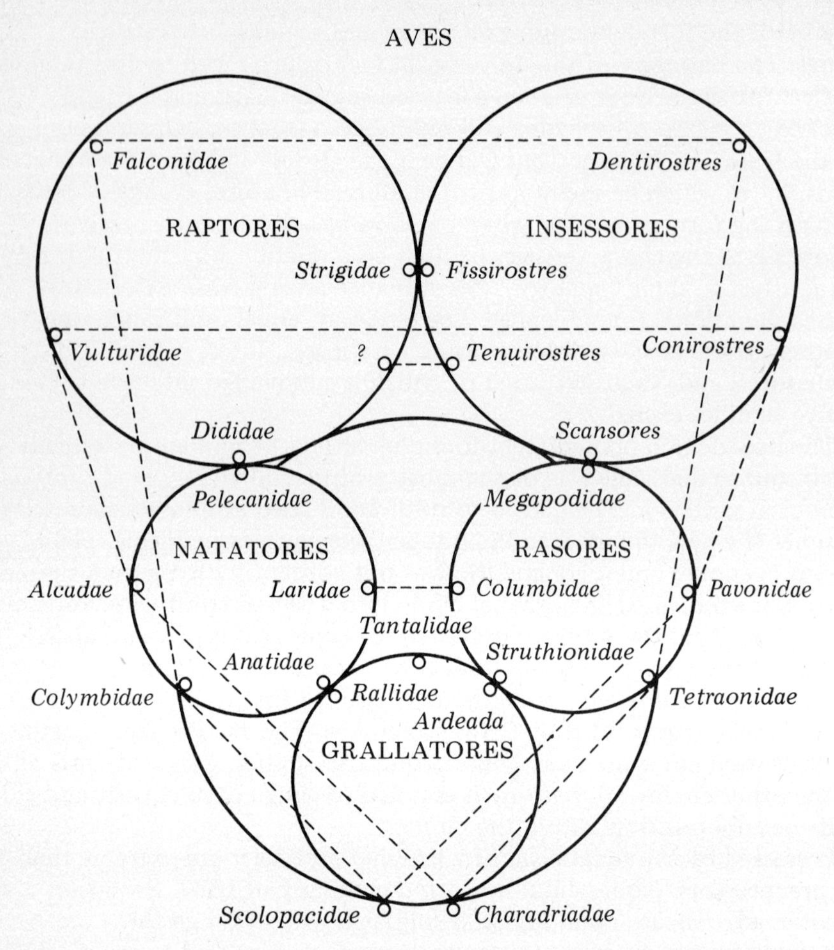

Figure 1. The classification of birds proposed by Swainson in 1837. The five orders are represented by three large circles, of which one, that for aberrations, is subdivided into three circles. Each circle includes five families, and family "affinities" join it to the two neighboring circles. The arrangement of the three remaining families is determined by "analogy" (broken lines).

circles, analogous to the five circles of mammals:

1.	Ordo	Insessores	Quadrumana
2.	"	Raptores	Ferae
3.	"	Natatores	Cetacea
4.	"	Rasores	Glires
6.	"	Grallatores	Ungulata

Because each circle has its analogies in all other circles, the horizontal rows in this example correspond by analogy:

Families of the Dentirostres	Tribes of the Insessores	Orders of Birds
1. Laniadae	Dentirostres	Raptores
2. Merulidae	Conirostres	Insessores
3. Sylviadae	Scansores	Rasores
4. Ampelidae	Tenuirostres	Grallatores
5. Muscicapidae	Fissirostres	Natatores

In each of these three columns the first and fifth constituents, which complete the circle, must have close affinities. According to Swainson, the Laniadae are related to the Muscicapidae through *Tyrannus*, the Dentirostres to the Fissirostres through the motmots (*Prionites*), and the Raptores to the Natatores through the frigate birds (*Fregata*).

Whereas the highest-ranking circles, for instance the classes, are connected with one another by analogy, the lower ranks (genera and subgenera), are connected by affinity. Great delicacy of perception is often needed to discover in which of two touching genus circles a subgenus belongs (Fig. 2). The decision is facilitated by the fact that no circle can enclose more than five subordinate ones.

When Swainson was working on the flycatchers in 1838 he thought he had discovered that even the species are subject to the general law: "The variation of the species is regulated by the same law as the variation of groups," that is, each type has its analogue (its representative) in the other groups. Therefore a natural subgenus must consist of five species, of which each must "represent" one of the five genera of its family. Many circles must temporarily remain incomplete because the missing species have not yet been discovered.[14]

William Swainson, born in Liverpool in 1789, was not only an idealist subject to enthusiasms but an experienced and extremely talented naturalist. The passion for all sorts of creatures that possessed him from childhood and drove him to collect and to draw could never establish itself on a firm foundation because he had had to leave school at 14, and soon thereafter set out on his adventures. From 1807 to 1809, as a member of the British Mediterranean Army, he was mostly in Sicily, where in 1809, shortly after he had published a small book called *Instructions for Collecting and Preserving Subjects of Natural History and Botany* (Liverpool, 1808), he became acquainted with Rafinesque-Schmaltz, a merchant passionately interested in zoology. Swainson had scarcely reached home before he began to dream of winning fame through a journey to the tropics, like his hero Levaillant. An Englishman who had already been exploring the newly accessible territory of Brazil chose him as a companion for a second Brazilian journey in 1816, from which they returned two years

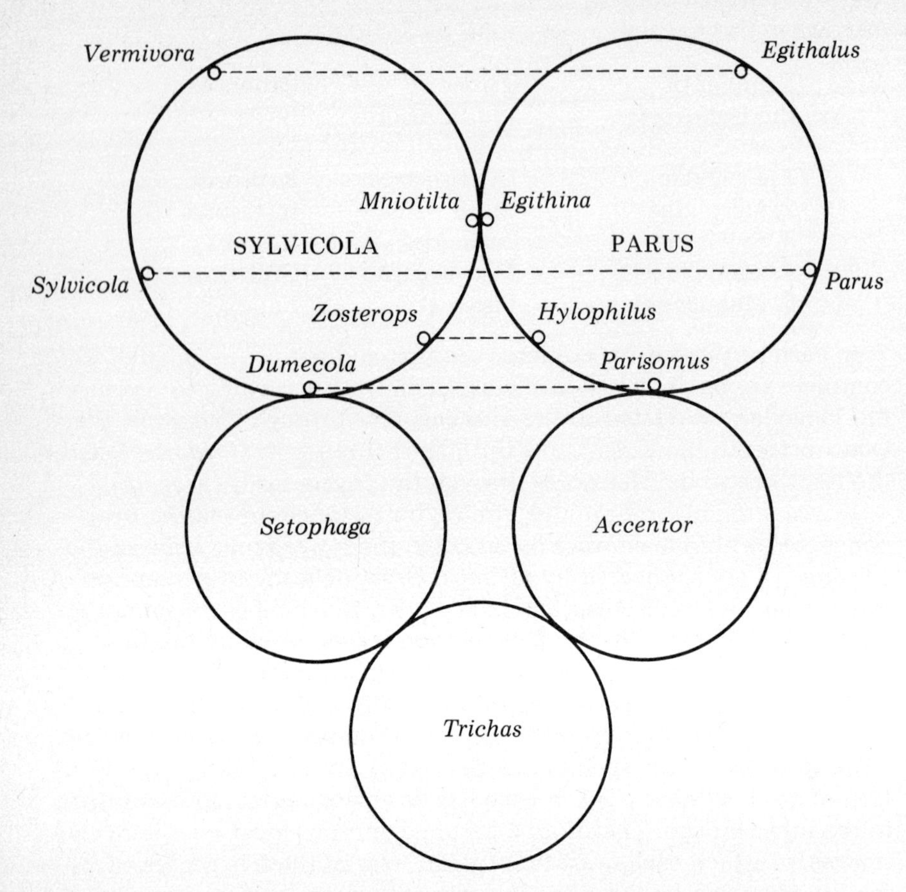

Figure 2. The taxonomic structure of the subfamily Parianae (tit-like birds), according to Swainson in 1837. The subfamily breaks down into five "typical" genera, of which the two large ones (namely the "typical" and the "subtypical") are to be divided into five subgenera.

later with large collections (including 760 bird skins). Swainson's career was now decided — he devoted himself entirely to zoology. First he worked on his Brazilian material, which contained many new species, and practiced lithography (recently introduced into England) so that he could illustrate his descriptions himself (*Zoological Illustrations, or Original Figures and Descriptions of New, Rare, or Interesting Animals*, first series, 3 vols., 1820–1823). In 1822, after he had competed in vain for the keepership of the Natural History Department of the British Museum, from which Leach had retired after becoming seriously ill, he realized that he would have to earn his living by zoological writing. As a member of the Zoological Club of the Linnean Society, founded by the Reverend William Kirby and others (and forerunner of the Zoological Society of London),[15] he took an active part in the scientific life of the capital for some time. With Audubon he went to visit the Paris Museum in the autumn of 1828, and there made the acquaintance of Lesson, with whom he established a real friendship. (He had alienated many earlier friends, including Vigors, by his fanaticism.) Finally he must have realized that in his urge for understanding he had been for years naïvely trusting in and following a will-o'-the-wisp, because in his last book, *Taxidermy; with the Biography of Zoologists* (London, 1840) — which includes several pages of revealing autobiography — he took leave of his readers with the words of Ecclesiastes: "Then I looked on all the works that my hands had wrought, and on the labour that I had laboured to do: and, behold, all was vanity and vexation of spirit." In 1840 he emigrated to New Zealand, where, retired from the world, he died in 1855.[16]

Vigors died in 1840, but both his ideas and those of Swainson continued to be influential for a long time. Even such a notable naturalist as Thomas Horsfield, who had earlier been driven into the natural philosophy camp by his friend Vigors, acknowledged his adherence to the quinary system as late as 1854 (*Catalogue of the Birds in the Museum of the Honourable East India Company*). In 1862 T. C. Jerdon (1811–1872), author of the pioneering book, *The Birds of India*, preferred the image of the universe provided by idealistic morphology to Darwin's theory of evolution:

> That species were created at hap-hazard, without any reference to others, either of the same group, or more distant ones, is a doctrine so opposed to all the affinities and analogies observed throughout the animated world, that the mind refuses to accept it, and intuitively acknowledges the evidence of design.[17]

Its unity was most clearly recognizable through the often unexpected analogies between widely separated groups and species.

In England, 1862 marked the end of contributions to "logical" systematics for birds, but not yet in Germany. The German trend in natural philosophy had absorbed speculative elements that had not affected the English. Its founder, Lorenz Oken (1779–1851), borrowed from Robinet's *De la nature* (1763) not only hylozoism, but also the idea that the final goal of creative nature was mankind. According to Robinet, "All beings were conceived and formed according to a single design, of which they form variants graduated to infinity. The prototype is the human form, whose metamorphoses are to be considered as progressive insofar as they approach the most excellent form of all beings (the human form)."* This was exactly the dogma proclaimed by Oken in his *Lehrbuch der Naturgeschichte: Zoologie* (1815): "Man is the measure and measurer of all creation; his body is thus measure and measurer of animal bodies; he gives the creatures rank and name." As a disciple of Robinet he taught that man, the ultimate goal of creation, came into existence only after many attempts that had to some degree failed — failed because in them the four organic systems of mankind (digestive, skeletal, muscular, and nervous) were arranged in false proportions. Therefore there must be as many principal classes of creatures as there are organic systems in man, that is, four; and from this it follows that the distinctions among the principal classes repeat themselves analogically within the subdivisions (whose number is always a multiple of four), namely, orders, families, and genera, in steadily diminishing degree, whereas forms immediately following each other in the same direction are linked by affinity.†

In 1835 C. J. Sundevall, a Swedish ornithologist whose later writings procured him wide recognition, worked on Oken's ideas in his first considerable publication to see whether they would serve for a new ornithological systematics, and failed in the attempt. Probably at the insistence of the philosopher C. G. Carus, Dr. F. A. L. Thienemann in Dresden tried again. He sketched his philosophical systematics in the second volume (1849) of the ornithological magazine *Rhea*, after having already stated in 1846 that

> the emergence of every organic being is always dependent on movement, but the only possible movements for a sphere are rotation and oscillation, both of which in progressing create a spiral. Therefore it must be the spiral course of the sphere that guides each individual being, each species, and so on up until the whole is connected . . . Only in this way will it be possible to acquire some closer understanding of the connection of earthly organisms. This proposition can be demonstrated without difficulty in an ornithological system.

But it was reserved for Ludwig Reichenbach (1793–1879), director of the Dresden Zoological Museum and an ornithologist with experience in many fields, to be the first really to master the problem.

In his world view, developed from Robinet's and Schelling's philoso-

*This idea is also related to earlier conceptions. Nehemiah Grew, in *Musaeum Regalis Societatis* (London, 1681), had already observed that the animals are "better placed according to the degree of their approximation to human shape, and one to another, and so other things, according to their nature. The very Scale of the creatures, is a matter of high speculation."

†Oken says, "Because vertebrates are divided into four classes, therefore each class can be broken down into only four orders, because these of necessity partly represent the three other classes, and partly their own natures."

phy, he agreed with his friend C. G. Carus (1789–1869): the divine Being, outside time, carries within it the seminal, but at the same time eternal and changeless, ideas that in the temporal universe, of unceasing becoming, appear always anew. By so doing, within a sequence of creations, they steadily approach man, the final goal. The animal creation is therefore the sum of all conceivable retarded forms along the path of development that has led to the organism of man. This development proceeds in *four parallel rows,* within which progression is constantly interrupted by new beginnings. All these beginnings start with species and are carried to their conclusion in species, because the species is that stage in classification which, as it is found in nature, appears as a distinct entity. The other eight categories of the system also belong to the plan of construction and take part in the progression, but without "appearing distinct." Therefore the animal world reveals construction in stages; each stage begins, develops, is completed. Each category has its type; the various species of the category are related to the type either in stagnating before completion or in attempting to reach a higher stage.[18]

> The first feeble attempts of nature at a bird, a sort of half bird, fish and bird at the same time, were the penguins . . . During the course of the era of stratified rocks they were followed by large marsh birds . . . Then with the development of vegetation came the arboreal birds, and prototypes of woodpecker, thrush, crow, and osprey were created to preserve a certain balance in the primeval world of worms, insects, and fish according to primeval law . . . The forms of the primitive ancestors of the bird world culminated in a mighty family of ostriches, at the opposite extreme from the penguins and again losing the power of flight . . . Thus we have seen every prototype of bird (probably all produced by individual primeval breeding from primeval eggs) develop in stages comparable to the age of an individual bird (penguin: stage of unfledged nestling; ostrich: an avian type grown to senescence). The later and more closely related intermediate forms perhaps arose only from the mating of a few ancestral pairs.[19]
>
> Thus in a *scientific* work we must always consciously follow the ascending stages of a type's development, so that we properly grasp the significance of the single members in relation to the whole, and so that we are in a position to indicate the point at which a type culminates and then again its relapse into heterogeneous forms, in always clearly evident repetition of what has already been. Thus science parallels art, which can perfect itself only by the harmony of its parts. The amateur plays casually with all his pieces lined up together, and the eclectic deals out singly those that he happens to have, without thinking of the relation to the whole.

> Those who like using nature for this sort of capricious exposition of their own personality, which gives birth to as many systems as there are natural scientists, should reflect that the presentation and demonstrable proof of *four* classes of vertebrates in nature, fish, amphibia, birds, and mammals — this quadripartition, as the brilliant and ever-immortal Oken discovered — has been dictated to us by Nature herself as the necessary and only true arrangement. Even if one were to forget that heaven and earth and all arts and sciences, even music, with its chords and systems of tuning, are quadripartite, just as everything living is quadripartite, one would have to admit that this *quarternary* division, whose sound is echoed vividly everywhere in Nature, is at least not *arbitrary*, and endlessly changeable, nor individual and violently *confused*, but is something that those who judge as experts can call composed and settled.[20]

> With the help of the law of relationships

> we shall be able to determine the relationship of birds algebraically, we shall see ourselves in a position to demonstrate, by mathematical equations, how large is the share of each genus in the characteristics of other preceding or succeeding genera, and thus, according to this calculation, accurately discover the place assigned by Nature herself to a problematic genus.

This comes from the introduction to *Das Natürliche System der Vögel* (1852, p. 10), in which Reichenbach logically (meaning in a strictly "quaternary" manner) followed his systematics of the orders "Natatores" and "Grallatores" through to the subgenera. In 1854 he treated the hummingbirds in the same way,[21] and in 1862 he published *Die vollständigste Naturgeschichte der Tauben und taubenartigen Vögel: Wallnister, Erdtauben, Baumtauben, Hocco's. Columbariae: Megapodinae, Peristerinae, Columbinae, Alectorinae* [The most complete natural history of pigeons and pigeon-like birds: wall-nesters, terrestrial pigeons, arboreal pigeons, curassows . . .], whose title alone is enough to reveal the eccentricity of the system.

Even more intricate, if that were possible, is the "natural system" of Johann Jakob Kaup (1803–1873), director of the Natural History Museum in Darmstadt, who after several preliminaries published his work of art as *Classification der Säugethiere und Vögel* [of mammals and birds] in 1844. Though Kaup, when establishing his chief principles of division, paid tribute to Oken as "the great master whose outstanding mind long ago indicated the path that every zoologist must travel," he later proved a heretic. The "inevitable result" of his studies was, in fact, not the number four but "the sacred number five," deduced from the five senses, because these are "to be regarded as the flowers of the five anatomical systems." In every class, order, suborder, family, genus, and so on, one of the five anatomical systems, one of the five senses, and one of the five areas of the body has been predominantly developed, according to the following scheme:

5 anatomical systems	5 sense organs	5 areas of the body
1. Brain	Eye	Head
2. Lungs	Ear	Chest
3. Bones	Nose	Trunk
4. Muscles	Tongue	Abdomen
5. Skin	Sexual organ	Pelvis

Man, the most perfect representative of the first class, as the most purely brain, eye, and head animal, is the last and greatest work of the Creator; birds belong to the second class of lung, ear, and chest animals, and this class in turn breaks down into five orders of Gallinae, as skin, sex, and pelvis birds; Ichthyornithes (= Rapaces + Natatores of other authors) as muscle, tongue, and abdomen birds, and so forth.

In 1854 Kaup tried out his basic principles on the classification of the Corvidae (and in 1862 on the Strigidae);[22] as "bones, nose, and trunk birds" the Corvidae occupy the third rank among the five families of the Dentirostres (which "self-evidently" form the fourth suborder of Passeres, the second order). A "pentagram" from this paper illustrates the position of the five subfamilies and their principal genera (Fig. 3). "Were the majority of the small groups discovered or determined, this diagram could be endlessly subdivided by converting each separate part into the same diagram, so that these small groups and the names of species could be entered."

In addition, Kaup thought that his method permitted an important calculation. According to the number of "hordes," that is, primeval types, recognized by him in 1844, he estimated that if all created forms of birds still existed they would make up 535 families, 2,675 genera, and 13,375 subgenera.

The two systems of German natural philosophers just cited were far too mystical to excite any emulation;* indeed, almost all contemporary

*Friedrich Berge sought to remedy this deficiency in 1855, with a "Versuch einer natürlichen Klassifikation der Vögel" (*Naumannia*, 5: 196–212), which was based on the Empedoclean quartet of "organic masses," water, earth, air, and fire, and the Oken quartet of "animal systems," bone, muscle, nerves, and senses. For him birds were "air or nerve animals."

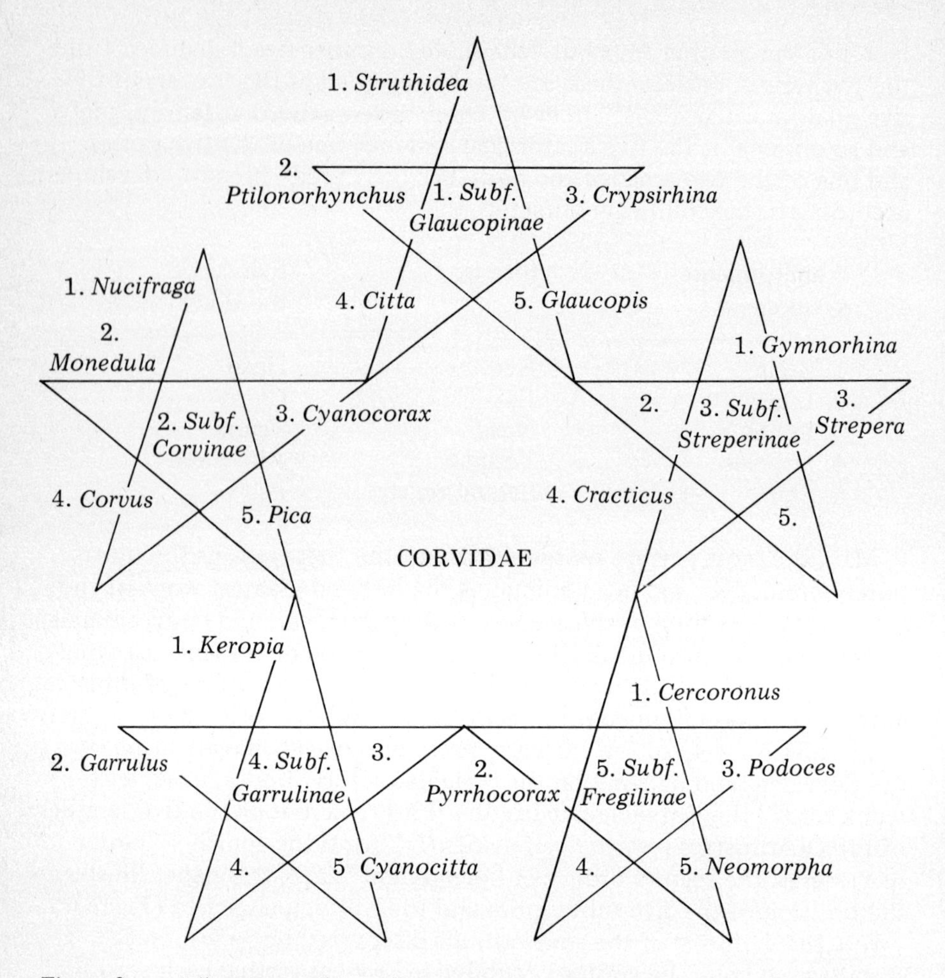

Figure 3. A "pentagram" that illustrates the supposed relationship of the genera of Corvidae. The blank triangles have been left for genera "still to be discovered." After Kaup, *Journal für Ornithologie*, 2 (1854), pl. II, fig. 10.

authorities laughed at them ("Kaup is still working at his monomania of the number five," wrote Hartlaub in 1852[23]). A notable exception was the Viennese zoologist Leopold Fitzinger (1802–1884), who, in a treatise written under the influence of "the ingenious Kaup," *Über das System and die Charakteristik der naturlichen Familien der Vögel*, published in three parts in 1856, 1862, and 1865, distributed all the classes of vertebrates, including birds, into five parallel rows, each of which, according to him, could be assigned to three orders, representing a continuous series from the lowest to the highest forms. Fitzinger thought that in this arrangement "the continuous and parallel relationships (affinity and analogy) of the individual groups appear most clearly," and only this procedure would lead "to a natural classification of the animal kingdom."

The English intellectual games were more popular than the German because they were less complicated. A number of amateurs were bewitched by them, although clear thinkers like Rennie (in 1831) soon attacked them and Macgillivray (in 1837) made fun of them as phantasmagoria. Even such a reserved scholar as J. F. Brandt considered it proper to join in the protest:

> The discerning scientist will make no effort to use geometrical or arithmetical schemes to set up a logical and continuous sequence of the different forms of animals from the highest powers to the lowest, or the other way round. The great differentiation and modification of the forms resist any such approach, which arises from narrow, one might say fantastic, preconceptions. The multiplex world of animals does not allow itself to be confined in cyclical groups returning upon themselves, or in rigidly subordinated ones. The birds . . . too . . . obey the great law of more coordinated than subordinated development of forms.[24]

In England it was chiefly Strickland (first in 1840 and again in 1845) who attacked the natural philosophers and, preparing the way for Darwin, led ornithologists back from deduction to induction. As a sober thinker he commented in his famous article, "Report on the Recent Progress and Present State of Ornithology" (1845):

> The researches of the comparative anatomist universally lead to this result — that all organized beings are examples of certain general types of structure, modified solely with reference to external circumstances, and consequently that the final purpose of each modification is to be sought for in the conditions under which each being is destined to exist. But these conditions result from the infinitely varied arrangements of unorganized matter, they are consequently devoid of any symmetry themselves, and the wild irregularity of the inorganic is thus transmitted to the organic creation.

Thus,

> irregularity, and not symmetry, may be expected to characterize the natural system, and to form, like the features of a luxuriant landscape, not a defect, but an element of beauty. Such a supposition

> appears . . . to be more consistent with the benevolence of an all-wise Creator, than the theory which would consider the final purpose for which certain groups of organic beings were created, to be the fulfillment of a fixed geometrical or numerical law.[25]

But the warning voice of reason had been heard even before Bonnet's principles of classification had been distorted into Swainson's and Kaup's cabalistics. Kant's counterarguments (1781) have already been mentioned. The great J. F. Blumenbach observed in the first edition of his *Handbuch der Naturgeschichte* (Göttingen, 1779) that he had always considered the idea of evolutionary stages in nature to be one of the most interesting speculations in natural philosophy, but "the most apt allegory can become stale, and can degenerate into fiddle-faddle, when it is pursued too far." And he warned that this was to be feared with the well-known images of chains, ladders, and networks that had been superimposed on nature.

Blumenbach was thus content to call the terminology of Bonnet and his successors allegorical. Pallas, on the other hand, had already objected in his *Elenchus zoophytorum* (The Hague, 1766) that the relationships they had constructed were "superficial and idealistic," and that Nature paid absolutely no attention to them when she produced the whole multitude of organic beings in the greatest order, both simultaneously and sequentially. Rather, Nature was concerned with structure, steadily diminishing size, methods of generation, and so forth.

> While beginning with the simplest things, she changes and creates organs gradually to suit their purpose, but she little by little introduces all sorts of new items even in the simplest structure and fits them together most appropriately. This is the origin of the great proof that is deduced from the similarity of forms, but that must not be employed without mature reflection.

By this splitting of the Aristotelian concept of resemblance into its two components, functional equivalence ("superficial and idealistic" resemblance), and morphological equivalence ("structural" resemblance), Pallas opened a way for an evolutionary treatment of the problem. Reasoning "only superficially," he continued in the same passage:

> Much the best among all the pictorial representations of the classification of creatures would be that of a tree arising from a root made up of the simplest plants and animals but at once forming two different trunks, an animalian and a vegetal, which, however, occasionally

come together. The first, beginning with invertebrates and reaching to fish, would, after producing a large side branch for insects, grow toward amphibia. And though this trunk must have quadrupeds in its uppermost tips, below these an equally large side branch would have to be indicated for birds.

In these words the young Pallas gave expression to a presentiment of the natural interrelation of living things, being the first to choose the image of a genealogical tree. Although he remained much attracted to the study of variability, he did not proceed to develop the momentous ideas of evolution because such a concept contradicted the systems of the philosophers, beginning with Aristotle's doctrine of entelechy, Leibniz's monadology, and the demonstration of John Ray, a devout thinker, that the immutability of species followed from the words of Genesis: after creating the animals, God rested from his work on the seventh day. It is therefore not surprising that no later naturalist dared to cross the threshold of the mystery, especially after Kant had called the theory of descent so monstrous "that reason shudders away from it." "A hypothesis of this kind," he wrote in 1790 in *Kritik der Urteilskraft*, "may be called a daring venture on the part of reason; and there are probably few even among the most acute scientists to whose minds it has not sometimes occurred." But no empirical proofs can be given for the transmutation of species, nor can theoretical considerations make it acceptable. The most important objection raised by Kant, however, rested on the Aristotelian hypothesis that organisms are created wholly for a purpose and that all their parts cohere in a structure whose harmony is related to the need for stability. Kant saw in this hypothesis a transcendental principle of judgment, and therefore warned:

> Once we depart from this principle we cannot know with certainty whether many constituents of the form presently found in a species may not be of equally contingent and purposeless origin, and the principle of teleology, that nothing in an organized being which is preserved in the propagation of the species should be estimated as devoid of finality would be very unreliable in its application.[26]

There were other reservations. In 1790 Blumenbach, with delicate irony, has the great Haller say:

> Such errors must be censured, because they will be greedily snapped up by atheists, only too glad to demonstrate the inconsistency of Nature from the development of new genera as well as from the presumed destruction of old species. And that must not be; if order in the physical world goes by the board, so does order in the moral world, and finally the whole of religion is jettisoned.[27]

The young Illiger also yielded to the force of such arguments when he wrote, in his *Terminologie* (1800):

> However much one would like to believe that the species belonging to one genus sprang from a common original stem, and are related to one another through this link, we must nevertheless abandon this thesis as unprovable, even contradictory to the nature of species, until at some point more observations give us the key to the secrets of Nature — how she generates species, and how, from accidental changes, caused by exterior effects, she creates hereditary varieties; and we must restrict ourselves to the subject as we find it at present. Because we really do find in nature that several species resemble one another considerably in their habitus, and differentiate themselves from others, and because we see that one particular species through this greater resemblance is more closely related to a number of species than to any others, it is entirely natural for us to conclude that Nature herself formed these aggregates.[28]

Accordingly, when Buffon's disciple Lamarck 9 years later brought out his *Philosophie zoologique*, "contradictory to the nature of species," he found hardly any support. Only after another 50 years had the study of variation and heredity developed far enough for the idea of evolution to break victoriously through the phalanx of its opponents. One is tempted to ascribe the slowness of progress to the corrupting effect of natural philosophy. Especially in Germany, it evoked in most of the empiricists familiar with the appearances of variation a hostility to any sort of employment of the deductive process. This hostility continued even after 1859 and obstructed the advance of Darwinism, thereby holding up scientific advance. Many people lumped the theory of descent together with the metaphysical constructions of Schelling and Oken, and branded them all as natural philosophy, as J. H. Blasius did in 1860 during the Stuttgart meeting of ornithologists. He distinguished two intellectual tendencies:

> The one proceeds from the unprejudiced investigation of particulars, and follows observation and experiment to the general result; the other preconceives a general idea as the foundation of all knowledge, and tries, by reconstructing reality from this general idea, to understand the particular. The one view is conscious that it will not be able to solve the riddle of the first creation; it is modestly content to study relationship as it stands at present; the other finds it necessary to leap across millennia in order arbitrarily to tear the veil from the first act of creation, and to illuminate the present in the light of the predetermined dogmas of creation. Both of them

> are attached by positive knowledge to the particulars seen at the present day; but the one sees in the present, despite all the variability of testimony, an immutable order persisting from the beginning of creation; the other sees a senseless, rapidly disappearing point of transition in the meaningless battle for existence . . . For the one view, study of particulars is an end in itself; for the other, it is simply a means to furnish evidence for the preconceived hypothesis. Such opposites cannot be scientifically reconciled.[29]

And with Blasius the majority of leading German ornithologists continued for decades to adhere to the group who find satisfaction in studying the particular. That was the reaction to the specter of the "immortal Oken."

Today we are still reminded of the period of natural philosophy by a mass of generic names that Swainson, Kaup, and Reichenbach had to devise in order to follow out logically their arithmetical constructs. These were the instruments used to force their victims into the Procrustean bed of the "natural system."

11

Hermann Schlegel (1804–1884)

In 1857 the careers of three famous ornithologists ended. Soon after Bonaparte died on July 29, the deaths of Johann Friedrich Naumann and Hinrich Lichtenstein were announced; both men were nearly 80. On January 30, 1858, Temminck died. Two or three decades earlier each of them, in his own way, had guided the development of ornithology on the European continent. But for a long time new leaders had been growing up in the shadow of these great authorities, men who could keep pace with the furious rate of scientific progress and who had been waiting impatiently for their moment.[1]

Hermann Schlegel, who now replaced Temminck as director of the Leiden Museum, had been obliged to spend 33 years in a dependent position and to sacrific his ability to the service of an often capricious superior, whose intellectual sphere he had long since outgrown. Now that he could at least imagine that he had reached his goal, he was in his fifty-fifth year. He had planned to create an institution to be scientifically valuable according to his own ideas; but would there be enough time left?

Like Bechstein, Naumann, and Brehm, the best-known ornithologists in Germany around 1820, Schlegel came from Thuringia. He grew up in Altenburg, where he was born on January 19, 1804, the first of eleven children of a brassfounder. His father, a well-educated and intellectually lively citizen of the town, had a small business supplying the Altenburg farmers with brass harness fittings. He took great pains with Hermann's education. Like many other boys, Hermann at first became interested in entomology, stimulated by his father's butterfly collection (the elder Schlegel had undertaken to look after a little museum founded by the Osterland Naturalists' Society) and, as his posthumously published autobiography[2] mentions, by the unusual interest in natural history taken by young people at that time. "There was widespread collecting . . . especially of butterflies, birds' eggs, plants, and rocks." A chance discovery of a buzzard's nest with eggs led him into ornithology; he learned taxidermy, made a collection, and kept birds, especially all sorts of birds of prey, in cages. Bechstein's *Naturgeschichte*, a present from his father, became an invaluable guide. In his thirteenth year he was already so experienced with the local birds that he was encouraged to visit Pastor Christian Ludwig Brehm in Renthendorf. Brehm had not yet become an author, but he had already made contact with the Osterland Naturalists' Society. "This strange and brilliant man kept me with him during the entire school holidays. Every day he walked with me through the woods and fields, and he was delighted to find how well I knew the flight and calls of the birds." But one particular experience during this time remained unforgettable for Schlegel; he proved to his instructor that the marsh warbler (*Acrocephalus palustris*) described by Bechstein was to be found near Renthendorf, "which Brehm himself refused to credit when I pointed out to him a bird of this species singing in alder bushes near his house, taking it to be an unskillful, as he expressed it, and not yet completely mature icterine warbler (*Hippolais*). The question had to be settled, and as it was already growing dark I undertook to shoot the singer the next morning, while the pastor was conducting church service. The sermon was not yet at an end when I entered through the open church door, bird in hand." After the service, in his study, Brehm examined the bird closely, and exclaimed: "You are right, I have never shot this species and it is not in my collection. Come to my arms — you have the making of a true naturalist!"[3] This was on June 4, 1816, according to the label on the marsh warbler still extant in the Brehm collection.*

When Hermann finished school he had to choose a profession. His father wanted to send him to the university, but with the stipulation that he should choose a safe breadwinning subject, and not perchance the natural sciences. But Hermann did not intend to give up his beloved work with animals, and considering that his father's trade would most readily provide him with leisure for his hobby, he gave up further academic study and learned brassfounding and bookkeeping at home. But soon he felt too cramped by his environment, and before he was 18 he quarreled with his father and set off on his travels. He went on foot to Dresden, where he worked for two years for a business friend of his father's and then at Easter 1824 took his savings and walked to

*Now in the Alexander Koenig Museum, Bonn, whither it was transferred on exchange from the American Museum of Natural History in 1960 (Dr. Lester L. Short, *in litteris,* November 29, 1973). For the Brehm collection see Chapter 14.

Prague and afterward to Vienna. Pastor Brehm had given him recommendations to several Viennese naturalists, including Joseph Natterer, the brother of the famous traveler, who at this time had been seven years in Brazil and had sent home copious collections. Natterer secured him a small post in the Imperial Cabinet of Natural History, then rapidly expanding, which he accepted gladly as the first step toward the lofty goal he had in mind. The ichthyologist Heckel, whom he helped with his work, initiated him into the study of fishes; and Fitzinger introduced him to reptiles and amphibia. Carl von Schreibers, director of the museum since 1806, often invited him to his friendly household, where he shared in the stimulating conversation of intellectually cultivated men. He formed a particularly warm friendship with a young Hungarian, Salomon von Petényi, who became so devoted that he compelled Schlegel to share board, lodging, and all necessities of life with him.*

But Schlegel's scientific inclinations were not the only ones to find plenty of encouragement in Vienna; his old love of music was most happily gratified. Indeed, he was able to hear the last performances given by the orchestra that Beethoven himself had trained, and Schlegel thought himself lucky when he and his friends could quietly and from a distance watch the great man taking a walk, either in Vienna or in Baden.

After Schlegel had enjoyed about a year of this life among the gaieties of Vienna, Schreibers received a letter from Temminck, whose acquaintance he had made during a journey to Hungary in 1818. Temminck asked "if perhaps in Vienna a young scientist can be found who not only is familiar with zoology but also possesses practical experience, no matter whether he has attended the university or not." Because Herr von Schreibers pointed out to Schlegel that, as a North German and a Protestant, he had little or no chance of promotion in Catholic Austria, he let himself be persuaded to give up his most attractive position in Vienna. He left after only a few days and, traveling by way of Altenburg and Renthendorf, arrived in Leiden on May 25, 1825, where Temminck was able to offer him only a modest and temporary job as preparator because all the other places were filled for the moment.

*Johann Salomon Petényi (1799–1855) is regarded as the founder of Hungarian ornithology. When he met Schlegel he was studying theology and had not yet made his début as an ornithologist. A particularly full letter written to his "dearest heart's friend" by Schlegel on October 1, 1853, and evoking a delightful reminiscences of youth, was published by Otto Herman in his biography of Petényi. It reads in part: "The instruction that you gave me on the moral condition of man greatly influenced my entire life afterward, and rooted itself the more deeply because it sprang from the pure fount of friendship, and was touched by the breath of poetry that surrounded our meetings and our life together. Those days were perfectly adapted to intensify pure and noble thoughts into enthusiasm, and to make their memory unforgettable."

At that time the rooms of the new Rijksmuseum, where the gigantic collections of Kuhl and van Hasselt were being studied, were full of activity. Boie, Macklot, and Müller, already busy with the final preparations for their great journey, had admitted J. J. Kaup and F. Cantraine, both students, into their group of friends, and now Schlegel was allowed to join too. He was equally at home in Temminck's hospitable house and with Professor Reinwardt.

Boie and his companions sailed for Java on December 21, 1825, and Schlegel took over their former jobs, intending to join the Natural History Commission after a few years. "But when the news arrived that Boie, designated by Temminck as his successor, had died, he begged me not to leave the museum. I gave up my plans for traveling because I realized that I could do more important work in such a famous seat of scientific learning." In order to achieve this aim properly, during the next years Schlegel strove chiefly to fill the many gaps in his knowledge. As a registered student at the University of Leiden in 1830–31 he attended not only the natural history lectures given by Reinwardt (who became his favorite teacher), but also courses on human anatomy, physiology, geology, mathematics, physics, and chemistry, and became more familiar with classical and modern languages. In 1832 the University of Jena gave him an honorary doctorate.

During his first years he worked primarily on reptiles, but soon sought to extend his research to other groups of animals. His wish was granted when he met Philipp Franz von Siebold, the famous explorer of Japan. Siebold was born in Würzburg on February 7, 1796. He took his medical degree there in 1820 and the following year was called to Holland as personal physician to King William I. Because he did not care for court life, he had himself transferred as "surgeon major" to the Dutch colonial army and landed in Batavia on September 22, 1822. There he attracted the attention of the governor general, Baron van der Capellen, who was greatly interested in the scientific exploration of Japan. He persuaded young Siebold to accompany as doctor J. W. Sturler, minister to Japan and recently nominated head of the Dutch trading station on the island of Decima, near Nagasaki. So Siebold reached Japan on August 8, 1823, and immediately began developing activities in many directions. Because doctors were in demand, he was allowed to take a country house near Nagasaki and to move about freely.[4] Thus after only two years he was able to send back to Java not only many crates with natural history and ethnographic specimens, but also tea seeds, which were the origin of the later flourishing tea plantations on that island. He married a Japanese and, together with his assistant and later successor, Heinrich Bürger, accompanied the minister on his visit to the Shogun (imperial regent) in the capital city of Yedo (now Tokyo). Because he had secretly made and kept maps of Japan he was accused of spying, in November 1828, and after long investigation was "permanently" banished from Japan. He left on December 30, 1829, and stopped at Batavia on his way to Holland, where he landed in July 1830 with his collections. He bought a large house in Leiden, which was partly arranged as a Japanese museum. His collections were acquired by the Dutch government in 1837. In 1859 he was able to revisit his beloved Japan for another two years, but sordid intrigues completely ruined Holland for him and he died on October 18, 1866, in Munich.

Soon after he arrived in Leiden, Siebold became a friend of Schlegel's. "Siebold suggested to me the publication of a *Fauna Japonica*, and we agreed that I should undertake the vertebrates, except for the terrestrial mammals, which Temminck reserved for himself." The first volume, dealing with reptiles and amphibians, appeared promptly in 1833, but that on birds only more than ten years later (1844–1850), after Schlegel had spent the interval on various other lengthy projects. Chief among these, after Salomon Müller's happy return in August 1837, was the joint study of the enormous collection of vertebrates that Müller had brought back from New Guinea, Timor, Sumatra, and Borneo (published 1839–1844). In 1838 Schlegel's friend Verster van Wulverhorst, inspector general of hunting, had interested him in the rapidly declining art of falconry, and for several years he studied most attentively the capture and training of hunting birds at Valkenswaard, the famous school for falconers.* He presented Verster, who wanted an illustrated book on the subject, with the *Traité de fauconnerie*, in imperial folio, published between 1844 and 1853 in Leiden. It is one of the most sought-after books on falconry, not least because of its magnificent plates. Schlegel's search for artists who understood how to paint birds of prey led to the discovery of a remarkably gifted young man, Joseph Wolf, who had been recommended to him by J. J. Kaup, himself a good artist and art critic.

Claus Nissen gives the following details concerning this still unsurpassed master of bird art:

> Joseph Wolf (1820–1899), son of a peasant who lived at Mörz (near Münstermaifeld) in the Eifel, was one of those passionately dedicated painters of animals whose highest aim is to represent as faithfully as possible the beloved and ever newly admired creatures.

*For a very long time trained falcons had been used to hunt herons on the royal preserve, "het Oude Loo," near Valkenswaard, south of Eindhoven. [The name seems to have been acquired from the trapping of migrant or passage hawks there, which continued for many years.] It became famous with the foundation in 1839 of an international falconers' club, the "Royal Loo Hawking Club," whose members hunted herons until about 1861 (J. E. Harting, *Bibliotheca accipitraria*, London, 1891, p. 264).

From his earliest years he pursued them with net and gun in order to capture their likenesses in his sketchbook. The extent of his progress, without any instruction at all (for a few years he had lithographed wine labels and business announcements at a small press in Coblenz) can be seen from the fact that Eduard Rüppell, whom he visited while on his travels, immediately entrusted him with the illustrations for his *Systematische Übersicht der Vögel Nordost-Afrikas* (Frankfurt, 1845). His fortune was made. Through Professor Kaup of the Darmstadt Natural History Museum, to whom he had been sent by Rüppell, Wolf made the acquaintance of Conrad and Eduard Susemihl and Hermann Schlegel, to whose guidance he owed much. The plates that he was commissioned to prepare for Schlegel's *Traité de fauconnerie* show him as already at the height of his powers and serve to render this still standard work the most beautiful that has ever been devoted to its attractive theme. This Darmstadt period was further of significance for him by bringing him into contact with the fine arts; he received regular instruction, copied zealously in the painting gallery, and produced hunting and animal scenes of his own invention for a rapidly growing circle of purchasers. But this rewarding activity, ample for many another, failed to satisfy him, and in 1847 he went to Antwerp, for further improvement. The stirrings of revolution again brought about a remove, and so he came a year later to England, where David William Mitchell, secretary of the Zoological Society, and then engaged in the illustration of G. R. Gray's *Genera of Birds*, greeted him with open arms.[5]

In London Wolf quickly settled down, becoming a favorite purveyor of oil paintings for country estates and city residences. His scientific work was to be seen in the *Proceedings* and *Transactions* of the Zoological Society, in the *Ibis*, in *Zoological Sketches* (1861–1867) and *The Life and Habits of Wild Animals* (1873), and in such large folio works as D. G. Elliot's monographs of the pheasants (1872) and birds of paradise (1873) and John Gould's *Birds of Great Britain* (1862–1873). "Wolf was never satisfied with a mere portrait, but always tried to reproduce the animal in full life and action, preferably in the midst of a movement peculiar to it . . . He rarely drew from nature, where he was simply an observer, and only at home did he attempt to give form to what he remembered."[6]

In Leiden, Wolf had the opportunity to display his great ability. Hardly any other ornithologist of those days was as well equipped to appreciate this as Schlegel who, starting as a boy, had practiced drawing and painting mammals and birds, and who had himself produced the plates for some of his own publications, including most of those in the *Fauna Japonica* (1844–1850) and *Neue grosse Edelfalken* (1843), among others. The most important function of the illustrations for him was the reconciliation of artistic and scientific demands, which requires a very detailed study of the living animal. The partly stiff, partly artificial manner of the older French school, still approved by Temminck, was odious to Schlegel. But even though he had already, by his own efforts, gone far in developing a new style for representing birds, he found his master in Wolf, who in turn had to admit that Schlegel outdid him in field observation. So the meeting of these two and their resulting friendship was of lasting benefit to their later artistic achievements.

Although officially Schlegel held only a subordinate position with a rather modest salary, even before 1850 he had become the intellectual representative of the Leiden Museum, and was left a free hand by the aging Temminck. Scientists who wished to sample the abundant material of the Rijksmuseum for their studies applied to Schlegel, not to the director, and anyone looking for cultivated society found it in Schlegel's house, just as Bonaparte did during his long stay (1849–1850). Several other systematists besides Bonaparte rated Schlegel as "le premier zoologiste de notre époque." Nevertheless, when Temminck's eyes closed forever in January 1858, Schlegel was nearly cheated of the reward for his long and patient efforts. Jan van der Hoeven, professor of comparative anatomy at Leiden University and representing the field of general zoology, quickly secured government approval of his appointment as "chief director" of the Rijksmuseum, superior to the future "director," Schlegel, to whom he could dictate. A real struggle began between "university zoology" and "museum zoology"; if Schlegel had lost, the Rijksmuseum's fate as a research institute would have been sealed. But he fought the exciting battle for the scientific inheritance entrusted to him with such skill and persistence that after two years he won, and in June 1860 van Hoeven had to resign as "chief director."

At last Schlegel had a free hand and could shape the museum according to his own ideas. It was high time, and indeed from his own point of view it was in many ways already too late. As he complained in 1866:

> It is sad that there is scarcely an institution with enough material to rebuild the structure of the sciences and to decide the fundamental question of the definition of species . . . What use has it been for me to have tried, ever since I arrived at the Rijksmuseum in 1825, to establish selected series of specimens belonging to each species, equipped with the data essential to a study in depth? Because my predecessor never found this sort of collection appealing, the Rijks-

museum contains only faint traces of what I had been intending to build up since that period.[7]

The "fundamental question of the definition of species" was of no concern to Temminck, who found any philosophic view of nature entirely foreign to him, but it must have troubled Schlegel after he became acquainted with the writings of his teacher C. L. Brehm, the many ideas in Friedrich Faber's book, and the frequent discussions of the subject from 1823 to 1831 in Oken's *Isis.*

Decades earlier Immanuel Kant had recognized that this development was demanded by logic, because in his *Critik der reinen Vernunft* (1781) he says:

> This logical law [of genera and species] would be without meaning and application if it did not rest upon a transcendental law of specification . . . which . . . imposes upon the understanding the obligation of seeking under every discoverable species for subspecies, and under every difference for yet smaller differences. For if there were no *lower* concepts, there could not be *higher* concepts . . . The knowledge of appearances . . . demands an endless progress in the specification of our concepts, and an advance to yet other remaining differences, from which we have made abstraction in the concept of the species, and still more so in that of the genus.[8]

Kant called this continuous transition from any species to any other, through the graduated increase of differences, the law of the affinity of all concepts.

The first ornithologists who, inspired by Buffon, studied the variability of species were not yet capable of harmonizing their results with Kant's postulate; Darwin was the first to succeed. Therefore they also energetically rejected a progressive "specification of concepts," although nature prompted it. Faber's statements in *Über das Leben der hochnordischen Vögel* (Leipzig, 1825–1826) were highly applauded in Germany:

> From the very beginning Nature created the different kinds of species, by assigning to one or more pairs of birds of each species the distinguishing marks that were to characterize them as a species, and she settled these primary individuals, the basic types, in one or more places on the earth, from which original places they spread, as from a center, over a certain area as individuals increased in number. As the progeny of the first individuals of the species moved farther and farther from the center, they underwent, particularly if they became resident in these distant places, gradual and separate changes in form and color, which were created by the conditions in

> their local environment; that is to say, they developed species according to climate and also transmitted to their offspring this specific change which, after being inherited by several generations, became constant. Thus arose the *climatic races* of many species. These races many ornithologists are inclined to establish as separate species, but without grounds, in my opinion, because, although they are specifically changed, they are genuine offspring of the first individuals of the type, and therefore belong absolutely to the primeval form of the species. The criterion by which these climatic varieties can be recognized as somewhat altered progeny of the primeval form is the obvious agreement in external and internal structure, in habits and history, with the basic types; and also the way in which, free from human interference and with no lack of neighbors belonging to the same variety, they pair as freely and impartially with individuals of the basic type as with those belonging to the variety, in the first instance breeding offspring that partly preserve the genuine, and partly the changed, form and color.
>
> We possess not a little evidence that individuals belonging to the same species strongly tend to adopt climatically constant changes in form and color, even within not very widely separated degrees of latitude and longitude. Among other examples, I can cite several from my own experience of aquatic species whose individuals occur in Greenland as well as Iceland; and certainly such climatic alterations will also be found between individuals of the same species in Iceland and in Siberia, Kamchatka, and elsewhere. If, therefore, we were to accept all these local varieties as the same number of genuine species, then no systematics would finally be able to include them, no ornithologist to classify them, and no memory to contain them; the theory of the laws of the geographic distribution of birds would be choked at its source, and science itself rocked to its foundations.

Schlegel was among those who accepted these views and developed them further. His significant research soon demonstrated to him, however, that the local changes of the typical species could not be ascribed to the effect of climate, as Buffon, Pallas, and Faber had thought, and so he gradually became convinced that geographic varieties had existed since the beginning of creation and were immutable. This corresponded to the theories long held by his fatherly friend Christian Ludwig Brehm.

The more the Renthendorf pastor's collection grew (in 1823 he had already amassed about 4000 skins from Europe), the more eagerly he devoted himself to the manifestations of variation. After he had discovered in his own Thuringian territory two species of tree creeper and two of goldcrest that had not previously been distinguished from each other, he noted in 1822, half in jest and half in earnest: "It really looks as if the Creator had made some of the species of animals in order to exercise the ingenuity of naturalists." Soon he began fervently to exercise his own ingenuity still further by describing new "species" by the dozen, and finally by the hundred, thus becoming the *enfant terrible* of the systematists and, in Faber's words, "rocking science to its foundations." For example, he named the representatives of all his newly discovered varieties binomially, even when their differences were minimal (or existed only in his imagination), and thereby made no distinction between individual and geographic variation. Only his explanatory notes, not his nomenclature, revealed whether he considered a discovery to be a "species" or a "subspecies." "A slight but constant difference in size, beak or skull form, or even in color"[9] sufficed him for a "subspecies." For instance, near Renthendorf he found breeding three subspecies of the redbacked shrike (*Lanius collurio, L. spinitorquus*, and *L. dumetorum*), and thought that "only examples of the same subspecies regularly pair together." According to Brehm, the subspecies did not develop from the "genuine" species by multiplication and spread of individuals; he was much more inclined to believe "that, because we accept the clear evidence of a marvelous and permanent law in the constancy and immutability of the structure of creatures . . . these subspecies were produced from the very beginning by God's unfathomable creative power and will remain as they are." And, just as if he had misunderstood Kant's "logical principle of the continuity of forms," according to which all varieties are related to one another, because all of them descend, through all further grades of classification, from a single highest genus, Brehm maintained that "the various subspecies appear as a sequence between the old species, linking them together so well that it is often hard to tell where one stops and another begins."

These views of his friend's became part and parcel of Schlegel's own. When he was trying to convert them into nomenclatural practice, he found the obvious solution in 1844 by following advice published long before, in 1828, by Carl Friedrich Bruch:

> I cannot deny that I am reluctant to see the system unnecessarily cluttered with names, and therefore I should like to put this question: supposing these new or secondary species prove to pair with their related species only very rarely or irregularly, is it therefore necessary to distinguish them with special names, and would it not be a relief for the already overburdened memory to introduce a *triple nomenclature*, by leaving the old name to the typical form and characterizing the varieties with a third word? The differences among these new species are often so slight that it may be difficult for many ex-

> perienced ornithologists to identify one of them exactly, if he does not have individuals of the related subspecies available for comparison. Also the relationship between one of these new species and its fellow subspecies is quite different from that between it and one of the old species: Brehm's *Platypus mollissimus* relates quite differently to his *Pl. leisleri* from the way it does to his *Pl. spectabilis;* the name characterizing this relationship must vary to the same degree.[10]

Thus trinomial nomenclature came into being; Schlegel created it in order to distinguish the "local" varieties (which he called "conspecies") from the "typical species," and adopted it in 1844 both in the *Fauna Japonica* and in the *Kritische Übersicht der europäischen Vögel*, where, for instance, one finds a *Falco tinnunculus japonicus* and an *Alcedo ispida bengalensis.* But, in contrast to Brehm, he left individual variations (those not geographically isolated) unnamed for practical reasons, even though it seemed very important to him that they should be known and described in detail, because they belong to "the nature of the species."

When in 1858 Schlegel began to collect systematically according to these principles, the English private collectors had already stolen a march on Leiden. And not in this respect alone; they had forced their way deep into the Rijksmuseum's most particular sphere of interest. After the Natural History Commission had been disbanded in 1850, the Dutch had done nothing more with zoological research in the Malay Archipelago, and in the meanwhile Alfred Russel Wallace, an Englishman, had won undying fame with a multitude of discoveries made in Schlegel's own province. No one else in a few years could so thoroughly have knocked the bottom out of the museum's resources as Wallace, all of whose projects seemed to succeed, because his precise and analytical intelligence, his gift for reliable synthesis, and his unceasing energy were faithfully accompanied by a luck denied to a Kuhl, a Boie, a Forsten.

This pioneer of evolution was a truly self-made man, who only by chance employed his genius for research into first causes to the benefit of natural history.[11] The child of poor parents, he was born on January 8, 1823, in Usk (Monmouthshire), attended elementary school till he was 14, and learned to become a surveyor. At this time he began to study botany as a hobby and about 1844, in Leicester, where he had become a teacher, he made the acquaintance of a young apprentice in the wool trade, who collected insects intensively. This was Henry Walter Bates, born in 1825. A fantastic scheme for a journey together to the tropics took concrete shape; hoping to cover their expenses out of their zoological material, the two friends set out for Pará in 1848. Wallace stayed in the Amazon area for four years, the last three of which he devoted to studying the fauna of the Rio Negro, until he fell seriously ill in 1852 and realized that he would be forced to go home (while Bates stayed until 1859). But in mid-ocean the ship caught fire. Everything that he had laboriously collected in the last two years, along with all his drawings, went up in smoke, and he could count himself lucky only in being able to leave the burning wreck in a lifeboat along with the other passengers. After ten days they were sighted by a passing ship and landed in England in October 1852. Here Wallace immediately wrote two books about his journey, and used the proceeds from the sale of the Amazon collections that he had sent home earlier for a new tropical expedition, this time to the Malay Archipelago. With a vivid consciousness of the many problems that challenge the mind during such a journey, he added to his knowledge by studying books and periodicals and, like Darwin at the same time, was particularly and lastingly impressed by Lyell's *Principles of Geology* and Malthus's *Essay on Population.* On November 1, 1854, Wallace reached Rajah Brooke's realm of Sarawak (by way of Singapore), and stayed there until January 1856. From then on he made a systematic investigation of the Dutch possessions, which ended only after six years, in January 1862, with a stay in Sumatra. During this time he worked over the islands of Bali and Lombok (1856), Celebes, the Moluccas (Ambon, Seram, Buru, Ternate, Batjan, and others), Waigeo (1860), Dore in northwestern New Guinea (1858), the Aru Islands, and Timor, while his assistant, Charles Allen, collected on Sulabesi, Morotai, Misool, Salawati, and Flores.

Since 1854 Wallace had been using the proceeds from the sale of duplicates in his zoological collection to finance the remainder of his journey (with the help of his London agent, Stevens), and the collection itself exceeded all expectations. It numbered some 125,000 items, including 8,000 bird skins, which were studied at the British Museum by G. R. Gray, though Wallace himself had already worked on a fairly large proportion of them. They greatly expanded the knowledge of species diversity.

Of still greater importance was the success of the journey from another point of view; it led to revolutionary insights into the laws governing the distribution of animals and the origin of species.

The increase of zoogeographic knowledge had long been a powerful motive for research expeditions; the final result believed to be attainable from such a compilation of scientific data was, however, the opposite of a "natural" systematics of fauna; the earth's surface was in fact to be divided into main "natural" biogeographic sections. In 1858 the English ornithologist P. L. Sclater published a study in which, basing his theory on avian distribution, he set up six zoological regions (which "are of course naturally separable into secondary divisions or provinces") with the following names: Palaearctic Region, Aethopian Region, Indian Region, Australasian Region, Nearctic Region, Neotropical Region. Not the least of the reasons for this was his continued belief in Creation. As he said at the time:

> I suppose few philosophical zoologists, who have paid attention to the general laws of the distribution of organic life, would now-a-days deny that, as a general rule, every species of animal must have been created within and over the geographic area which it now occupies. Such being the case, if it can be shown that the areas occupied by the primary varieties of mankind correspond with the primary zoological provinces of the globe, it would be an inevitable deduction, that these varieties of Man had their origin in the different parts of the world where they are now found, and the awkward necessity of supposing the introduction of the red man into America by Behring's Straits, and of colonizing Polynesia by stray pairs of Malays floating over the water like cocoa-nuts and all similar hypotheses, would be avoided.[12]

While Sclater in 1858 was continuing to regard zoogeographic problems in the old way, Wallace had already gone deeply into the matter several years earlier. In his article written during February 1855, in Sarawak, "On the Law Which Has Regulated the Introduction of New Species," he already stated

> that the present condition of the earth and of the organisms now inhabiting it is only the latest step in a long and unbroken procession of changes to which they have been subjected, and accordingly any attempt to explain and interpret present conditions without considering these changes must lead to very incomplete and erroneous conclusions.[13]

Therefore a true classification can be achieved only by paying heed to the geological and geographical facts.

> If a group is limited to one district and is rich in species, then almost without exception the nearest related species is found in the same locality or nearby, and therefore the natural sequence of species formed by relationship is also a geographical sequence . . . It is evidently possible that two or three species could have had the same prototype, and that all of these in turn became prototypes from which other and related species were developed . . . The shallow sea between the Malaccan peninsula, Java, Sumatra, and Borneo

> was probably a continent or a large island in an earlier epoch, and perhaps sank when the chain of Javanese and Sumatran volcanoes rose. We see the effects of this in a very notable number of animal species that are common to several or all of these areas, while at the same time there exists a number of representative species, belonging particularly to each area, proving that a significant length of time has elapsed since their separation.

Thus Wallace, stimulated by Robert Chambers's *Vestiges of the Natural History of Creation* (1844), discovered the signs of a natural law, whose clarification occupied him steadily during his further travels through the Malay Archipelago. Learning that Charles Darwin was working on similar ideas, he started corresponding with him in 1857. In February 1858, while an attack of malaria forced him to a period of inactivity on Ternate, he wrote his famous article, "On the Tendency of Varieties to Depart Indefinitely from the Original Type," which transformed the biological sciences. For Wallace sent his manuscript to Darwin, to be passed on to Lyell. He, along with Hooker, persuaded Darwin to present a selection from his manuscripts, as well as Wallace's treatise, to the meeting of the Linnean Society on July 1, 1858. The combined papers were then published in August in the *Journal of the Proceedings of the Linnean Society.*

In this treatise Wallace first attacked the general belief that something like "permanent or true" varieties exist, races of animals that steadily breed true to type, and then sought to prove by examples "that there is a general principle in nature which will cause many *varieties* to survive the parent species, and to give rise to successive variations departing further and further from the original type." This is effected by the struggle for existence, in the course of which useful variants have a tendency to increase, useless and harmful ones to disappear. Surviving variants would finally produce the extinction of the original species.

In *On the Origin of Species*, published the following year, Darwin brilliantly expressed all these theories, which were also his own — and had been so partly ever since 1844 — and thereby provided systematists and zoogeographers with many most important matters to reflect upon. To avoid the apparently insuperable difficulty of supposing that, in the course of long periods, the individuals of one species and those related to it all arose from a common source, he presented the hypothesis that all the principal phenomena of geographic distribution could be explained by the theory of migration and the resultant changes and multiplication of new forms. Then great importance must be attributed to natural barriers — land or water — between different botanical and zoological regions.

For the first time it thereby became clear how much more nearly close cooperation between faunal systematics and faunal geography had come to solving the basic problem of biology than had any other attempts. An unprejudiced analyst of the new line of thought was forced to recognize that a real understanding of the reason for the multiplicity of living forms was beginning to take shape. Museum zoology, occasionally derided by the anatomists and physiologists as a withering or dead branch, had achieved a success with far-reaching consequences. At the time, it is true, few people were aware of it, and those who closed their minds most tightly against this reasoning were the old and experienced systematists. They resisted the evidence of the mutability of species and varieties, and doubted that natural selection, which was prerequisite to the theory, could create such effects as man produces in domestic animals by artificial selection.

Wherever the valid answer to the question of species might lie, Wallace's and Darwin's publications only confirmed Schlegel in his old opinion, inherited from Brehm, that the study of individual and geographic variation was particularly promising. On March 29, 1859, he wrote to his "chief director" (for further communication to the trustees of the University of Leiden) that in the present state of science a "duplicate" meant something quite different from what it had formerly meant. "If one wishes to know a species thoroughly, one must possess specimens of different sexes, different ages, from different seasons, and from the most important localities."

These guidelines were also contained in the instructions that Schlegel gave to his 17-year-old son Gustav in 1857, when he allowed him to travel to China, and hoped that the unexplored Middle Kingdom could still supply much new material for his museum. But here also he was too late in the field, and again it was an Englishman who arrived just before him: Robert Swinhoe (1836–1877), at that time British vice-consul in Amoy and an ornithologist as gifted as he was devoted. Indeed, Swinhoe and Gustav Schlegel quickly became friends, as we are reminded by Swinhoe's *Anthus gustavi* (Pechora pipit), but Gustav completely abandoned his father's profession, returning home not as a famous naturalist, but as an experienced scholar in Chinese, which he had chosen as the subject of his life's work when he was still a boy.*

Particularly dear to Schlegel's heart was further exploration of the Malayan islands, especially the Moluccas and New Guinea, where he could still hope to win the race against Wallace. "Every day it becomes clearer," he wrote in 1861,

*Gustav Schlegel (1840–1903), the elder of Hermann Schlegel's highly gifted sons, later became a famous sinologist at the University of Leiden. His father wrote to Baldamus in December 1853: "Gustav, a boy of 13, [is] now in the Gymnasium and nicknamed the Chinese, because he has already been working at the language hard for three years, and wants to specialize in the study of East Asian languages." (Letter in the archives of the Deutsche Ornithologen-Besellschaft.) The younger son, Leander, was very musical and became a composer.

> that the difference among species and races is nowhere greater than in the islands of the East Indian Archipelago and the South Seas, and that here investigations have to be set up in a manner completely different from that hitherto employed, if reasonably useful results are still to be obtained for modern science.[14]

At that time none of these islands promised the naturalist such a rich yield as New Guinea, the mysterious home of rare birds of paradise for which every traveler (including Wallace) had searched in vain since Sonnerat had brought some battered skins back to Europe in 1773. Schlegel knew of two naturalists, Dr. H. A. Bernstein and H. von Rosenberg, already in government service there, who could do well by the Leiden Museum if they found proper support. His first choice was Bernstein, who had already applied to Temminck from Java in 1856 for a recommendation to join a New Guinea expedition planned for the Dutch naval vessel *Etna.*

Heinrich Agathon Bernstein, born at Breslau on September 22, 1828, the son of an orientalist at the university, studied medicine there and obtained his doctorate in November 1853 with "De anatomia corvorum. Osteologia," a dissertation produced while working in Barkow's Institute. He had attended lectures on natural history as well as medicine with special interest, and had found the old professor of zoology, Gravenhorst (1777–1857), particularly stimulating. As a boy he was already eager to get to know the tropics; after visiting the Leiden Museum he decided in 1854 to emigrate to Java, where he had been offered a post as director of the sanatorium in Gadok, not far from Buitenzorg. Here he began to concentrate intensively on the study of birds. The extent of his progress in a short time is demonstrated by a series of articles in the *Journal für Ornithologie* (1859–1861), "Über Nester und Eier javascher Vögel," which almost to the present day held a leading place among contributions to the biology of Javanese birds. His thorough anatomical training enabled him in 1856 to solve the problem of the building material used in edible birds' nests, by research into the salivary glands of the cave swiftlets (*Collocalia*), and led him to discover that cuckoos belonging to the genus *Centropus* have only one developed testis.[15]

In March 1859, Schlegel proposed this highly gifted and excellently trained ornithologist to the government as an official naturalist, and recommended that he be assigned scientific research missions in New Guinea and the surrounding islands. Schlegel was able to support his proposal with reference to a royal decision in 1850 concerning the future employment of occasional traveling scientists, to be named by the Rijksmuseum, to supplant the disbanded Natural History Commission. Bernstein was the first to benefit from this decision. After approval by Rochussen, the cabinet minister, Bernstein left Batavia for Ternate in November 1860. From 1861 to 1863 he made it his base for trips to many islands of the northern Moluccas, as well as Waigeo and Obi. With Schlegel's agreement, he began by thoroughly exploring this archipelago, partly because its investigation promised particularly instructive material on geographic variation and zoogeography, but also because he wanted to train his team for a more difficult enterprise. After some months of leave in Java for recuperation, he returned to Ternate in May 1864. On October 18 he set off with two proas for the land of the birds of paradise, but the coasts of the Papuan islands of Gebe, Batanta, and Salawati attracted him for a fairly long stay, and he landed in the village of Sorong, New Guinea, only at the end of November. Here his great hopes were quickly dashed as he fell seriously ill, and in addition found himself in a bloody conflict with the natives (mid-January 1865), which forced him to return to the coast of Salawati. Three months later, on April 19, 1865, he died of a liver abscess on board his sailing vessel as it lay at anchor in the harbor of a small island near Salawati.

Though nothing could have been more painful for Schlegel than the news of Bernstein's death, he had not staked everything on the one throw. In 1861 he had been visited by a young German ornithologist called Otto Finsch. A Silesian like Bernstein, he seemed to Schlegel's experienced eye to possess the right equipment for a major scientific undertaking. Schlegel gave him the chance to learn more by studying the Leiden bird collection, and before a year was out employed him as an assistant. Finsch received a well-thought-out training for New Guinea, where according to Schlegel's ideas he was not merely to collect for Leiden but to found a biological station as well. But the government refused to sanction the plan.

By 1862, however, Schlegel had succeeded in acquiring financial support for two other collectors, though neither of them had had a scientific or practical training and therefore could hardly justify any such high expectations as had attended all of Bernstein's undertakings. Both of them belonged to the administration of the governor of the Moluccas. One was D. S. Hoedt, an "Indo-European" who had been born in Ambon, and the other was Hermann von Rosenberg. They were to explore independently, first the islands of the eastern Malay Archipelago, and then New Guinea.

Hoedt began his work on Seram and Buru (1863), and during the next three years included other islands in his collecting range: Sulabesi, the Sangihe, Kai, and Southwest Islands (Leti, Wetar, Kisar). After the governor general had named him to succeed Dr. Bernstein on October 27, 1866, Hoedt went to Salawati and Misool in 1867 and to Banda in April 1868. A quarrel with the resident of Ambon caused his resignation in that year. Thus his work for the Rijksmuseum ended before he had been able to visit New Guinea.

Hermann von Rosenberg, born on April 7, 1817, at Darmstadt, served for a short time in the Hessian army before transferring to the Dutch colonial army. Soon after his arrival in Java he joined Dr. Franz Junghuhn, who had been commissioned to explore the Batak lands, and arrived in Sumatra with him in October 1840. He spent the next 16 years there as a subaltern. Rosenberg does not seem to have started collecting birds until he was assigned in 1858 as topographical draftsman to the expedition of the naval vessel *Etna* to the southwest and north coasts of New Guinea (as far as Humboldt Bay). Through such activity he attracted the attention of Schlegel, who succeeded on June 4, 1862, in getting him a government grant to work for the Leiden Museum. In his new capacity as an official explorer, Rosenberg landed on the northern peninsula of Celebes in April 1863, and collected intensively until August 1864. He would have been less successful then and later if he had not been able to profit from Bernstein's travel experiences, as transmitted by Schlegel. In fact, in 1863 Schlegel advised him to explore every islet, no matter how small, because there are species restricted to a very small area or with very scattered distribution. Schlegel also called his attention to Wallace's failure to measure birds and mammals before skinning them, or to provide them with an exact locality and a date, "which means that his observations can be regarded only as provisional and do not advance the cause of science in the way our museum considers proper."[16] Schlegel himself had evidently taken his example from C. L. Brehm and Johann Natterer, both of whom had begun about 1817 to attach a tag to one leg of each bird skin, on which sex and age (determined by dissection), locality, date, and (for Natterer) the colors of the iris, the beak, and the feet were exactly described.

After he returned from Celebes to Ambon, Rosenberg undertook for Schlegel a journey to the Aru Islands, in January 1865, and then pushed on to the Kai and Southeast Islands, from the end of July until October of the same year. Because his health had suffered on his travels, he took a two-year leave in Europe, meeting Kaup in Darmstadt, getting to know Alfred Brehm, and visiting his "fatherly friend" Schlegel in Leiden. In May 1868 he was back in Batavia and ready to carry out Schlegel's new assignments. On January 1, 1869, he reached Dore by way of Ternate in a rented schooner, and, because the season seemed unsuitable for climbing the towering Arfak Mountains, decided to content himself with visiting the entirely unexplored islands in Geelvink Bay: Numfor, the Schouten Islands (Biak-Mysore), Japen (Jobi), and Mios Num. Hardly ever leaving the safe quarters of the schooner, he employed his native hunters to shoot birds for him. The campaign for the year, very successful zoologically, ended there, and Rosenberg returned to his base at Ternate, sailing for Dore again only in January 1870. On February 18, 1870, he arrived at the foot of the Arfak Mountains and sent into them his native hunters, who were also experienced collectors. This turned out well, because they came back with a really sensational bag of mountain animals, including the much-prized birds of paradise *Lophorina superba* and *Parotia sefilata*, as well as a new though unspectacular species, *Amblyornis inornatus.* Rosenberg's usefulness ended here; no longer active enough at 53 for such enterprises, he asked in 1871 to be pensioned off and returned to Europe. Later he described all his travels in a widely read book *Der Malayische Archipel* (Leipzig, 1878), which, however, cannot be even faintly compared with Wallace's *Malay Archipelago* (London, 1869).

Delighted with the quantity of surprising novelties that Rosenberg had brought back from the Arfak Mountains, Schlegel immediately attempted to win government support for further research in New Guinea. But his appeal to national prestige fell on deaf ears, and in the future he had to look on helpless while his province became a field of discovery for foreign explorers. The Italian d'Albertis and the German A. B. Meyer acquired fame in western New Guinea and the surrounding islands, and sent their new finds not to him but to P. L. Sclater in London for description. He had to look on while the elderly John Gould, enchanted by these exotic discoveries, added another to his series of magnificent volumes, *The Birds of New Guinea and the Adjacent Papuan Islands*, which started appearing in parts in December 1875, and an Italian, Count Tommaso Salvadori, prepared a large work on the birds of New Guinea and the Moluccas, *Ornitologia della Papuasia e delle Molucche*, published in Turin (1880–1882).* Matters

*Soon after the various parts of Italy had decided in 1861 to unite into a kingdom, a period of great activity began for Italian ornithologists. Marchese Orazio Antinori (1811–1882), who, like his old friend Charles Bonaparte, had to flee abroad because of his revolutionary sympathies after the seizure of Rome, supported himself with difficulty at first (1849) in the Levant as a dealer in natural history specimens. From 1858 to 1861 he took part in Piaggia's expedition to Bahr el Ghazal, returned home in 1861, and decided to live in Turin, then the capital of the unified kingdom. In 1863 he joined a fellow enthusiast (a pupil of Paolo Savi who was continuing his ornithological studies in the Zoological Museum of the University of Turin) on a journey to Sardinia. This new friend and energetic bird collector was Tommaso Salvadori (1835–1923), who three years earlier had been a young doctor with Garibaldi's victorious army in Sicily. He was destined to become Italy's most famous ornithologist, and he soon had an opportunity to prove himself. Professor Filippo de Filippi, who had headed the Turin Museum since 1848 and had gone with the Marchese Giacomo Doria as part of an embassy from the king of Italy to Persia in 1862, faced another long journey in 1865, this time around the world in the Italian frigate *Magenta*, which was deputed to show the flag of Savoy in all seven seas. (He was accompanied by the young zoologist, E. H. Giglioli, as assistant.) Before he left, he entrusted the Turin bird collection to the care of Salvadori, who, when he heard of de Filippi's death in Hongkong in 1867, suddenly found himself confronted by heavy duties. He already had such a reputation as a painstakingly exact expert on exotic birds that nearly all Italian travelers entrusted him with their collections to be determined: the Marchese Giacomo Doria and Dr. Odoardo Beccari, who in 1865 had made a joint expedition to Sarawak, from which Beccari did not return till 1868; Antinori, who traveled through the Bogo lands of Abyssinia from 1870 to 1872; Beccari again, who (at first with d'Albertis) explored western New Guinea had gone so far that Leiden was not even leading the field in material from the Dutch possessions, and the museum's great role, for which Temminck and Schlegel had struggled for 50 years, was finished.

In the 1860s Schlegel, as systematist and zoogeographer, was still more highly regarded than most others, but he had actually passed the peak of his reputation, even though he remained as active as a young man and in his late seventies kept on introducing innovations, such as the foundation of a new periodical, *Notes from the Leyden Museum* (1879). The task of maintaining the steadily growing apparatus of the Rijksmuseum, which already had four curators, left him no time for lengthy research, and unfortunately for him he did not share Temminck's luck in training up a highly useful assistant in his special field, that of the higher vertebrates. Thus several new bird collections that arrived after Bernstein's death were looked over only briefly, though this superficial treatment did not affect the material from New Guinea and its surrounding islands.

About 1863, while Bernstein's ship was still sailing among the Moluccas and his exciting reports were arriving by mail, a more lively spirit reigned in the Museum. At that time Schlegel had taken on the young Otto Finsch as assistant, and had himself attacked with undiminished energy the writing of a major work called *Muséum d'histoire naturelle des Pays-Bas.*[17] It was intended to catalogue the entire enormous bird collection in a series of monographic surveys of families and orders, with detailed descriptions also of all "local forms." At the same time he had gladly followed the proposal of the governor general of the Dutch East Indies, Sloet van de Beele, and, to interest wider groups in research into the avifauna of the area, had begun to write an outstandingly informative book called *De Vogels van Nederlandsch-Indië, beschreven en afgebeeld* [The birds of the Netherlands Indies, described and illustrated] (Leiden, 1863–1866), for which he himself painted a large number of plates. "To make the book useful also to people of limited means, I reproduced the figures only on the smallest scale, but added all the variations necessary for proper recognition of species, cospecies, and aberrations," so that here better information might exist than anywhere else about forms of color variation. At the same time he enjoyed the satisfaction of being able to educate and employ in the museum a remarkably gifted young painter of birds, J. G. Keulemans.

Jan Gerard Keulemans, who delighted ornithologists for over 40 years with an endless series of admirable plates, was born in Rotterdam in 1842. From his youth up he drew and painted birds with enthusiasm,

---

from 1871 to 1875 and then the islands west of it, coming home with a very important collection; and finally d'Albertis, famous for his explorations (1874–1877) of the northernmost parts of Australia and southern New Guinea.

and in 1861 he came to the attention of Schlegel, who, enchanted by his special talent, soon attached him to the Museum and trained him further. In 1864 he was engaged as a draftsman, but was granted a long leave in 1865 to go on an expedition with Dr. H. Dohrn (of Stettin) to Principe and the Cape Verdes. After he had returned to Leiden he set to work with Schlegel to illustrate the latter's work on the fauna of Madagascar (1868). About 1867 there arrived in Leiden a young Englishman, R. B. Sharpe, who, although only 20 and absolutely penniless, had conceived the bold idea of writing a richly illustrated monograph on the Alcediniidae, and wanted to learn more about kingfishers by studying the treasures in the Museum. He was surprised to find a painter there who was exactly to his taste, and therefore commissioned Keulemans to make the plates, from Leiden specimens, for the first part of the monograph (published in July 1868). Thus was decided Keulemans' future, for in 1868 he was persuaded to leave Leiden and emigrate to England, where he splendidly illustrated not only Sharpe's monograph but also that on the Capitonidae (1870–1871) by the Marshall brothers, and the enormous work by Sharpe and Dresser, *A History of the Birds of Europe* (1871–1881). Starting in 1869, Keulemans undertook most of the plates for the *Ibis*, as a worthy successor to Joseph Wolf who, already famous, was glad to leave such a commission to another protégé of his friend Schlegel.

Another extremely able animal painter trained by Schlegel was Joseph Smit, born in 1837 in Lisse, where Temminck lived. After an apprenticeship in the Leiden Museum (where he painted most of the pictures of predators illustrating Schlegel's *Vogels van Nederlandsch-Indië*), he moved to London in 1866 and entered the service of the Zoological Society, illustrating their *Proceedings* for many years.

As if it were not enough that three explorers (Bernstein, Rosenberg, and Hoedt) were all collecting for Schlegel at state expense, a fourth scientific enterprise was in fact under way. A young Dutchman from a wealthy family, François Pollen (1842–1886), who had been attracted by Schlegel's works on Dutch birds, among others *De Vogels van Nederland* (Leiden, 1854–1858), with 362 plates by Schlegel himself, had approached the famous author and declared that he was ready to devote himself entirely to natural history. Schlegel prompted him to undertake an expedition to Madagascar, the "land of wonder and promise," as it had been christened a little earlier by Hartlaub, who had written the first monograph on its avifauna (1861).*

*Accompanied by his friend D. C. van Dam, Pollen visited Réunion, Mayotte, and northwest Madagascar from 1863 to 1866, bringing back important collections, especially of mammals and birds. They provided the material for a collaboration between him, Schlegel, and Keulemans on *Recherches sur la faune de Madagascar et de ses dépendances* (Leiden, 1867–1868).

But a succession of troubles followed. Finsch left him in 1864 because the Bremen Museum offered him better terms, and in December of the same year Schlegel's wife, the daughter of a Dutch clergyman, with whom he had lived in harmony for 27 years, suddenly died.* In 1865 came the news of Bernstein's death. In 1868 Keulemans went to seek a more rewarding field in England. In 1866 lack of sales had stopped work on *De Vogels van Nederlandsch-Indië* after only the monographs on pittas, kingfishers, and diurnal birds of prey had been published. And his large catalogue, the *Muséum d'historie naturelle des Pays-Bas*, seemed likely to remain a mere torso, because the monographs on the song birds, half of the planned total, had not been written.

To further "higher points of view" resulting from his frequently expressed conviction that systematics and physical geography were closely related, Schlegel had collected a gigantic quantity of material in Leiden over the years. From some 12,500 stuffed birds, which he had taken over when the directorship changed in 1858, the number had grown to 51,000 toward the end of his life. Most of them were set on stands because all his life, like his colleague Cabanis in Berlin and the Parisian ornithologists, he considered only mounted specimens useful. As late as 1878 he energetically opposed the "technique so prized by some English scholars" of having only the most important species mounted and preserving all other birds in the form of skins kept in drawers. "In my opinion this method was invented simply for the greater comfort of scholars, while sacrificing general usefulness. To introduce this sytem is therefore to rob the whole nation, which owns the institution." However, since British ornithologists remained just as firmly attached to their method (which had already been recommended by Swainson†) as he did to his, he was forced to watch the London collection grow much faster than that in Leiden after 1872, and, under Sharpe's vigorous directorship, soon outstrip it, though until that time Leiden's had been the largest in Europe.

In 1860 Eduard Baldamus had been conducted by Schlegel through the Museum, then undergoing renovation, and had written a most en-

*In 1869 Schlegel married for the second time, taking as his wife Mrs. Albertine Pfeiffer, the beautiful daughter of his dead friend Professor van Lidth de Jeude. Her name is borne by the rare starling, *Streptocitta albertinae* Schlegel 1866, and her father's by *Garrulus lidthi* Bonaparte 1851; the professor had a large zoological collection.

†Swainson mentioned the matter in 1836: "The last and best method of arranging birds is by leaving them, as it is technically called, 'in their skins.' It is now nearly twenty years since we began this plan, to the great surprise and disapprobation of our scientific friends; but the practice has now become almost general. It is simply arranging the birds in drawers, as shells or minerals are kept; the specimens of each genus being placed together, and laid in rows upon carded cotton," Swainson, *On the Natural History and Classification of Birds* (London, 1836), vol. 1, pp. 276–277.

thusiastic report about it.[18] Certainly anyone not looking at the collection with a systematist's eyes would have thought he had got into an attic, as did Philipp Leopold Martin, one of the creators of modern lifelike taxidermy. His visit to Leiden in 1868 gave him the opportunity to satirize, without the slightest reticence,

> the gloomy pedantry of the whole style of exhibition . . . White cupboards and whitened stands and shelves blurred into one another in the not over-illuminated galleries. Everything overflowed, as nowadays in our shops. When I reached the woodpeckers, I thought that my eyes must be confused by the dim light from the windows, because I saw dozens of cracks in the crooked stands. When I looked more closely I saw that I was right, because pieces of tree trunk, bark and all, had actually been taken and — painted white! But that was not all, because in another cupboard were drawers full of freshly stuffed exotic sandpipers. Like old-fashioned tin soldiers from one's childhood, 30 to 40 specimens of a species stood there in exactly identical positions, with one outspread wing, as if they had been called out on zoological maneuvers. I went away speechless.[19]

We know how Leiden looked in 1867 from a report by Baron Ferdinand von Droste-Hülshoff, who visited it with his friend Altum.

> The staircase leads us in one sweep . . . to a little anteroom containing nothing but birds of paradise. A whole cupboard full of *apoda*, 46 specimens, from the drab to the most splendid plumage in the most wonderful gradations. And these are not all of the group. In the skin storerooms there are masses of skins, 21 of the wonderful *rubra*, all collected by Bernstein; 15 *P. speciosa*, 9 *wilsoni*, 25 *regia*, 25 *wallacei*, nearly all collected by him too; then 16 *papuana* and 1 *sexpennis*, supplied by Müller. All these species prove that a great many so-called "good species" are not valid. Most of the examples of these species were collected in widely separated places so that no suspicion can arise of having to deal with more than one species. These show a gradual change in plumage color and a slow sprouting or even a complete alteration of the plumes . . . From here we turn left and reach the long corridors and galleries, which interconnect and lead through various wings until we finally get back to the birds of paradise. All these extensive rooms are stuffed to bursting with birds and mammals. One grows dizzy at the endless rows of falcons, owls, song birds, etc. . . . And when one has reached the end of all this, there are still the most immense skin storerooms, holding countless skins. This museum is the place to form a valid judgment on the independence of species and the many varieties that have been

elevated into species . . . It becomes obvious that many birds . . . from different islands, often in the same archipelago, show constant differences. Not uncommonly at the first glance one can assign individuals of the same species to one island or another, on account of the greater or lesser intensity of color, or the dimensions, of a particular spot or band.[20]

Schlegel aroused a great deal of interest among his contemporaries with his theory of color change, which he published several times from 1852 on, without knowing that he had long been anticipated by English and American ornithologists. He maintained nothing less than that all birds molt only once a year and that the nuptial plumage appearing in many species arises from a change in the colors of an originally drab plumage.

During the development of the sexual urge because of an excess of juices, by which it is caused, the blood vessels in the feathers, which had seemed completely dried out, are filled with new life and are thereby enabled to resist external attrition produced by chemical and mechanical causes. Now, as before, comes a second period for the full development of the feathers. Here one should note that

a) the faded and rubbed feathers are replaced by new barbs and barbules . . .

b) at this time a fairly large quantity of pigment enters the feathers . . .

To the honor of German ornithologists be it said that this gross error was immediately corrected by E. F. von Homeyer, Altum, Mewes, and Schlegel's former teacher Christian Ludwig Brehm, which did not prevent Schlegel from persisting firmly in his opinion and winning over others (like H. Gätke, C. L. Gloger, and Otto Finsch), so that even at the beginning of this century it was still maintained by a few people (P. Kollibay, for instance). Schlegel's hostility to the use of the microscope and his occasional fits of obstinacy (as in the matter of evolution) damaged his scientific reputation.

Schlegel, who had been tormented by the problem of species even in his youth, was an inveterate opponent of the theory of evolution, and considered Darwin's ideas to be ingenious speculation. A man with an enormous collection of factual material under his eye could easily be tempted to bolster his skepticism with knowledge of phenomena that the Darwinians had not been able to explain satisfactorily. Those who knew only the Darwinian show pieces could devoutly accept the new teaching if they liked; for Schlegel it was incontrovertible that the problem of species was not yet solved. As he liked to say, he would gladly be satisfied if only one well-attested example of a transitional form in nature could be produced from which a species breeding true had been developed. In this he agreed with J. H. Blasius, who, as one of the spokesmen for German ornithologists, had said in 1860:

> Darwin sets up an arbitrary hypothesis and tries to make it plausible with acute reflections, not entirely apt analogies, and unproved possibilities. Direct evidence for this hypothesis is impossible; no one has even tried to demonstrate the development of any particular species. In my firm opinion, Darwin has not taken the definition of the concept of species one step further, to say nothing of making it a convincing certainty, either theoretically or practically.[21]

The Berlin ornithologist Cabanis agreed. "The valleys and plateaus of the cordilleras appear to be particularly favorable territory for so-called local races or aberrations. The establishment of . . . the reasons for this must be left to a future decision by a perfected science. Premature and 'ingenious' fantasies do not promote knowledge."[22] Schlegel did not see in such "local races" or geographic variations any developing germs of new species, perhaps stimulated by climatic influences. Rather, they were for him "conspecies," that is, local and — as he still held with C. L. Brehm — immutable manifestations of the type of the immutable species. Though, as we have seen above, he was the first zoologist to employ trinomial nomenclature, and to persist in it without any concern for the protests of his contemporaries,* he was also the man who from the very beginning opposed the designation of geographic forms as subspecies (Brehm's term for all variations), because this expression could all too easily imply the idea that the variation in question was, though not a true species, on the way to becoming one. This idea seemed to him to threaten the purpose of his entire life's work; in fact, if the future should confirm Darwin's view, then (in the words of his friend Blasius in 1860) "every investigation of species would become useless and impracticable."[23]

In an article published in 1866 on the kingfisher genus *Tanysiptera*, very productive of local forms, Schlegel developed his views most clearly, and was able to support them with the study of no fewer than 120 specimens from the area of distribution, explored by Bernstein. He wrote then:

> To what conclusions do the phenomena represented by these briefly discussed birds lead us?

*"Here we must repeat our objection to the use of trinomial nomenclature for subspecies or races — conspecies"; Hartlaub in a review of the first fascicle (Leiden, 1862) of Schlegel's *Muséum d'histoire naturelle des Pays-Bas: Revue . . . de la collection des oiseaux*, *Journal für Ornithologie*, 11 (1863): 134.

> Naturalists who are partisans of the inventive Darwin's fantastic doctrines will unhesitatingly explain these facts by the supposititious theory that the forms of life are mutable. But in so doing they will err, because no proof is available to support such speculative assertions. The ornithologists who adopt the latest fashionable procedure have distinguished these birds from one another with a certain number of names, and this practice would be correct if, true to their own principle, they had further multiplied the distinctions and were in a position to provide exact and explicit signs for the recognition of what they call species. As for me, I do not deny that a naturalist dealing with such facts becomes perplexed when he seeks some way of converting his concepts into a complete, simple, and clear expression of the law of nature. And it is all the more difficult to do this because, research on these birds having only just begun, there can be no question yet of an end to the debate. Then let us temporarily regard the creatures here dealt with as so many isolated stems ['souches'] of the same type, which can be included in the general name of *Tanysiptera dea*,* and restrict ourselves simply to entering the new items in the scientific ledger as we come across them[24] [that is, without giving all these geographic variants a scientific name].

Until the end of his life, therefore, Schlegel was to be, in Darwin's words, "incessantly haunted by the shadowy doubt whether this or that form be in essence a species."[25] Because he never succeeded in hitting on a usable (for instance, a geographic) definition of species, he considered that many differences later recognized by systematics as distinguishing species were generally unimportant, and thought they could not justify any promotion to a new name. In his *Tanysiptera* article he wrote:

> It is no longer appropriate to our times to imitate (not so much to advance science as to satisfy one's own vanity) certain examples of an endless splitting of species, genera, and groups, examples of childish exaggeration that are damaging to science and cause the names of their authors to grate upon the ear.

This rebuke was clearly addressed to Wallace and especially to Cabanis, who had explained his procedure as follows:

> In many . . . cases certain consistent variations among individuals

*Wallace, on the other hand (*Ibis*, 4, 1862: 348–350), described a new form belonging to this group as *Tanysiptera doris* (on the island of Morotai, off Halmahera), and added that after this volcanic island appeared it was probably colonized by birds flying across the sea from Halmahera. *Tanysiptera* seems "peculiarly subject to modification," and has undergone such a great change on Morotai that its representatives there already display "well-marked specific characters."

> from Mexico, New Granada, Cayenne, Brazil, etc., do not escape the practiced eye, and certainly the advancement of science is better served by calling attention to such noticeable climatic aberrations by specific distinctions than by referring them back, like Buffon, to one species or another. In many cases the definitive answer to the question of species or subspecies must be left to later and better knowledge of all the contributory factors.[26]

Schlegel's admonitions went unheeded. The new generation preferred Darwin, who in 1859 had written:

> Systematists will have only to decide (not that this will be easy) whether any form be sufficiently constant and distinct from other forms, to be capable of definition; and if definable, whether the differences be sufficiently important to deserve a specific name . . . Hereafter we shall be compelled to acknowledge that the only distinction between species and well-marked varieties is, that the latter are known, or believed, to be connected at the present day by intermediate gradations, whereas species were formerly thus connected . . . We shall have to treat species in the same manner as those naturalists treat genera, who admit that genera are merely artificial combinations made for convenience. This may not be a cheering prospect; but we shall at least be freed from the vain search for the undiscovered and undiscoverable essence of the term species.[27]

Thus young scientists, led by Sharpe, Salvador, and Severzov, could always cite this passage as authority for their zealous persistence in describing even slight geographic variations from the "type" as separate species. But the process more and more quickly brought systematic ornithology into a critical situation. What the experienced Schlegel had sought to avoid, although with impractical methods and an untenable theory, had become reality.

His brusque rejection of Darwinism was incomprehensible to his younger contemporaries chiefly because they had learned to believe in the omnipotence of natural selection. Only a later generation, which had drawn new conclusions from the abundance of data and had been able to provide a more solid basis for the theory of evolution, was milder in its judgment of his skepticism. Until that point was reached, countless minutely detailed preliminary studies on variation and geographical distribution were needed, studies such as men of Schlegel's stamp had produced or had at least made possible by their enormous collections of material.

"Conservez ce jeune homme, il ira loin [Keep that young man, he'll go far]," wrote Heinrich Boie to Temminck from Java in 1826, emphatically repeating that he should cherish Schlegel. Anyone who knew these prophetic words may well have remembered them when the gray-haired naturalist was carried to his grave on January 21, 1884.[28]

12

Otto Finsch (1839-1917)

On October 17, 1884, just nine months after Schlegel's death, the German flag was raised on the coast of New Guinea by none other than Dr. Otto Finsch, the same man whom Schlegel had taken under his wing 23 years earlier, in order to concentrate his adolescent yearning for adventure on a single major aim, New Guinea.

For Schlegel, New Guinea meant birds of paradise, cassowaries, and marsupials, and for decades after Salomon Müller had told him about it he worked to learn more. Who at that time could have suspected that the seed would so spring up, or that Schlegel's own disciple was destined some day to plant the national emblem in that far land?

It had been a long and tortuous journey from the bird rooms of the Rijksmuseum to the glittering shores of Konstantinhafen (near Madang), for Otto Finsch's restless spirit and ceaseless activity had driven him far. They conditioned him perfectly for his own period, that of boundless opportunity.

He was born on August 8, 1839, in Warmbrunn, a little spa at the foot of the Riesengebirge. His father, a tradesman and painter on glass, could not afford to send him to secondary school and wanted to make him a painter like himself. The boy would not listen, however, because he was passionately devoted to collecting and observing birds, and had thus attracted the attention of Dr. Ernst Luchs, the physician at the spa, who owned a collection himself and further strengthened the young man's inclinations. At 18 he decided to seek his fortune in foreign parts, and in 1857 he went to Budapest. Shortly thereafter he managed to reach Bulgaria, a Turkish dominion which attracted him powerfully because it was one of the European countries whose fauna was almost unknown. He stayed there all through 1858 and into the following summer, before returning home. In Bulgaria he had partly earned his keep as a tutor in Ruschuk (Ruse), and had traveled as much as he could to watch birds, getting as far as Shumen (Kolarovgrad) and the eastern Balkan Mountains. He was so delighted with what he found that he wrote it up in an article, "Beiträge zur ornithologischen Fauna von Bulgarien," and sent it to Cabanis, who published it in his periodical, *Journal für Ornithologie*, in September 1859. That was the point at which Finsch decided to become an ornithologist.

Luck was with him. Schlegel, who was busy at the time reconstructing and improving the Rijksmuseum and who needed help desperately, encouraged Finsch to come to Leiden in 1861, and found such pleasure in his great enthusiasm and quick understanding that he employed him as an assistant the following year. Finsch, whose knowledge was still modest, began assiduously absorbing everything Leiden offered, and concentrated particularly on studying the East Indies, because when he was a boy Cook's Pacific voyages had interested him much more than Levaillant's South African travels. While his new friend, the Dutch lieutenant A. Goffin, was writing the catalogue of "Buccones" for Schlegel[1] and preparing for a journey to West Africa, Finsch was being trained by Schlegel for New Guinea.* But that meant more than mere acquaintance with everything known about the fauna, because Schlegel demanded that his envoys should be as well informed as possible about the geography, climatology, and ethnology of the countries they were to travel in. To Finsch's bitter disappointment the New Guinea expedition planned for 1862 was abandoned, but to avoid wasting his efforts he published a brochure in Bremen in 1865, *Neu Guinea und seine Bewohner. Mit einem Anhang: Liste der jetzt bekannten Säugethiere und Vögel Neuguineas und der Molukken* [New Guinea and its inhabitants, with an appended list of the mammals and birds so far discovered in New Guinea and the Moluccas]. It was the first German book to call attention to New Guinea.

He also developed a special enthusiasm for parrots. As early as 1863 he described some new psittacids (including *Lophochroa* [*Cacatua*] *goffini*, to commemorate his friend, who had died young) from the Amsterdam Zoological Garden, which had developed magnificently under the direction of Schlegel's close friend G. F. Westerman,† and which at that

*On May 23, 1862, Schlegel wrote from Leiden to Baldamus: "Finsch is a really outstanding person, tirelessly busy and very quiet, which is good for the country we have in mind. I hope that I shall soon succeed in shipping him to the Indies." (Letter in the archives of the Deutsche Ornithologen-Gesellschaft.)

†Gerardus Frederik Westerman (1807–1890) was one of three men from Amsterdam who in 1836 founded the Royal Zoological Society "Natura Artis Magistra," together with a zoological garden, of which Westerman became director. The Society began issuing its own periodical in 1848, the admirable *Bijdragen tot de Dierkunde* [Contributions to zoology], in which Schlegel was glad to publish. The magnificent book *De Toerako's afgebeeld en beschreven* [The turacos

time contained no fewer than 70 parrot species. This was the prelude to his great monograph, *Die Papageien* (two volumes; Leiden, 1867, 1868), with which Finsch immediately took his place among outstanding ornithologists, as Temminck had done earlier with his *Pigeons*. Finsch's extraordinarily wide reading, his accuracy and care, his acuteness, and the diversity of his interests emerge clearly from this smoothly written book, particularly from the first part, the "General Natural History of Parrots," which describes in detail the history of scientific knowledge about them, their biology, distribution, "intellectual endowments," "appearance and exterior appendages," plumage, anatomy, and systematics. This achievement immediately received an appropriate reward: in 1868 the University of Bonn conferred on the author an honorary doctorate.

But when *Die Papageien* appeared, Finsch was no longer in Leiden. In 1864 he had been called to Bremen as curator of the natural history collections belonging to the Museum Society, because he had come to the attention of Dr. Gustav Hartlaub, who had long wanted a well-trained colleague.

These two outstanding scientists formed an alliance that benefited ornithology for 14 years, though they differed fundamentally in their backgrounds and in many of their characteristics.

Gustav Hartlaub, who for almost 50 years was Germany's most distinguished representative in the field of general ornithology, was already, in 1864, at the zenith of his international reputation. Traditional and social ties bound him to Bremen, where he had been born on November 8, 1814, the son of a wholesale merchant and senator, and which as an adult he never left for long. But his connections covered the whole world, which indeed lay at the gates of the old port city.

In his exuberant youth he had felt quite differently. Under the stimulus of Karl Ritter's lectures, three students of general zoology in Berlin enthusiastically decided to explore the High Tatra and the central Carpathians on a scientific walking holiday, and to climb its highest peak, the Lomnitz, in one rash attempt. This all took place in August 1835, the three being Count Alexander Keyserling from Courland (1815–1891), J. H. Blasius from Brunswick (1809–1870), then a student in the philosophical faculty, and the medical student Gustav Hartlaub, all of whom were well known a decade later.* Hartlaub

illustrated and described] (Amsterdam, 1860), by Schlegel and Westerman and illustrated by Schlegel, also appeared under the Society's auspices.

*Keyserling and Blasius together wrote a book, *Die Wirbelthiere Europas* [European vertebrates] (Brunswick, 1840), which was particularly prized for its information on mammals. The former made a name for himself later in Russian geology, and the latter made a large collection of European birds in Brunswick, and helped European ornithology by introducing new methods of diagnosis and by being "a vigilant and critical censor" (Hartlaub). After Naumann's death in 1857 he became the intellectual leader of the Deutsche Ornithologen-Gesellschaft.

took his degree at Göttingen in 1838; since his family was wealthy, he was in no hurry to start practicing, and joined a collecting trip to Lake Balaton and Croatia, organized by the Viennese ichthyologist Heckel and Joseph Natterer's two sons, which was also joined at Budapest by Schlegel's old bosom friend, J. S. von Petényi. Even later, when Hartlaub was practicing medicine at home, he treated it rather as a hobby; he was much more interested in cultivating his mind. "Geography, travel, history, literature, and art were all more or less thoroughly familiar to him, and he followed all developments with lively interest. Always excellently informed, he could permit himself to make judgments even in fields far removed from ornithology."[2] But in that he quickly became an expert.

When he made up his mind in 1840 to take up exotic ornithology, probably stimulated by the impressions received when he was a student in the Berlin Zoological Museum, this branch of learning interested few scholars in Germany, although it was cultivated by many in France and England. Illiger, Kuhl, Boie, and Wagler, who had all seemed destined to instill new life into general ornithology, had died young. Apart from Rüppell in Frankfurt, only Lichtenstein in Berlin could be regarded as a real expert; but the former had concentrated for 20 years almost entirely on his northeast African birds, while the latter, a species collector of Temminck's stamp, had published hardly anything for a long time. Hartlaub, however, did not intend to remain an idle spectator of foreign progress. Because he could give his wishes free rein, he provided himself amply with ornithological literature and, at his own expense, bought specimens from dealers to enlarge the originally insignificant bird collection belonging to the museum of the Bremen Natural History Society. He made such striking progress in a few years that it had grown to 2000 species by 1844 and could easily stand comparison with similar collections. Strickland, who had visited Hartlaub in the summer of 1845, noted in his diary:

> A highly scientific ornithologist, and most obliging person . . . excellent library, and a very admirably arranged collection of natural history . . . It is really refreshing, after seeing the masses of antiquated confusion which commonly go by the name of museums, to find a collection in which every specimen is put in its right place, and marked with its right name, according to the latest and most approved zoological principles.[3]

Hartlaub soon discovered undescribed species among his acquisitions, and began to study them in detail. Because he could not hope to find an audience in Germany, he began by publishing his articles, of which the first appeared in 1841, exclusively in the French *Revue zoologique*, patronized by Baron de Lafresnaye and other French collector-ornithologists. Even in his earliest publications (such as revisions of barbets and of Mexican fringillids, 1842), Hartlaub was concerned to provide a general survey and to fit scattered building blocks together into a definite structure; in so doing he demonstrated his detailed knowledge of the literature. In 1846 he was commissioned to write the yearly reports for Troschel's *Archiv für Naturgeschichte* on the progress of ornithology, a job which from then until 1871 made it possible for him, as an intelligent judge, not only to observe in detail the development of ornithology, but also to guide it. Leverkühn aptly remarked in 1901: "The character of this wonderful man is reflected in the bibliographic surveys; he lets no special considerations deter him from pillorying inaccuracy or superficiality, while excellence and value he praises with the enthusiastic pleasure of a young man. He was a critic to be feared, but he himself was seldom, if ever, attacked."[4]

Cabanis, who in 1841 had become Lichtenstein's right-hand man, could not have found a more powerful ally than Hartlaub, because both of them wished for a rebirth of ornithology in Germany, and they agreed in their conviction that this aim could not be reached by leaving the leadership to men like Naumann and Baldamus, who were familiar only with European or even solely German birds, and had almost no connections with the rest of the world. As a protest against the spirit of *Naumannia*, with Hartlaub's powerful support a new periodical was founded in 1852, the *Journal für Ornithologie*, whose editor, Jean Cabanis (1816–1906), intentionally called it "a central organ for general ornithology," and which splendidly fulfilled its purpose from the beginning.

At the outset, Hartlaub found birds from all parts of the world of equal interest. However, the arrival in the Hamburg Museum of the zoological collections of Carl Weiss, who had been traveling on the Gold Coast and the islands of São Tomé and Principe, prompted him to specialize in detail in West African ornithology and to assemble what was known about it. The result first appeared in 1850 and then, considerably enlarged, in a book published in 1857, *System der Ornithologie Westafrica's*, which, more than any other work, promoted the flourishing development of African ornithology for years to come. To achieve the greatest accuracy and completeness Hartlaub had spent the interval traveling from one museum to another on the Continent for purposes of study. In 1854 he passed some time in Leiden, to use "the vastly rich treasures of the Dutch royal museum"[5] with Temminck's permission, and particularly the large collections that H. G. Pel had sent back from the Gold Coast. In 1856 he went to Paris, where he enjoyed the helpful collaboration of Jules Verreaux, "freely quaffing from the brimming spring of his practical experience and unrivaled specialized knowledge." What he still needed to complete his material was to cross the Channel. For almost 20 years he had watched from a distance the magnificent revival of ornithology in England, where Vigors, Jardine, Gould, Strickland, Blyth, Eyton, Fraser, and Sclater had appeared in quick succession and seized the primacy from France, which had been leading until 1840. His contacts with English ornithologists gradually grew closer. To start with, he met Strickland,* who visited him in Bremen in 1845 and saw to it that his very promising pupil, Philip Lutley Sclater (1829–1913), when he visited the Continent about 1849, paid his respects first of all to the doctor in Bremen. Meanwhile Sclater had become one of the leading ornithologists in England, indeed in the world, an authority particularly in the field of neotropical speciation and of zoogeography, and the moving spirit in a group of friends who in 1858 were preparing to found an English ornithological journal, the *Ibis.* In the same year Hartlaub finally decided to repay Sclater's visit, and to study the excellent British Museum material under the care of G. R. Gray. Delighted by what he saw, from then on Hartlaub became a dedicated intermediary between Continental and English ornithologists. He published many of his contributions in English periodicals, principally in the *Proceedings of the Zoological Society*, so that he could let the rest of the world know of the progress being made in Germany, and could be certain of an understanding audience. When he returned to Oxford in 1860, he was chosen as the first honorary member of the newly founded British Ornithologists' Union.

Finsch, who had already been well trained by Schlegel in basic systematics, began his higher education with Hartlaub, whose influence is noticeable in the production of the *Papageien.* In many points he rapidly equaled his master. Hartlaub persuaded him not to complete the monograph without having studied the parrots in the most important European museums, as Heinrich Kuhl had done so profitably in 1818 and 1819 when he was similarly engaged. Therefore, at Hartlaub's "urgent

*Hugh E. Strickland is distinguished for having given the decisive direction to ornithological activity in England, converting it from a "scientia amabilis" into a serious science. He created a clear overall picture, rejected the last traces of old-fashioned arbitrariness, and showed the systematists important subjects to investigate. He was born on March 2, 1811, to well-to-do parents at Righton, Yorkshire, and studied the natural sciences at Oxford. After taking his M.A. in June 1835, he accompanied the geologist W. J. Hamilton on a scientific journey to the Near East, going by way of Paris and Trieste. They returned the next summer. In July 1845, Strickland married a skilled painter of birds, the daughter of his fatherly friend Sir William Jardine, the ornithologist. On their honeymoon the Stricklands went to Holland (where he noted a picture of a dodo by Roelant Savery, in The Hague), and then to places where he could meet ornithologists — Bremen for Hartlaub, Copenhagen for Professor J. T. Reinhardt — before reaching his destination, Lund. After exotic ornithology he was principally interested in geology, whose study may well have influenced his taxonomic views. In 1850 he became reader in geology at Oxford. On September 13, 1853, he met with a fatal accident when, while examining a railroad cut carrying a double track, he avoided a freight train on one track only to be run over by a passenger train speeding round a curve on the other.

insistence," Finsch, who had already visited the museums in Berlin, Dresden, and Darmstadt, and the Museum Heineanum in Halberstadt,* went to London in 1864. There he studied parrots not only in the British Museum but also in the large private collections of Wallace and Sclater, and made personal contacts that he carefully cultivated later.

At the same time he began to work on Polynesian ornithology because the spirit of adventure that was still lively in the Hanseatic cities, among them Bremen, had recently opened up a new area for science in the South Seas. Few of the island groups had yet been visited by zoologists, though Reinhold Forster (1773–74) and William Anderson (1776–77), who accompanied Cook on his voyages of exploration, had early provided a reasonably good knowledge of the birds of the Society Islands and the Tongas. Since that period the South Seas had often been crossed by voyages of discovery none of which contributed so much to ornithology as did Dumont d'Urville's to the South Pole, 1837–1840, with the naturalists Hombron and Jacquinot,[6] but they had only sampled the bird life of some of the islands. The first change came with the United States Exploring Expedition into the South Seas between 1839 and 1841, under the command of Charles Wilkes. One of the two naturalists on the expedition was Titian R. Peale from Philadelphia, whom we encountered before when he set out to collect in Florida for Bonaparte in the autumn of 1824. He made many interesting discoveries in Samoa and the Fiji Islands. There had been no further research, and many groups of islands, including almost the whole of Micronesia, remained unknown.

While affairs were in this state, described by Hartlaub in 1854 in an article, "Zur Ornithologie Oceaniens,"[7] Johann Cesar Godeffroy, the head of a large Hamburg trading firm with principal interests in the South Seas, decided in 1861 to found a South Sea museum. Directed after 1863 by the ethnologist J. D. E. Schmeltz, it was intended to augment the knowledge of the ethnology and the zoology of Oceania. Godeffroy ordered the captains of his merchantmen to provide mate-

*About the middle of the last century, the Museum Heineanum was the largest private collection in Germany, with approximately 10,000 mounted specimens of 5,000 species. In 1843 its owner, Ferdinand Heine (1809–1894), tenant of monastic lands and a high bailiff in Halberstadt, had got to know Cabanis, who energetically went to work to enlarge the collection and in 1850 began a descriptive catalogue of it, with important critical comments. *Museum Heineanum* remained incomplete, though the first and second parts appeared in 1850 and 1851. The third and fourth parts were written by Cabanis together with the owner's son, Ferdinand Heine, Jr., who had acquired a considerable knowledge of exotic ornithology, and were published in 1859 and 1863. Later this museum, which is still interesting because of its many types and occasional rarities, was taken over by the city.

The other famous German private collection during the first half of the nineteenth century was that of Prince Maximilian von Wied, which at his death in 1867 consisted of 4,000 mounted birds and 600 mounted mammals. With all its important types (mostly Brazilian), it left the country when J. A. Allen succeeded in acquiring it in 1870 for the recently founded American Museum of Natural History in New York.

rial for the museum, and also employed collectors. The first of these, and the only one with a thorough scientific education, was Dr. Eduard Gräffe, a Swiss from Zürich, who undertook an ornithological reconnaissance of the Tonga, Samoa, Fiji, and Ellice islands, of the Wallis and Phoenix groups, and others. In 1865 Alfred Tetens, who had returned to his native Hamburg, called Godeffroy's attention to the commercial possibilities of the Carolines. This gave a final impetus to the rapidly expanding South Sea trade of the firm of J. C. Godeffroy and Co., acquired by the German Empire as it took over the Bismarck Archipelago, the Marshalls, the Carolines, the Marianas, and other islands. Tetens was sent out to Palau in 1865 as captain of the brig *Vesta*, to found a trading station and at the same time to make a natural history collection, which he assiduously did for two years, not only there but on Yap, the Ninigo group, and the Hermit islands. About 1871 Godeffroy's captains Heinsohn and Peters took over the job. From 1869 on an American, Andrew Garrett, sent birds to Hamburg from Rarotonga, the Marquesas, and other islands. Additional collectors for the Godeffroy museum (which in 1871 began to issue its own expensively produced periodical) were Theodor Kleinschmidt and Franz Hübner.* But none of them was so persistent a bird collector as Johann Stanislaus Kubary, who had arrived in Hamburg as a political refugee from Warsaw, and had been sent by Cesar Godeffroy to collect in the Pacific in April 1869. He began on Savai'i, Western Samoa, where he discovered the (later extinct) rail, *Pareudiastes pacificus.* Then he went to Yap, and in 1871 settled in Palau. He must be credited with the first thorough ornithological exploration of Palau, Yap, and Ponape.†

When Finsch moved to Bremen in 1864, Dr. Gräffe's first consignments of birds from the larger islands in the Samoa and Fiji groups had just arrived in the Godeffroy museum. In order to study them as thoroughly as they deserved, Hartlaub joined forces with Finsch, and together, after long and detailed preliminary work, they published in 1867 *Ornithologie der Viti* [Fiji]-, *Samoa- und Tonga-Inseln.* A large number of important articles followed during the next 12 years (until 1879), dealing with further bird shipments to the Godeffroy museum (first of all Tetens's collection from Palau, Yap, Ninigo, and Hermit Islands).

*Theodor Kleinschmidt (an uncle of the ornithologist, Dr. Otto Kleinschmidt) worked with great success from 1872 to 1878 in the Fiji Islands, where he discovered *Lamprolia victoriae, Drymochaera badiceps [Vitia ruficapilla badiceps]*, and other startling endemics. In April 1881 he and his two companions were killed on the little island of Utuan (off Duke of York Island) during a reprisal action. Franz Hübner was on Samoa, Tongatapu, and Niuafo'ou from 1875 to 1876; he died of sunstroke in 1877 at the Godeffroy trading station of Mioko, off Duke of York Island.

†Cesar Godeffroy was interested not only in the islands of the Pacific but in the then little-known territory of Queensland, where Frau Amalie Dietrich collected plants and animals for his museum (principally between Rockhampton and Bowen) from 1863 till about 1872. For Amalie Dietrich, see Charitas Bischoff, *Amalie Dietrich. Ein Leben* (Berlin, 1909); translated by A. L. Geddie as *The Hard Road: The Life Story of Amalie Dietrich, Naturalist, 1821–1891* (London, 1931).

The articles, written partly together and partly separately, and published in various periodicals, including English ones, named over 50 new South Sea species.

No wonder that Finsch yearned more and more to see with his own eyes the islands that so constantly occupied his imagination. But before he could realize his desire, his urge to travel was temporarily diverted. From August to December 1872 he accompanied a geographer from Bremen, Dr. M. Lindeman, across the United States to San Francisco. In the summer of 1873 he crossed Lapland from Alta as far as the Varanger Fjord, and in 1876, commissioned by the Society for a German Expedition to the North Pole and accompanied by Dr. Alfred E. Brehm and Count Karl von Waldburg-Zeil-Trauchberg, he went to western Siberia. They traveled south as far as Alakol and the Dzungarian Alatau, and from there they turned north and, crossing the Altay, penetrated as far as the bay of the Kara Sea.

> In the short time between spring and autumn they covered almost 7000 kilometers in sleds, carts, and on horseback, so that their work could only be a sort of superficial reconnaissance, and not many skins were taken. But it is astonishing how many species could nevertheless be identified and how many accurate individual observations could be made.[8]

During the intervening years Finsch had by no means limited himself to studying South Sea birds, because, like Hartlaub, he was interested in birds from all over the world. A special opportunity for the two men to concern themselves with African birds was provided by the collections of the famous traveler Baron Claus von der Decken, who had spent five years studying the geography and natural history of tropical East Africa before he was murdered at Bardera, Somalia, on September 19, 1865. The birds collected by the expedition had indeed been transmitted to the Berlin Museum, where Cabanis described the 20 new species, but a major contribution to ornithology was made only when Hartlaub, with Finsch's help, was able to employ his thoroughness and encyclopedic knowledge in the fourth section of the work on von der Decken's travels. This stout volume, published in January 1870 as *Die Vögel Ost-Afrikas*, assembled everything previously known about East African birds, thereby laying a foundation for further research, which was soon carried on with great success by other German travelers (J. M. Hildebrandt, 1872–1877; G. A. Fisher, 1877–1886; R. Böhm, 1880–1884).

But, as we previously noted, Finsch when young had not been so stirred by Levaillant's accounts of his travels as by Cook's report of his voyages, and that was how he continued to feel. He had hardly finished working up the material from his west Siberian journey before he decided on a trip to the South Seas. Cesar Godeffroy could no longer help him, because his formerly powerful firm, which had been supported solely by its coconut plantations, had got into financial difficulties and could do nothing more for science.* So Finsch turned to the Academy of Sciences in Berlin, which in January 1879 granted him 30,000 marks from the Humboldt Foundation on condition that he give all the material collected to the Royal Museums. However, the Bremen senate disliked Finsch's frequent absences, and when he applied yet again for leave, this time for two years, he was told to choose between his journey and his directorship of the Natural History Museum. Without a moment's hesitation he left.†

Finsch intended to explore the atolls, rarely visited and poor in fauna, when he left on April 21, 1879, with Ernst Rehse, a preparator. They traveled by way of New York, San Francisco, and Honolulu to visit the Marshall and Gilbert Islands (September to November 1879), and then Kusaie and Ponape. Years earlier Finsch had described many new species of birds from the skins sent by Kubary from Ponape; now, under the guidance of the remarkable expert himself,‡ he was able to study the living examples (March 1880). His journey took him next to New Britain and Duke of York Island. Not much was left for a visiting ornithologist there, because George Brown, the head of the Wesleyan Mission Station, had already skimmed off the cream. The alert P. L. Sclater had barely got wind of the missionary's intended departure for Duke of York Island when he commissioned him to collect birds. So when Brown arrived, on August 15, 1875, he was already accompanied by an Australian preparator, James Cockerell, and went busily to work to satisfy the famous zoologist. In fact, Sclater was enabled to describe a number of remarkable new species in the *Proceedings of the Zoologi-*

*Publication of the *Journal des Museum Godeffroy* was halted in 1879, and in 1882 the museum's curator, Schmeltz, accepted a call to the Ethnographical Museum in Leiden. When Cesar Godeffroy died in 1885, the contents of his museum, valued at a million marks, were auctioned off and scattered all over the world. Many of the birds, including numbers of Finsch's and Hartlaub's types (that of *Pareudiastes pacificus* among them), were acquired in 1886 by the Hamburg Zoological Museum, and were destroyed by bombs in 1943.

†Finsch's departure from Bremen marked the end of his long collaboration with Hartlaub, who then returned to his first love, the birds of Africa. At just that time Dr. Emin Bey (Arthur Schnitzler, 1840–1892), who had been governor of the Egyptian Equatorial Provinces since 1878, wrote from Lado to the famous author of *Die Vögel Ost-Afrikas* for ornithological advice, which Hartlaub promptly supplied. Success was not long in coming; Emin, excited by the many new species Hartlaub had been able to describe from the collection of birds sent to Bremen from Khartoum, distinguished himself greatly in inner African ornithology until his violent death.

On November 20, 1900, Hartlaub, the Nestor of German ornithologists, who worked for the universal advancement of his beloved subject almost till his last breath, died at Bremen like the sort of wise man who, as he wrote Leverkühn in 1890, "really lived on only in Goethe."

‡Johann Kubary, son of a Hungarian father and a German mother, was born in Warsaw in 1846. After six years in the South Seas he visited Germany in 1875, though he returned to Ponape. After many ups and downs (from 1885 to 1891 he ran the Konstantinhafen trading station of the New Guinea Company), he died there in October 1896. (Obituary by Schmeltz in *Internationales Archiv für Ethnographie*, 20, 1897.)

cal Society for 1877, leaving almost nothing for Finsch, who therefore soon went on to Sydney. Here there was an ornithologist to visit, E. P. Ramsay, the respected curator of the Australian Museum.

The real goal of the journey south, however, was New Zealand, where Finsch arrived in October 1881. The event had a long preliminary history, for as a young beginner he had been keenly interested in New Zealand birds, especially parrots, and in 1867 he translated into German and published Walter Buller's first work, *Essay on the Ornithology of New-Zealand.*[9] This interest had been much heightened by a visit to the Vienna Museum, also in 1867, and had led to a lively correspondence with outstanding scientists resident in New Zealand, Dr. J. Hector and Julius von Haast, the latter of whom from 1869 on sent many New Zealand birds to him in Bremen. In 1871 Walter Buller and Captain F. W. Hutton also began to consult Finsch, who was thereby enabled to publish a "Revision der Vögel Neuseelands,"[10] and to return to the subject several times later. No wonder that he was very keen to meet these friends, who welcomed him warmly and accompanied him as far as the glaciers of the Southern Alps.

From New Zealand Finsch returned to the northernmost tip of Australia, visited the islands in the Torres Straits and, finally realizing his boyhood's intense desire, landed near Port Moresby in the southeastern corner of New Guinea. There he was pleasantly surprised by meeting another German, formerly a gold prospector but at that time a collector of natural history specimens, Carl Hunstein, an expert on the country and its inhabitants, who guided him some distance into the interior and helped him to add rapidly to his bird collection.

Finsch came home in November 1882, by way of Java, and delivered to the Berlin Museums his large and important collections, particularly valuable for ethnography and anthropology. A few months later a malign star brought him into contact with the *Geheime Legationsrat* [privy councilor] von Kusserow, of the Foreign Office, and the *Geheime Kommerzienrat* Adolph von Hansemann, who directed the Discount Credit Bank. The latter had founded a "Syndicate to Prepare and Establish a South Sea Island Company," and attached great importance to a discussion of his secret plans with such an expert on the area as Finsch. A meeting with them led ultimately to a secret mission, commanded by Finsch, to annex for Germany as much of eastern New Guinea as possible, together with New Britain and neighboring islands, thereby forestalling action in this direction by Australia.[11] The flag raising at Konstantinhafen on October 17, 1884, was succeeded by similar demonstrations in New Britain and back in New Guinea at the newly christened Finschhafen. Imperial sanction for an accomplished fact was now in order, and on December 19, 1884, Bismarck notified the powers of a German protectorate in eastern New Guinea, New Britain, New Ireland, and other island areas. More flag raising ensued in the New Guinea portion, with the British joining the race, until by governmental compromise in April 1885 a border was established between Kaiser Wilhelm Land (so named by Finsch and his companions) and Papua (in the strict sense) that endured to the end of the German colonial empire in 1918.

Finsch's disillusionment came quickly. Returning to Berlin on September 2, 1885, he found that the financial moguls who had initiated and supported his adventure were guided by principles very different from those of the scientists whose company he had forsaken when dazzled by ambition. He was kept on until June 1886 as an "adviser" whose advice was not followed, and then fobbed off with a contract so glaring in its tactlessness and unscrupulous avarice that Finsch "as a scholar could not contemplate such an offer."

Seduced by the rustling of palm trees, he had abandoned all the solid foundations of his existence for dubious rewards; now he had to make his way as a freelance writer and try to earn some money from his travel experiences.

In his hasty journeys as the "Conqueror of Kaiser Wilhelm Land" he had almost completely given up collecting animals, but as luck would have it he did not come home empty handed. In December 1884, while he was at Cooktown, where his steamer the *Samoa* had gone to send off urgent telegrams to Berlin, he met his old acquaintance Carl Hunstein, just returned from New Guinea, where he had collected and made a daring expedition from Port Moresby into the Hufeisen Mountains, penetrating for the first time into the rhododendron zone.

Carl Hunstein had led an adventurous life. Born about 1843 in Friedberg (Hesse), he had emigrated to America as a young man and then had gone on to Queensland to try his luck as a gold prospector. When the Australian papers announced in 1877 that the suitably named Andrew Goldie had discovered gold near Port Moresby, Hunstein was one of the first to cross over in 1878 and prospect as far as the Hufeisen mountains — always in vain. So there was nothing left for him but to collect natural history and ethnographic specimens, at first as a companion to Andrew Goldie during his further travels to Milne Bay and the d'Entrecasteaux Islands. According to Finsch, Hunstein "should be credited with the discovery of most of the new species of birds sent by Goldie to Australia and England."[12] In 1882 Finsch met him at Port Moresby, as we have seen. The next year Hunstein made himself independent; he undertook his successful trip into the Hufeisen mountains in 1884 entirely at his own expense. Finsch, finding him in Cooktown, immediately took him aboard the *Samoa* and established him as head of a trading station that he had just set up on the northeast coast of New Guinea. Soon afterward Hunstein entered the service of the New Guinea Company, founded by the Hansemann syndicate and sovereign owner of Kaiser Wilhelm Land by imperial letters patent. On March 13,

1888, he was drowned on the west coast of New Pomerania (New Britain) by a powerful tidal wave, created by a submarine eruption, which destroyed an entire expedition sent out by the New Guinea Company to establish plantations.

When they met, Hunstein had his bird collection with him, and Finsch, who immediately noticed magnificent new species, bought it on the spot. When he was back in Germany he approached Dr. A. B. Meyer,* the only person there who understood anything of New Guinea birds, who was glad to obtain these sensational specimens for the Dresden Museum. In the same year (1885) Finsch and Meyer wrote up the new birds of paradise in the *Zeitschrift für die gesammte Ornithologie*, edited by Julius von Madarász.[13] The prize items were outstandingly beautiful, the blue bird of paradise (*Paradisaea rudolphi*) and the Princess Stephanie bird of paradise (*Astrapia stephaniae*), both named by Finsch for the crown prince and princess of Austria.

But he worked chiefly on the geography and ethnology of the areas that he had traveled in, especially New Guinea, writing a book, *Samoafahrten* (1888), with an "Ethnological Atlas" after his own drawings appended to it, as well as a large number of articles for newspapers and periodicals, because he had to earn money at all costs.[14]

He had completely deserted ornithology, yet it suddenly provided him with a lifeline. What he had least expected happened, and in 1899 he was offered the post of curator in the bird department of the Leiden Museum, which had become vacant when Schlegel's last assistant, Dr. Johann Büttikofer, a Swiss, left.

Finsch came back an old man, widely traveled, highly experienced, and bitterly disappointed in society, to the same rooms that he had left so lightheartedly in 1864, full of energy and the unsatisfied urge to achieve something. He now looked backward instead of forward; he was happiest when occupied with his memories, and the laborious verification of long-past events had grown more important to him than competition in the arena of the present day. Young men could go and chase new phantoms there. Many of his articles were now entitled "Toward a Catalogue of the Ornithological Department,"[15] and were intended finally, though much too late, to demonstrate the full value of the achievements of Reinwardt, Kuhl, van Hasselt, Boie, Macklot, Müller, and Forsten.

And yet he was once more to taste the pleasure of enthusiasm for something new. In 1901 a young German, Max Bartels, came to his

*Adolf Bernhard Meyer, born at Hamburg in 1840, traveled twice to the Malay Archipelago (1870–1873), on the second occasion going as far as northwestern New Guinea. In 1874 he was named to succeed Ludwig Reichenbach as director of the Dresden Zoological Museum. Distinguished for his expert knowledge and great organizational gifts, he helped the museum to become world famous, particularly in the fields of the zoology and ethnography of the Malay Archipelago and the New Guinea area. Dismissed in 1905 because of his unstable character, he died in obscurity on February 5, 1911, in Berlin.

attention. Bartels had emigrated to Java in 1895 and, as a worthy successor to Boie and Bernstein, had elected to do research on the Javanese avifauna and its habits. Finsch, happy at the signs of new and hopeful exploration, was at his most cheerful when studying Bartels's reports and shipments from Java, and his last ornithological contribution, written in 1910 when he was 71, was "Über die Geschlechtsverchiedenheit von [On the sexual differences in] *Caprimulgus Bartelsi*."[16]

At that time Finsch was no longer in Leiden. Homesickness, stronger than his devotion to ornithology, had drawn him back to Germany, where in 1904 he had been offered a modest post as head of the ethnological department in the Brunswick city museum. Here he wrote a book much esteemed by ethnologists, *Südseearbeiten* (1914). Then came the war. Fortunately for him, before he realized that Germany had lost her huge colonial empire, including his own Kaiser Wilhelm Land, he died on January 31, 1917.

Arnold Jacobi, who knew Finsch in his last years, described him as a stately figure even then, upright and proud, content simply to know the value of his own achievements, even though the world withheld well-merited recognition.[17] And this it had most signally done! The *Journal für Ornithologie* said of him in the year of his death: "Finsch belonged to a vanished past. The present generation of ornithologists, following other paths to other goals . . . has almost forgotten the man who, in his own time, did so much so well for the development of ornithology in Germany. They have found themselves new leaders."[18]

Even at the turn of the century, when the battle over the definition of species flared up again, Finsch was regarded as an archreactionary by the younger generation, who paid no attention to his earlier achievements. Uninterested in new ideas, defending the point of view that he had adopted 40 years before from Schlegel and Hartlaub, he grimly drew his sword against all innovators who, he was convinced, threatened to bring nothing but catastrophe to ornithology. In the *Ibis* (1904) he warned:

> With this load of always increasing subspecies, even our Palaearctic birds are now likely to be overwhelmed, and how much more the exotic forms, if the splitting into "subspecies" goes on at the rate now practised by the modern school . . . The members of the "old school" may therefore continue on the way that they think to be correct.[19]

Once more, just as when Darwin's *Origin of Species* first appeared, a new battle cry had split the ornithological ranks into two camps. One was for and the other against trinomial nomenclature and the distinguishing by name of slight variations from the "typical species." Under cover of a dispute about forms, the quarrel between ideologies continued as doggedly as ever.

13

The Effect of the Theory of Evolution

By 1900, European ornithologists were aroused to take part in the battle for the last positions to which the opponents of the theory of evolution and their sympathizers had withdrawn. Once the systematics of the higher categories had been revised according to the historical point of view, there was no choice but to treat the concept of species in the same way; the revolution had to be carried into the peaceful regions where specialists were working on actual specimens and were studying primarily the differences among closely related "forms."

Fewer and less-impassioned warriors had struggled to rearrange the higher groups, orders and families, soon after Darwin's work came out. The natural philosophers had themselves reduced their preconceived ideas to absurdity and could provide no serious opposition. Next came those systematists who had long given themselves to the devoted task of patient induction, and had succeeded in organizing the cornucopia of forms into a number of clearly defined bundles of widely varying size. After 1859, most of these abstract groups could be placed without any difficulty in the hypothetical scheme of the historical relations of organisms, and could even serve as a basic outline for it.

Before Darwin, the efforts to discover the plan behind the diversity of phenomena had been futile; even quite recently there had been authors who followed Buffon in stigmatizing any attempt at systematics as arbitrary. Chief among them was the intelligent and witty Scot, William MacGillivray, equally experienced as anatomist and biologist. In the introduction to his *History of British Birds* (1837) he said:

> To present affinities, species would require to be fixed in empty space, and represented by forms bearing no resemblance to birds,

> but having parts more or less elongated, to meet parts of other near and distant species; so that such an arrangement, were it made by a being who understood all the relations of the species, would, by exhibiting such multitude of reticulations, be just as unintelligible to us as is the order of things studied in the economy of nature such as we see it. When objects, then, are placed in linear series, whether directly extended, circularly bent, or otherwise disposed, they cannot manifest any other relations than those that refer to the structure of a single organ;* and thus, all arrangements must be merely artificial, as we have no means of disposing our descriptions of species otherwise than in elongated series. The only remedy is to indicate in words all the relations that have been observed, while some principal organ is chosen as a general medium of connection.[1]

But such skepticism could hardly be brought into line with experience; the morphologists therefore did not allow themselves to be distracted in their search for a truly natural system. Particularly outstanding among the ornithologists of this persuasion was Strickland, as proponent of an idealistic morphology based on induction. In 1840, in Charlesworth's *Magazine of Natural History*, he attempted, like Buffon, to explain the existence of related groups by the hypothesis that, comparatively speaking, the first Creation had produced only a few important types of structure or centers of creativity, from whose various modifications are derived the characteristics of the innumerable species now in existence. Therefore species were created not absolute, but relative, in regard not only to their particular prescribed manner of living but also to other species destined for similar though not identical conditions. In 1844, at a meeting of the British Association for the Advancement of Science, Strickland showed a diagram purporting to represent the relationship of all the groups of birds.[2] Categories of equal rank were represented as figures differing from one another in form and size, and they were placed closer together or further apart according to the degree of their supposed relationship. These groups could perhaps be regarded as so many realizations of an abstract idea, which the Creator did not abandon even in the most unusual and eccentric modifications of the form. Clearly (as Strickland said in *The Dodo and Its Kindred*, London, 1848), we stand here before unknown hieroglyphs, which certainly have a meaning, but one that we are only just beginning to be able to decipher.

*This conception can be found as early as Johann Hermann: "The arrangements of genera will differ one from another, depending on whatever resemblance you start with, and whatever characteristic you adopt as the basis of your system" (*Tabulae affinitatum animalium*, Strassburg, 1783, p. 34). He in turn must have taken it from Félix Vicq d'Azyr, *Table pour servir à l'histoire anatomique et naturelle des corps vivants* (Paris?, 1774).

As the morphologist-systematists of the period learned more and more to distinguish the essentials (constants) from the inessentials (variables), they were slowly nearing their goal. As early as 1829, Nitzsch, outstanding both for his minutely detailed research and for his acute reasoning, aligned the swifts with the hummingbirds and removed them a long way from the swallows; Blyth (1838) and Nitsch (1840) recognized the close relationship of hornbills (Bucerotidae) and and hoopoes (Upupidae); and L'Herminier (1827) and Nitzsch (1840) discovered relationships between the gulls and the shorebirds. Even the dodo (*Didus ineptus*) was assigned in 1848 to its appropriate slot, because Strickland declared it to be a large flightless pigeon, after it had earlier been defined by Vigors (1823) as a link between *Struthio* and *Crax*, by Blainville (1835) as a relative of the birds of prey, by Owen (1845) as a relative of the vultures; while J. F. Brandt, who still clung to J. Hermann's concept of networks of affinities, decided in 1847 that the dodo was a pigeonlike wader, linking ostriches to pigeons, and belonging to the same rank as cranes, storks, snipes, ibises, and rails. Real progress certainly took a good deal of time to overcome opposition; for decades most ornithologists continued to struggle against the separation of the swifts from the swallows, and in 1854 Schlegel maintained that he could see no affinity between the dodo and the pigeons; it was much nearer to the "ostrichlike" group (where Willughby had already placed it in 1676). The systematics of authors without anatomical training for a long time lacked any basic scientific principles. They grouped "birds into families just as they pleased, and it is not surprising if their families have almost no character and are simply irrational heaps, or if these heaps are differently arranged by different authors." Johannes Müller, who wrote this in 1845,[3] tried through research into the vocal organs to make definite progress in dividing the large agglomeration of "passerines" into natural sections. His success stimulated Cabanis, in 1847, to look for external characteristics corresponding to the internal ones. Bonaparte, as an evolutionist, revised the work of the pre-Darwin period and set up the dividing lines between orders and families with delicate discretion in systematics; by separating them into "parallel rows" according to the natural data, he took into account a greater number of factors than his predecessors had done.

The existence of more or less clearly defined groups helped Darwin in 1859 to understand their causal relations. By observing the world of organisms historically, he succeeded in harmonizing Leibniz's law of continuity with the results of research into natural history. The reality of his success in deciphering the hieroglyphics before which Strickland had stood helpless was acknowledged within a few years by all comparative anatomists and paleontologists. The proof provided by *Archaeopteryx*, discovered in the Solnhofen quarries in Bavaria in 1861, also contributed greatly. Andreas Wagner, who promptly announced this extremely significant find in the reports of the meetings of the Bavarian Academy of Sciences, warned in vain against using it to support erroneous conclusions (that is, as a proof of the theory of evolution). A. Oppel and Haeckel employed it to demonstrate the evolutionary relationship of reptiles and birds; Huxley classified them together as "Sauropsidae," and in 1867 declared: "Birds are greatly modified reptiles."

From that time on the systematists knew that their efforts were aimed at discovering a real evolutionary connection among organisms, instead of an abstract "logical" one. Historical morphology joined formal morphology, and the difference between *homology* (affinity) and *analogy* developed a new meaning.

As a result of this new mode of thought, Thomas Huxley, stimulated by Haeckel's example (*Generelle Morphologie*, 1866), drew up the first avian genealogical tree (*Proceedings of the Zoological Society*, 1868): a "phylum" of "gallo-columbine" birds, whose groups were represented as branches of a trunk with roots in the "carinates." Fürbringer subsequently observed: "Genealogical trees at the same time supersede the essential significance of the concept of parallel rows. Insofar as the principal concern is genetic descent from common ancestors . . . the parallelism of the rows becomes secondary and their common origin primary."[4]

At first, however, the methods of comparative morphology did not change, and the success of the new efforts toward a natural system did not meet expectations. Huxley's attempt at reform ("On the Classification of Birds," *Proceedings of the Zoological Society*, 1867), was soon revealed as a failure, because the author had assigned to an arbitrarily chosen characteristic, the structure of the bony palate, a taxonomic value (a degree of phylogenetic persistence) that it did not possess to such an extent. By underestimating the fact that every structure is influenced by its function, and that the study of form must proceed together with the study of function, the young English anatomist Garrod fell into the same error in 1874, and his alterations of the system (according to the development of the carotids, the *musculus ambiens*, and the adductors of femur and tibia) were in general no improvement. So the experienced Swedish ornithologist Sundevall* was quite right when he

*Carl J. Sundevall (1801–1875) was one of the most highly regarded ornithologists of his century. After working in 1827 and 1828 as a doctor on a merchant ship trading to Calcutta, he became much interested in exotic ornithology. In 1838 he became head of the vertebrate section in the Royal Museum, Stockholm. His learned curator was Friedrich Wilhelm Meves (born at Delligsen, near Brunswick, in 1814), who worked with him from 1842 until he retired in 1877. During Sundevall's administration Johan August Wahlberg, a Swedish ornithologist, made the two journeys to South Africa (1839–1845, 1853–1856) that resulted in the unmasking of Levaillant (Chapter 5).

wrote (in *Methodi naturalis avium disponendarum tentamen*, Stockholm, 1872–1873): "It is a fairly common, but absolutely false, belief that anatomical, or internal, characteristics are always better or more reliable than external ones." In setting up his own avian systematics he was therefore eclectic and tried to rely on a large number of external as well as internal characteristics, but because years of habit induced him to trust the former more than the latter, his arrangement was also far from perfect. Like Huxley, Sundevall assumed that relationship must be equated with common inheritance, so that species are indefinitely changeable, a view that "probably most natural scientists share."

Certainly this was true in Scandinavia and in England (where, however, the Marshall brothers, as late as 1871, in their handsome *Monograph of the Capitonidae*, devoted many folio pages to the "refutation" of the theory of evolution). Among leading German ornithologists the opponents of Darwin still had the upper hand; Cabanis, Schlegel, Heuglin, E. F. von Homeyer, B. Altum, and W. von Nathusius constituted a variegated coalition that believed it could block the progress of the "world-shaking" doctrine. The proponents of this view saw homology and analogy as concepts equal in importance, because they regarded all kinds of similarityas merely formal data, and did not hesitate to confound systems of natural philosophy with inductive systems. Their sole aim was an arrangement of categories to facilitate quick identification. This appears most crudely in the two volumes on birds in *Thierleben* [Animal life] (1866, 1867) by Alfred Brehm. This skillful popularizer of zoology was so helpless when faced with the results of earlier endeavors that he decided on the categories of crackers (parrots, finches, and corvids), catchers, searchers, runners, and swimmers, and in so doing unconsciously approached the sixteenth-century ornithologists Belon and Aldrovandi. Because he relied principally on the natural philosopher Reichenbach for details, we read in Brehm that the penguins "appear to provide the transition between birds and fish." Astonishingly enough, he acquired disciples among ornithological specialists; Reichenow followed him when he wrote, in 1893:

> An ornithological system must be primarily and above all a help in learning to know the kinds of birds now in existence. It must provide a general survey of them by a comprehensible arrangement of the forms into groups, bringing order out of chaos. The purpose of systematics is therefore chiefly practical.[5]

This sort of system, "corresponding most nearly to our thought processes," he called "logical."*

*To the end of his days Reichenow clung to his "logical" system, which arranged birds in six entirely arbitrary categories: "shortwings, swimmers, stiltbirds, skinbills, yoketoes, and treebirds." (See his *Handbuch der systematischen Ornithologie*, Stuttgart, 1913–1914.)

Even so, this Berlin ornithologist made an effort to arrange at least the genera, if not the orders and families, according to their probable historical relationships. But the declared enemies of evolution were not prepared to concede even this much, and were guided only by easy external characteristics. Their most influential spokesman was Eugen Ferdinand von Homeyer, whose *Verzeichniss der Vögel Deutschlands* was published in Vienna in 1885 by the "Permanent International Ornithological Committee" (meaning Dr. R. Blasius and Dr. G. von Hayek) and recommended for general use, although it was 50 years behind the times and went in for such combinations as "V. Order, Scansores: Families Picidae, Sittidae, Certhiidae, and Upupidae," a completely heterogeneous assemblage that derived from Willughby in 1676 (!). But this anachronism was outdone at the same period in France by Léon Olphe-Galliard's (1825–1893) four-volume work, *Contributions à la faune ornithologique de l'Europe occidentale*, published between 1884 and 1891. Carus's natural philosophy is exhaled from its pages like the chill vapors of the tomb. The author used the classification that he had outlined in 1857 in his "Versuch eines natürlichen Systems der Vögel,"[6] with only minor alterations.

Germany's reputation as the nursery of ornithological systematics, still surviving with the memory of Illiger's, Nitzsch's, Johannes Müller's, and Cabanis's earlier achievements, would have ended if such products of intellectual rigidity or feeble resignation had not been offset, at the right moment, by a German scholar's book that completely outstripped everything previously done in this field: Max Fürbringer's *Untersuchungen zur Morphologie und Systematik der Vögel.* No more lasting commemoration of a fiftieth anniversary could have been produced by the Amsterdam Zoological Society, Natura Artis Magistra, than these two thick folio volumes, which appeared in 1888 as part of its *Bijdragen tot de Dierkunde.*

Max Fürbringer was born at Wittenberg in 1846, and after finishing school at Gera went to the University of Jena to study mathematics and natural history. He was persuaded by Carl Gegenbaur to change to medicine, and was trained by him in comparative anatomy. After taking his medical degree at Jena in 1869, he followed his teacher to Heidelberg in 1874 as assistant professor of anatomy. In 1879 he accepted a call to Amsterdam as professor of anatomy; in 1888 he succeeded Oskar Hertwig at Jena, and in 1901 Gegenbaur at Heidelberg, where he died on March 6, 1920.

Fürbringer's interest in ornithology can be traced back directly to the influence of C. L. Brehm. The intermediary was Karl Theodor Liebe (1828–1894), who in 1878 (1876) founded the Deutscher Verein zum Schutze der Vogelwelt [German Society for the Protection of Birds], and edited its monthly periodical circumspectly from 1884 until his death. His ornithological interest had been aroused when, as a boy, he left his village of Moderwitz, near Neustadt on the Orla, to visit Brehm, the "bird parson," in Renthendorf. In 1861 Liebe began to teach mathematics and natural history in the Gera Gymnasium, Ruthenum Illustre, where young Fürbringer was among his pupils, and was soon allowed to help in looking after the large aviary belonging to his much admired teacher. This won him over to ornithology.

When Fürbringer began his labors, he was most thoroughly equipped with all the specialized knowledge necessary to evaluate the previous attempts at avian classification, because he was himself a comparative anatomist trained in the best school (that of Gegenbaur), his judgment was refined by long experience, and he had carried out a large number of research projects involving tedious and minutely detailed work, particularly those that seemed necessary to clarify genealogical problems. No one before him — and one can risk saying no one after him — had so completely marshaled the characteristics useful for classification, and so painstakingly evaluated their phylogenetic worth. He constantly kept in mind the connections between form and function, and therefore was able prudently to avoid the traps provided for systematists by the isomorphs (that is, morphologic forms that look alike but have different ancestry, the resemblance arising secondarily from the same functional influence); he was also already aware of the phenomenon described today as the "allometric growth of parts." By introducing four higher categories (Ordo, Subordo, Gens, Familia) into his system, he achieved a classification more finely articulated than that of his predecessors, because the category of "Gentes" corresponds to the orders of other authors. He recognized no fewer than 45 Gentes, whereas Linnaeus and Cuvier had 6, Sclater (1872) 21, Sclater (1880) 26, and Reichenow (1882) 17. More hypothetical were his groupings of Gentes into 24 suborders, and of these into 7 orders. One of Fürbringer's many important reforms deserves to be particularly noted here; he abandoned the traditional division of birds into Ratites and Carinates (deriving from Merrem in 1812), and postulated that the "keelless birds" make up an entirely heterogeneous complex of species that secondarily became flightless. This group he split up among several orders: Struthiornithes, Rheornithes, Hippalectryornithes, and in part Alecterornithes (kiwi).

Almost contemporaneously another pupil of Gegenbaur's, Hans Gadow (1855–1928), was working on another system for birds, because in his new post as curator of the Strickland Collection at the University of Cambridge he had begun to write the section "Vögel" for Bronn's *Classen und Ordnungen des Thierreichs* [Classes and orders of the the animal kingdom], and was therefore carrying out detailed research on the morphology of birds. His results, published in 1893, coincided in many points with Fürbringer's, because over the years these two friends had been exchanging opinions constantly. Only a few of Gadow's changes in Fürbringer's system can be described as improvements.

In fact, most ornithologists were hardly concerned with the methods of arranging the higher groups; they were interested in the "real" units, the species. So naturally the systematics of species became the arena where opinions clashed most violently during the advance of Darwinism.

Hartlaub tried to impress on ornithologists that "in systematics everything should be avoided that threatens to weaken the stability of the concept of species" (*Archiv für Naturgeschichte*, 1852), and we have already seen (Chapter 11) how eagerly Schlegel and J. H. Blasius followed the advice and passed it on, even after 1859. The idea that, if the theory of evolution were victorious, "every investigation of species would become useless and impracticable"[7] was particularly widespread among the older generation of German ornithologists, and may have contributed to the extreme unpopularity of Darwinism among them. One of their loudest spokesmen, the oologist Wilhelm von Nathusius (1821–1899), maintained that "for Darwinians the concept [of species] has lost any kind of meaning, so that their employment of it in any way whatever is simply illogical."[8] Incensed against the oologist Friedrich Kutter (1834–1891), who in 1877 and 1878 published "Betrachtungen über Systematik und Oologie vom Standpunkt der Selectionstheorie" [Observations on systematics and oology from the point of view of the theory of natural selection],[9] this epigone of the natural philosophers' century and defender of Oken's holism* thundered in 1879: "For all practical purposes Darwin's views on the personal relations between man and his Maker are exactly the same as those of the atheists . . . The denial of planned development involves the denial that the development of presumptive inferior beings into men was designed by the Creator."[10] Homeyer,† who had grown up during the

*One of the results of his premises was that Nathusius, though a painstaking researcher into the structure of eggshells, wrote a peculiar description of the development of the egg. "When a bird's egg leaves the oviduct it is a homogeneous organism, a fully organized individual, a complete bird, merely in an earlier developmental stage, that of the egg" ("Die Structur des Vogel-Eies und deren Beziehungen zur Systematik," *Journal für Ornithologie*, 19, 1871: 242). A further result of his theoretical position was that (like L. A. F. Thienemann, the founder of "scientific" oology) he was convinced of his ability to distinguish eggs even of closely related species (like carrion and hooded crows) from each other under all circumstances, because their progenitors (and thereby the eggs themselves) had been independently created and therefore could not be related by transitional stages.

†Eugen Ferdinand von Homeyer (1809–1889), a Pomeranian landowner, had worked on local ornithology from his boyhood (*Systematische Übersicht der Vögel Pommerns* [Systematic survey of the birds of Pomerania], Anklam, 1837). With the help of A. Tancré, an Anklam dealer in natural history specimens, and of many others, he made a collection of palaeartic birds that finally contained 7000 specimens and many nests and eggs, and assisted him in making an exact study of "local forms." After his death the collection went to the Natural History Museum in Brunswick. In his old age he became a passionate polemicist, far too often blind to other people's merits, and in his book *Die Wanderungen der Vögel* [The migrations of birds] (1881) he tried not only to refute Palmén's theory of migration routes by

period of eccentric speculation by the natural philosophers, and was therefore hostile to any theories at all, expecting salvation from blind induction alone, used even more forceful language. In *Die Wanderungen der Vögel* (Leipzig, 1881) he opposed to the theory of evolution Cuvier's theory of catastrophe, advancing the hypothesis

> that, at the time when our world was being created, an enormous number of different beings came into existence, that these multiplied so long as they found new means to subsist, that with changed conditions on the earth new species arose and those disappeared that no longer found conditions suitable for their survival. This explanation seems far more natural and more in accordance with the facts than the hypothesis that all later forms have developed out of the species that appeared originally.

For the opposition, then, the fixity of species remained a sacred dogma. "Those of us who are not Darwinians still cannot escape from the definition given by Linneaus, that 'species is the form created by God in the beginning,' " explained Nathusius in 1874.[11] His opponent Kutter, on the other hand, viewed "species" as "a collective term, serving essentially practical purposes."[13] Sundevall had already expressed a similar opinion in 1872; Nicolai Severzov, the great Russian explorer, agreed with him when he acknowledged in 1873 that earlier he had been "a fanatic for the most rigorous justification of species, an inquisitor into bad species,"[13] but that later he had discovered that Nature has no sharp distinctions between species and variety. Another supporter of this view was the Finnish zoologist J. A. Palmén.[14]

But at first the supporters of the theory of evolution in Europe could progress no further than these general observations. Their conversion into a new, solidly based nomenclatural practice was reserved for American ornithologists.

After Bonaparte had returned to Rome from the United States in 1828, transatlantic ornithology started a tremendous boom, noticeable first in Philadelphia. There the Academy of Natural Sciences, founded in 1812, began to construct a new building in 1829, to keep pace with the rapid increase in the library and collections. The bird collection proliferated remarkably after John Cassin (1813–1869) became curator of the Academy in 1842, because he intended to create a museum of

---

referring to his own wide experience, but also took the opportunity for a zealous attack on Darwinism. He was an old friend of the Blasius family, and in 1875, when the Deutsche Ornithologen-Gesellschaft was united with the Deutsche Ornithologische Gesellschaft zu Berlin as the Allgemeine Deutsche Ornithologische Gesellschaft, he was elected president, an office which he resigned in 1883 after violent quarrels.

world importance on the European model. In September 1846 a rich patron of the museum, Dr. Thomas B. Wilson, bought for 50,000 francs in Paris the famous bird collection belonging to Victor Masséna, Duc de Rivoli, which consisted of 12,500 mounted specimens, and gave it to Philadelphia; not satisfied with that, in 1847 he also bought in London, for the high price of £1000, John Gould's Australian birds (almost 1900 specimens), including all the types illustrated in *The Birds of Australia;* and in September 1850 he obtained one of the most important European egg collections, that of Oeillet des Murs. Through these and other acquisitions, the bird department of the Philadelphia Museum became in 1856 the largest and richest in the world, with 29,000 specimens (23,000 in glass cases), far surpassing even the Leiden Rijksmuseum. Cassin was therefore able to become one of the greatest experts on exotic ornithology and to determine incoming collections from all over the world, particularly those from West Africa made from 1855 to 1859 by the Frenchman P. B. de Chaillu.

But Washington, not Philadelphia, provided the new stimuli for which ornithology must be grateful to America.

The organizing ability and farsightedness of Spencer Fullerton Baird (1823–1887), who in 1850 had become assistant secretary (scientific head) of the Smithsonian Institution, founded four years before, ensured that in a short time science derived great benefit from James Smithson's vast bequest. Baird's plans concentrated on the detailed exploration of the North American continent, with particular attention to mammals and birds — a task comparable only to that which Pallas had set himself when he began to study the fauna of the Russian Empire. Well-planned expeditions accompanied the Pacific Railroad explorations, and soon furnished the United States National Museum in Washington, which Baird had created, with large numbers of bird skins from many parts of the United States. While working on them Baird became aware as early as 1859 of certain laws governing geographic variation, and especially one phenomenon known today as Bergmann's Law.* He noticed that as a rule, in bird species that are widely distributed along the meridians, their bodies become larger as they are nearer the north and smaller nearer the south. "The same principle applies, though in less marked degree, to an increasing altitude in the same latitude."[15]

*So called after the physiologist Carl Bergmann, who in his treatise *Über die Verhältnisse der Wärmeökonomie der Thiere zu ihrer Grösse* [On the relation of temperature to size in animals] (1847), after considering physiological evidence, arrived at the theory: "If there were genera whose species were differentiated as widely as possible only in size, then the small species would all need a warmer climate, to a degree exactly corresponding to the difference in size." The first person to observe this and attempt to explain it physiologically was not, however, Bergmann but the Emperor Frederick II (about 1240); see Chapter 1.

From that time on Baird paid particular attention to the study of such laws. Because they could be clearly deduced only from a large quantity of material, he arranged for the collection of long series of the same species from many localities on the North American continent. The importance of this undertaking was recognized during the same period by Hermann Schlegel, though he could produce no general conclusions because he was pursuing it in an area of tropical islands and islets. Baird, on the other hand, was led from one discovery to another. First of all he established that in the most southerly part of the range of some species

> there appears to be a tendency . . . to absolute increase of the size of the bill, even with diminution in general bulk . . . Another fact which may be mentioned in reference to birds of the different provinces, is that specimens from the Pacific coast are apt to be darker in color than those from the interior, the latter frequently exhibiting a bleached or weather-beaten appearance . . . In a careful study of large series of birds of any two representative species collected near the line of junction of their respective provinces, a combination of characters of both species will often be met with, explicable only on the supposition of the hybridization of the two.[16]

Further contributions to the subject followed closely and facilitated important theoretical studies. Robert Ridgway (1850–1929), a student of Baird's, who was a zoologist attached to the United States Geological Survey of the Fortieth Parallel, tackled the problem of the dividing line between species and variety. In 1856 Bonaparte had taught that "the transitions between the different races and their type are the best evidence we can supply to set aside putative species, which are to be relegated to races."[17] Ridgway, on the other hand, stressed in 1869 that

> it does not follow . . . [that,] because we have a series connecting by a gradual transition two extremes, we are to consider the whole as one species, the discrepancies indicating different varieties. The difference between these extreme examples is often too great to admit of this; and when we have traced a species through all its variations to a certain point where the discrepancy from the typical style is too great and uniform to be accounted for by any physical cause, it becomes us as naturalists to assign to such extreme conditions a specific rank.[18]

Another young ornithologist, Joel Asaph Allen (1838–1921), who like Ridgway was destined soon to be highly regarded and very influential, published in 1871 an important treatise "On the Mammals and

Winter Birds of East Florida."[19] In a section of 64 pages entitled "On Individual and Geographical Variation among Birds, Considered in Respect to Its Bearing upon the Value of Certain Assumed Specific Characters," following Baird's 1858 pattern, he published detailed statistics on a large quantity of material, to provide evidence for the individual and geographic variation in size of all measurable exterior characteristics. He found that Baird's "climatic laws" were justified and tried to explain their causes. For example, the westward increase in intensity of color "seems to be also coincident with the increased humidity to the westward, the darker representatives of any species occurring where the annual rainfall is greatest, and the palest where it is least." If the systematists were to name all these local variants, the number of distinct species would become limitless. Would it not be better to refer, in the description of each species, to the known laws of climatic variation, and to give an indication of the tendencies to variation and the degree to which they are developed? "The average characters being given, a line or two would suffice for a statement of [the species'] variations, both geographical and individual."

Here Allen had reached the position already indicated by Gloger to the ornithological systematists in 1833, when he devoted a small book to *Abändern der Vögel durch Einfluss des Klimas* [Change in birds produced by climatic influence]. However, the work of this author, who was full of ideas but lacking in experience, had absolutely no effect because he attempted to prove his audacious hypotheses by unsuitable examples.

Following the path already cleared, Allen published a second important piece of work in 1877, "The Influence of Physical Conditions in the Genesis of Species."[20] In it he attacked the theory of natural selection (or the deductive components of Darwin's theory of evolution), and tried to prove from examples of geographic variation in North American birds and mammals that other influences than selection are powerfully at work in the differentiation of species and subspecies, and that the effect of environmental factors ("geographic causes") is therefore greater than had previously been admitted. Among these immediate influences he included not only climate (temperature, humidity), but also "habits and food, and other conditions of life," as well as the "use and disuse of organs, through changes of habit resulting from changed conditions of environment." In short, Allen declared himself a decided Lamarckian. Following this thesis, he contradicted Baird's attribution (1866) of transitions between subspecies to the crossing of previously isolated populations, and argued, "Let . . . some of the connecting links become extinct, and these now intergrading forms would be resolved into distinct species," an opinion that long held sway among American ornithologists.

Meanwhile the question of nomenclature, raised by Allen in 1871, had been solved by other American ornithologists, but not along Allen's lines. Elliott Coues (1842–1899) recommended as early as 1871 that "geographical and some other differentiations" of species should be named,[21] and in his *Key to North American Birds* (1872) described all stages between two extremes as "subspecies or geographical races," adding their names, prefixed by "var.," to the species name. This practice was followed by Baird, Brewer, and Ridgway in their *History of North American Birds* (1874), and remained valid until Ridgway (*Nomenclature of North American Birds*, 1881), by dropping the "var.," took the decisive step toward true trinomial nomenclature. As we have noticed above, Schlegel had already resorted to trinomial nomenclature, but his principles were fundamentally different. He adhered to the theory of fixity, regarding his "conspecies" as stable variations from the "type" of the species, without letting his judgment be affected by the geographic distribution of the forms that he had compared. The Americans, on the other hand, were disciples of the theory of evolution; for them the third name served to designate the species that were coming into being ("nascent species"). Ridgway in 1881 set up the following rules:

> Every form whose characteristics bear unmistakably the impress of climatic or local influences, gradually less marked toward the habitat of another form, with which it thus intergrades, and all forms which certainly intergrade, no matter how widely distinct the opposite extremes may appear . . . together with intergrading forms whose peculiarities are not explained by any known "law" of variation, have been reduced to subspecific rank. On the other hand, where the difference between allied forms is slight, but at the same time apparently constant, and not necessarily coincident with a difference of habitat . . . specific rank is upheld.[22]

These rules were adopted, with exemplary unanimity, in 1885 by the committee on nomenclature of the American Ornithologists' Union (founded in 1883). The guiding American principle, reduced to a slogan, which long remained in force, was: "Intergradation is the touchstone of trinomialism."[23]

Although American ornithologists, under intelligent leadership, had united in adopting nomenclatural procedures to which the recognition of the theory of evolution must inevitably lead, in Europe everything remained in the old confusion. Some scholars, like Finsch, Gadow, and Radde (who proudly declared that in this matter he was Pallas's disciple), kept rigidly to the traditional concept of species and anxiously avoided including "bad" species, that is, slight variations from the "type," in their systems. Others, like Sharpe, Cabanis, Severzov, and Salvadori, gave a name to every "constant" variation that they discovered, even the smallest. But all of them were unanimous on one point; as a matter of principle, they did not disturb Linnaeus's binomial practice. The result was real chaos; the species of the two feuding groups were no longer at all the same, and it was doubtful which method was worse — the "old," which considered variations of a "species" unworthy of a name, but nevertheless made it easier to survey the whole; or the "new," which in trying to name every difference occurring in nature demolished the concept of species, and threw the unorganized and unrelated remains onto a trash heap.

Not until 1887 or thereabouts did a European ornithologist appear who was able to argue forcefully against this state of affairs, and who recommended the adoption of American principles as a certain cure. This was Henry Seebohm, a widely traveled and highly experienced English private collector. He was born in 1832 at Bradford, Yorkshire, into a Quaker family. As a Sheffield manufacturer he acquired a considerable fortune, and then devoted himself more and more to his hobby of palaearctic ornithology. He himself, on extensive journeys, laid the foundation for his large collection of bird skins, which he considerably augmented by purchase (the collections of Swinhoe, Harting, Shelley, Pryer, and many others). The first journey took him, in 1870 or 1871, to the Mediterranean, where he traveled in Greece and Asia Minor with the curator of the Athens Museum, Dr. Theobald Krüper (1829–1921), who came from Ückermünde. In 1875 he went with J. A. Harvie-Brown to the Pechora, and in 1877 through the Yenisey Valley from Yeniseysk to Gol'chikha, describing these experiences in two books, *Siberia in Europe* (London, 1880) and *Siberia in Asia* (London, 1882), which he later combined, revised, and reissued in one volume as *The Birds of Siberia* (London, 1901). The importance of geographic isolation for the differentiation of subspecies, and the fact of their interbreeding in secondary contact zones, became clear to him especially in the Yenisey area (from the examples of black and gray-headed goldfinches, hooded and carrion crows, and so on). In many of Seehohm's publications we note his efforts to find the right solution, unconcerned with the "hereditary conservatism" and "inherited superstitions" of his fellow countrymen. For instance, he combatted (first in the fifth volume, which he compiled, of the *Catalogue of the Birds in the British Museum*, 1881) the prevailing opinion that structural and external characteristics (forms of wing, tail, and beak) are phylogenetically older than characteristic patterns of plumage, maintaining that the opposite is true. This thesis, violently attacked at first, but used by Oates by 1890 as the basis for a new systematics of songbirds (according to the coloring of the immature plumage), revealed itself later as helpful; in other instances, however, his fertile ideas led him astray and provided his opponents, especially Alfred Newton, with welcome ammunition. He died while engaged in voluminous writings (including the *Monograph of the Turdidae*, completed by R. Bowdler Sharpe, 1898–1902), on November 26, 1895, in South Kensington, and bequeathed his famous collection of almost 17,000 skins to the British Museum.

In his illustrated book on *The Geographical Distribution of the Family Charadriidae* (1887), Seebohm complained that, though English ornithologists acknowledged the truth of the theory of evolution, they ignored it as a working hypothesis. It was now finally time to distinguish species *in statu nascendi* from sharply differentiated species by giving them names, and no better method could be found than the trinomial nomenclature that had been proving its value for years in America. Then, specifying more exactly his earlier observations in the *Ibis* (1882), Seebohm went on to explain, by examples, the many identifiable stages in the relationship of two closely related forms, through the degree of ability to crossbreed and the success of crossbreeding, from the first beginning of their differentiation to the complete differentiation into two clearly distinguishable species.

Seebohm, influenced by the debates carried on among Moriz Wagner (*Die Darwinische Theorie und das Migrationsgesetz der Organismen* [The Darwinian theory and the law of migration of organisms], 1868), August Weismann (*Über den Einfluss der Isolierung auf die Artbildung* [On the influence of isolation on the genesis of species], 1872), and A. R. Wallace (*Island Life*, 1880), but also particularly guided by his own experience, was the first ornithologist who fully recognized the significance of isolation for the development of species. He insisted that the spatial separation of populations was the *conditio sine qua non* for the division of species. Like many of his contemporaries, who still lacked the help provided by the discovery of gene mutation, he found it difficult to imagine the development of isolated populations. One separate variation existing alongside others would surely be rapidly exterminated by the "swamping effect"; therefore he considered that the same variations in the same localities appear in many individuals at the same time, which cannot be by chance, but must either be effected by the conditions of the environment or arise from an innate tendency in the species itself. Like Wagner, he rejected selection as a corrective factor in evolution and took refuge in Lamarckian and vitalistic ideas.

According to Seebohm, the transition from one subspecies to its neighbor is created chiefly by their crossbreeding in the contact zone:

> The final stages of subspecific forms are represented by species so closely allied that they habitually interbreed whenever the areas of their distribution overlap, and habitually produce offspring of various

> degrees of fertility — a condition of things which, if the fertility continue for a sufficient number of generations, must inevitably produce an unbroken series of intermediate forms. Intermediate forms generally occur in the middle of the area of distribution of a species which has a very wide range, and which has been modified in different directions at each extremity of its range, sometimes to such an extent that the individuals of the two extremes have become specifically distinct from each other, though remaining only subspecifically distinct from the intermediate forms which connect them . . . In practice it will be found that the most convenient line that can be drawn between a species and a subspecies is to regard two forms as specifically distinct, however near they may be to each other, whenever they are *not* connected by intermediate forms; and to regard two forms as only subspecifically distinct, however wide may be the distance between the extremes, whenever they *are* connected by a series of intermediate forms — without reference in the one case to *how* the intermediate forms are produced, or in the other to *why* they are *not* produced.[24]

Only those variations restricted to an isolated area should be called subspecies. "Whatever individual variation be found within the range of a species, if it be not capable of being defined geographically I do not regard it of subspecific value."

Such was the situation of ornithological systematics in Europe when Ernst Hartert began to study its problems and to look for his own path through the tangle of theories and systems.

14

Ernst Hartert (1859–1933)

In 1884 a spark flew over the ocean and kindled a fire in Europe; immediately a general alarm was sounded to preserve the venerable structure of Linnaean nomenclature and, if possible, to smother the conflagration entirely.

In that year Elliot Coues arrived in England as a representative of the American Ornithologists' Union, to awaken interest in the idea of trinomialism and the new American rules for its use. He was warmly received. On Sharpe's recommendation William Flower arranged a meeting of England's leading zoologists in the lecture room of the new Natural History Museum in South Kensington on July 1, at which Coues spoke. But the only person present who sponsored the adoption of the American practice was Seebohm, "with an extraordinary display of acute logic and brilliant dialectic." It was useless, for he did not succeed in dispelling the doubts of the meeting.

In September, Hartlaub, who had read a report of the event,[1] asked the ninth annual meeting of the Deutsche Ornithologische Gesellschaft, of which he was then president, to express its views. Those present approved "unanimously" a resolution drafted by Reichenow

> that the local varieties of many species of birds, showing often slight but constant differences, cannot be regarded as species in the traditional sense, and must be distinguished as subspecies or local races, and therefore trinomial nomenclature cannot be avoided. The meeting believes, however, that the *trinomina* should be employed with the greatest prudence, because they may easily be misused by less critical authors.[2]

Encouraged by this permission, a few German writers (Berlepsch, Hartert, and Reichenow) ventured on a modest trial of the novelty. The time seemed ripe to work out binding definitions for the Deutsche Ornithologische Gesellschaft on the American model, so Reichenow raised the issue at the fourteenth meeting, held at Münster (Westphalia) in 1889; he and the Freiherr von Berlepsch were commissioned to work out proposals and lay them before the next meeting.

How many blind alleys the systematists would have avoided, and how much better the wasted energy could have been used, if the proposal that Berlepsch finally thought of making had been actually offered, "simply to accept the American roles on nomenclature,"[3] because he could not agree with Reichenow. On May 12, 1890, the full assembly, "after hours of often very lively debate, which permitted hardly any hope of a successful resolution of the difficult problem," achieved a compromise and assigned the drafting of it to a five-man committee.[4] It was Berlepsch who revised it, and he was not exactly overjoyed to receive from Berlin the papers on nomenclature a few days before the meeting scheduled for May 12, 1891, at Frankfurt am Main, "with hair-raising corrections by Reichenow, Moebius, and A. B. Meyer." But there was no help for it; he was outvoted and, as reporter, had to serve up the watered-down version himself. It read in part:

> The trinomial nomenclature of subspecies may be permitted in certain instances to facilitate study . . . Whenever a third name is allowed, it will, for practical reasons, permit the particular characterization of those local forms which, because of variations in living requirements, differ very slightly but constantly from others, and whose separation as a species either appears unjustified or would not be acceptable to all authors. We shall thus avoid burdening the system unnecessarily with binomially named species that are difficult to identify precisely, though as a rule binomial description is completely sufficient.[5]

This was hardly a scientific definition of subspecies. The practicing ornithologists, carefully protecting their narrow little sphere of ideas, had won. The ragout they made out of a lot of leftovers was supposed to appeal to everyone's taste, whether he were friend or foe of the theory of evolution, whether he considered subspecies to be "nascent species" or "debased forms, produced by climate," of the fixed species. But finally all those sitting in judgment on the proposals seemed satisfied with this diplomatic contrivance — Freiherr von Berlepsch from Hanover-Münden, Professor Blasius from Brunswick, and Paul Matschie and Dr. Reichenow from Berlin. Only one man held out, the secretary of the meeting, Ernst Hartert. Influenced by Seebohm's writings, he had defined subspecies (in his *Katalog der Vogelsammlung im Museum der Senckenbergischen Naturforschenden Gesellschaft*, which had been published a few months earlier) as "forms not sufficiently distant from others to entitle them to the rank of species." And among the guests one was shaking his head, though for quite different reasons, ready to call attention to the danger of an outbreak of frivolous names given by scribblers who were only eager to publish and had no critical judgment, if a general introduction of trinomialism took place. This was Richard Bowdler Sharpe, Europe's most experienced ornithologist, who had followed Hartert's advice and stopped in Frankfurt for the meeting, on his way to Budapest for the Second International Ornithological Congress. Now he sat there in tense excitement and damned all the modernistic troublemakers surrounding him.

Sharpe, one of the most tireless workers ever known in the development of ornithology, was then at the height of his powers, after a rapid rise to fame. Born in London on November 22, 1847, he had as a boy annoyed his father, the publisher of a small newspaper, by neglecting to follow in his footsteps, preferring to watch animals in the woods and fields, and to add to the collection of birds that he stuffed himself. To cure him of this fad, his father took him out of grammar school when he was 15 and apprenticed him to a London publishing house. But there was no cooling the boy's enthusiasm for birds. Just after changing masters and starting to work for the well-known book dealer Bernard Quaritch in 1866, he began to visit the British Museum in Bloomsbury in his occasional spare time. There he studied kingfishers and consulted books on birds, because he had conceived the idea of a monograph on the Alcedinidae! These activities soon attracted the attention of several visitors to the bird room. Among them was Dr. P. L. Sclater, who in 1867 helped the industrious young man to a job as library assistant to the Zoological Society. Now his fortune was made. Constantly surrounded by all the books that he needed, he worked away feverishly at his kingfisher monograph. In so doing he came to know the most important ornithologists in England, and was so encouraged by their praise that he went to Leiden the same year, to improve his specialized knowledge by studying with Schlegel. Here he made the acquaintance of the bird painter Keulemans, and succeeded in acquiring his masterly talent to illustrate the monograph. The very first parts, published by the author in 1868, made a great stir, so that a rich London businessman, Henry Eeles Dresser, joined him in 1870 in a splendid undertaking, the *History of the Birds of Europe*, planned for several quarto volumes, which began appearing in parts in 1871. But only a year later Sharpe had to end his collaboration. George Robert Gray, in charge of the British Museum's bird collection since 1831, died on May 6, 1872, and Sharpe, on the recommendation of the keeper of the Zoological Department, Dr. Albert Günther, was chosen to succeed him. He was only 24 when he realized his most audacious dream.

His predecessor, a dry, pedantic civil servant, had not known how (and indeed had not even tried) to make the crowded bird collections in the cellars of the old museum in Bloomsbury into a center of ornithological research. Those not forced to go there avoided it and consulted the private collections, which in England were steadily increasing in number and splendor. The whole situation changed overnight. Sharpe, helped by the goodwill and trust of many reputable ornithologists, rearranged the birds completely and made the vaults as attractive as possible. He did not have to wait long for results. In 1873 he acquired A. R. Wallace's famous collection from the Malay Archipelago, and then in rapid succession one donor followed another in entrusting his treasures to Sharpe's care. He had taken over in 1872 at most 30,000 birds (of which 10,000 were mounted and on exhibition), but by 1890 he had no fewer than 230,000. Though at the beginning of his era the British Museum's bird collection ranked level with or behind those of the Berlin and Paris Museums, its rapid development left all competition hopelessly far behind. The most imposing contribution occurred in 1885, when Allan O. Hume presented to the British Museum his famous Indian birds, at that time the largest private collection in the world, on condition that Sharpe should fetch it personally from India. It contained 60,000 skins and 16,000 eggs.[6]

Merely the cataloguing, arranging, and checking of the innumerable acquisitions from all over the world would have completely swamped the powers of a normally competent person. Sharpe, however, made a thorough job of all that and much more besides. Certainly it was not simply his daemon that drove him to pile burden upon burden; it was also the anxiety about providing for the ten lively daughters with whom he had been blessed after his marriage at the age of 19. For their sake he completed the magnificent books that the elderly Gould had left unfinished, *The Birds of Asia* (1850–1883), *The Birds of New Guinea* (1875–1888), and the *Monograph of the Trochilidae* (1880–1887), and for their sake also he wrote the text to accompany C. W. Wyatt's fine pictures of swallows (*Monograph of the Hirundinidae*, 1885–1894). And all this could only be done in the evenings at home!

But Sharpe's really gigantic performance was his work on the *Catalogue of the Birds in the British Museum* (27 volumes, 1874–1898). Soon after he had entered his new position he decided on an undertaking, on the model of Albert Günther's *Catalogue of the Fishes in the British Museum*, which would include all the birds of the world, with citations of all synonyms and literature, exact descriptions of plumage, and complete details of distribution. He began with diurnal birds of prey, and had the manuscript of the first volume ready on June 1, 1874. In 1875 the second followed, on owls, and by 1890 Sharpe had written eleven volumes with his own hand. What that meant he described himself at the time:

> By working very hard for 300 out of 365 days in a year, and by burning much midnight oil, it is just possible to describe and to work out the synonymy of *one* species a day — not more. That is the average which a man can attain to by very close application; and if he can describe and collect together the synonymy of 300 species every year, it will be good work indeed . . . This is the result of my calculations — that a man can hope to acquire some practical knowledge of species and their literature by unswerving application to work for forty years![7]

And under what conditions the work was carried out! Sclater portrayed them in 1877:

> If the visitor to the British Museum will pause at the foot of the staircase leading up to the Paleontological gallery and look carefully into the obscurity in the right-hand corner he will perceive a door with a brass plate on one side of it. On entering this door and descending (with care) a flight of darkened steps, he will find himself in the cellar, which has for many years constituted the workshop of our national zoologists. Two small studies partitioned off to the left are assigned to the keeper of the department and his first assistant. The remaining naturalists are herded together in one apartment commonly called the "Insect-room," along with artists, messengers, and servants. Into this room is shown everybody who has business in the Zoological Department of the British Museum, whether he comes as a student to examine the collections, or as a tradesman to settle an account. Amid the perpetual interruptions thus caused, our national zoologist has to pursue his work. Some of the specimens are here, some in the galleries overhead, and some are stored away in cellars at a still lower depth than that in which he sits at work. The library attached to the department contains merely some of the most obvious books of reference, all others have to be obtained on loan from the great national depository of books in the centre of the building. No lights are allowed, and when the fogs of winter set in, the obscurity is such that it is difficult to see any object requiring minute examination.[8]

Only in 1883, after Sharpe had written six volumes of his *Catalogue* "in the gloom of this underground dungeon," did the move into the spacious and splendid building of the Natural History Museum in South Kensington take place.

Hartert must have been oppressed by dismal thoughts while he was sitting next to the great Sharpe in the Senckenberg Museum on that May 12, 1891. How different was his own situation! Born in Hamburg on October 29, 1859, the son of an army officer, he was already approaching 32 and still could not think of marrying. What good was it that he had traveled more than any contemporary German zoologist, or that nobody could hold a candle to him as an expert on birds and mammals? As long as his abilities carried no academic certificate, he was regarded by everyone as merely an industrious collector. Trusting blindly in his luck, from boyhood up he had done nothing but what he enjoyed. He had begun by collecting eggs when he was in the Breslau Gymnasium. Then came the free and easy years in East Prussia, where he learned taxidermy in Königsberg and for months stalked the woods with his gun, first to make his own bird collection, and then, in the early summers of 1882 to 1884, to oblige the elderly Eugen von Homeyer. Other boys were sitting in lecture halls after graduating from the Gymnasium, but in summer he was to be found in the marshes and on the North Sea beaches, and in winter among his eggs and bird books. Those were his apprentice years, but then came his years of travel, which took him into the tropics of two hemispheres. When he was 24 he joined the Flegel expedition, 1885–86, to collect in the Hausa country of northern Nigeria, and barely a year later he went to southern Asia, to earn some money again by collecting insects, birds, and mammals. From 1887 to 1889 he led exactly the sort of life he liked in the woods of Sumatra and Perak, in the wilderness of Upper Assam, and in the dry wastes of Rajputana. While he learned more from the observation of nature than he could have done from all the books, jobs and honors were being awarded to others. This could not continue indefinitely. Was he to go on living off his father, who in the meanwhile had retired as a major general and had moved to Marburg? Was he to continue with his lectures to geographic societies and colonists' clubs, which yielded hardly anything more than applause? Finally he pinned his hopes on Frankfurt, which is not far from Marburg. There, in November 1889, six months after he returned from India, he was commissioned to work through and catalogue the 10,000 mounted birds in the Senckenberg Museum — a notable collection, which had been founded in 1818 when a motheaten assortment was purchased for 6000 guilders from Dr. Bernhard Meyer in Offenbach. Eduard Rüppell had made it famous and looked after it for three decades, until in 1850 he lost interest. It was Hartert's task to rescue it from oblivion. He completed the job in nine and a half months, and at the beginning of February 1891 he was able to surprise those who had helped on difficult questions with the printed catalogue. Two men were particularly close to him: Dr. F. Kutter, Germany's most expert oologist, who had recently (following Hartlaub's resignation) become president of the Deutsche Ornithologische Gesellschaft; and the Freiherr Hans von Berlepsch, world famous as a sound authority on neotropical birds and a first-class systematist. "By your suitably professional work on such an important collection you have permanently established your reputation as a scientific ornithologist," wrote Berlepsch in a letter of thanks — but who was there to employ a scientific ornithologist? In November 1890 the Senckenbergische Gesellschaft had, for financial reasons, rejected Hartert's application to become curator of its collections. Berlin offered no hope, because the hardworking Reichenow had been encamped there since 1874, waiting for the shortly expected retirement of Cabanis. And Hartert's last card was trumped when not he, but the entomologist Seitz, was chosen to be director of the Frankfurt zoological garden after W. Haacke (the famed Richard Schomburgk's successor) went to Adelaide. There was only one glimmer of hope left: perhaps an ornithologist of his stamp could be used in England. With that in mind he went to London for a fortnight in February 1891, with Berlepsch's letter of recommendation to Sclater in his pocket. Sclater kindly introduced him to his circle, and arranged a meeting with Sharpe in the British Museum, but foreign specialists were not needed. Hartert could not even interest a rich private collector in an expedition to Venezuela and Colombia, which he had thought up in desperation and with the encouragement of his friend Berlepsch. The London journey was a complete failure.

Now he was back in Frankfurt among his German colleagues, writing the minutes of the argument about nomenclature and trying to forget his anxieties, while his fiancée was still waiting to get married. He could not have suspected that the great change in his fortunes was just around the corner, and that none other than Sharpe would produce it.

Hartert had become Sharpe's friend from the first moment of looking into his merry eyes. Sharpe was not at all like the "authorities" from Berlin, Brunswick, or Dresden, who were very conscious of their own dignity and emphasized it with every gesture. The "immense and almost boyish enthusiasm" commented on by a biographer had not been dampened in the slightest by age and honors. He went to work with cheerful humor, and whenever this seemed to desert him, his beloved red wine helped its recall. Although he was always ready to admire someone else's real achievements without a trace of envy, and to honor the efforts of a scientific opponent, he kept a watchful eye on the failings of his fellowmen and knew how to depict them in an inexhaustible flow of anecdotes, to everyone's delight. But there were many and varied anecdotes about him, too. His temperament was illustrated by the story of his having furiously seized an inkwell, like a second Martin Luther, and thrown it at a negligent assistant who had not been quick enough in bringing the book he wanted. The assistant ducked only just in time, letting the missile smash on the wall of the bird room. "But Mr. Sharpe!" was all he could say as he slowly got up off the floor, streaming with ink, whereupon Sharpe rushed over to him and hugged him warmly to his broad chest.

That was the sort of man he was, and the sort he would have had to be to win complete affection from Hartert, whose own warm heart and gay humor were always looking for a response. Lucid intelligence alone, which he himself possessed abundantly, was never enough for him in other people; he demanded "real men."

The moment the Frankfurt meeting was over, seven of the participants, including Sharpe and Hartert, took the train to Budapest, because the International Congress of Ornithologists was to open there on May 18, and Sharpe was to give the principal lecture, "A Review of the Recent Attempts to Classify Birds,"[9] which demonstrated most clearly his vast knowledge and astute judgment. Also on the program was a report by Reichenow on the rules for nomenclature just accepted by the Deutsche Ornithologische Gesellschaft, which were to be recommended to the Congress for international use.

Five weeks later Hartert received a letter from the British Museum, commissioning him to come to London to work over the nightjars and swifts for the sixteenth volume of the *Catalogue of Birds*, which had originally been allotted to Shelley. This was an earnest of Sharpe's friendship.

Some time earlier Dr. Albert Günther (1830–1914), born at Möhringen near Stuttgart, keeper of the zoological department since 1875, had realized that even a Sharpe would find it impossible to write a gigantic work, estimated at 27 volumes, all by himself, and had provided him with a group of assistants, mostly excellent: Seebohm, Gadow, Salvin, Sclater, Hargitt, Shelley, and Salvadori. But now Shelley, to Hartert's good fortune, had suddenly withdrawn after completing only half his assignment.

A few days after the letter arrived Hartert married Claudia Endres and they moved to England, without imagining that it would be their home for the next 38 years, because there was nothing to suggest that the settled position they wanted would be attained so soon. After Hartert polished off his commission for the British Museum in the startlingly brief time of eight months, he would still have been without a definite job if his new friends had not got to work in good time.

Through Dresser he made the acquaintance of young Walter Rothschild (1868–1937),[10] who two years earlier had built himself a small zoological museum at Tring, an hour's train ride from London. Rothschild's wealth quickly made it famous and respected. Hartert tried out again the Venezuelan plans he had already aired to so many others, and this time he was lucky. Rothschild financed the undertaking, and on May 1, 1892, the Harterts left for South America to collect birds and insects. Because political upheavals kept them from landing in Venezuela, they combed the islands in the Caribbean until, at the beginning of August 1892, they were surprised by a telegram from Rothschild summoning them back in a hurry, because Hartert was to become director of the Tring Museum.

He had been lucky again, because the English curator had been dismissed by Rothschild for gross abuse of confidence, and Dr. Günther had advised him to telegraph to Curaçao.

The unparalleled rise of the Tring Museum began on the day when Hartert first entered the little English town in its green setting. After two decades the two huge departments of birds and lepidoptera eclipsed those of all museums on the Continent and rivaled those of the British Museum. Sharpe had helped his most dangerous competitor into the saddle with his own hand, but he never regretted it. Only death, which unexpectedly claimed him on Christmas Day, 1909, could loosen the bonds of friendship between the two most important ornithologists of their generation. The conservative and the radical reformer loved each other, resented each other, and competed against each other for the great traveling collectors and the most precious rarities, the more enthusiastically because triumph sweetened the prize. When Sharpe reached 400,000 birds and eggs in 1908, Hartert was already on his heels. He never quite recovered from the all too early death of his only congenial rival and lively companion.

Sharpe had shown special ability in acquiring gifts of private collections to enrich the British Museum, but the Tring Museum, too, profited from the zeal of earlier collectors. On Kleinschmidt's advice, Rothschild bought in 1897 the once famous collection of Christian Ludwig Brehm,[11] which after the "bird pastor's" death in 1864 had been slumbering in the attic of the old parsonage in Renthendorf.* The 9,000 skins were now finally taken out of their wooden coffins. In 1912 an enormous collection of Australian birds (more than 45,000, and more than all the bird skins owned at the time by the Berlin Museum) was obtained from Gregory M. Mathews. But the most important acquisitions of the Tring

*In 1932 the Brehm collection went to the American Museum of Natural History, along with the rest of the birds from Tring (see reference 29), but in 1960 the bulk of the Brehm specimens returned to Germany through an exchange with the Alexander Koenig Museum in Bonn.

Museum came from the collectors whom Rothschild and Hartert sent out to all the areas marked "unexplored" on the map. They concentrated particularly on the Indo-Australian islands and on New Guinea, because Wallace and the collectors for the museums in Leiden and Genoa had been able to complete only half their job, and anyone who wanted to make a thorough study of the phenomena of geographic variation could hardly find a more rewarding field.

From just such work Hartert had hoped for a reform of systematics, ever since he had catalogued the Senckenberg collection. But when Seebohm died in 1895, Hartert was the only remaining European who wished to arrange the classification of forms according to the theory of evolution. Everyone in England totally rejected the use of trinomial nomenclature; in Germany it had begun to be employed eccentrically in opposition to Hartert's main ideas.

These he had defined in detail in 1891, when he wrote:

> I agree with Seebohm, who views subspecies as forms in the process of development. Seebohm wishes to have classed as species all forms that are not interrelated with other forms, and as subspecies all those that are related by a series of intermediate forms. That this expresses the nature of subspecies in a scientific manner I do not doubt, but I object to taking it as a definition . . . I believe that it is right to regard as subspecies forms that differ only in a small variation in size, lighter or darker coloring, or small variations in pattern, even though one does not have the intermediate forms at hand. This type of nomenclature shows the closeness of the relationship, whereas the simple specific name gives no indication whether the species are poles apart or very nearly related.[12]

The bold and significant step had been taken; Hartert had dropped the American leading strings. "Intergradation" had been replaced by generalizing induction, the "judgment trained by practice." But at first there was no sign of agreement. There was still no sign some years later, when Hartert's conception of the practical use of his recommendations became clear to everyone. In the new periodical *Novitates Zoologicae*, which the Tring Museum had been issuing since 1894, he followed his method with growing determination, but instead of support, there arose in Germany a hydra of opposition. To strangle it, he went home in 1897 and (after buying Brehm's collection in Renthendorf) took part in the twenty-first annual meeting of the Deutsche Ornithologische Gesellschaft at Dresden, where Berlepsch (now a count) was to give a lecture on May 29 about the concept of subspecies.

Three speakers appeared against Hartert — Berlepsch, Wiglesworth, and Kleinschmidt — each representing a different opinion.

Berlepsch objected that Hartert's procedure would burden nomenclature with hypotheses. "We are still completely uncertain about the development of species, and therefore should not introduce our guesses into nomenclature." Hartert, using trinomial nomenclature, had recently been lumping species together. But Nature herself knows no lumping, and produces species only side by side. Therefore the rules laid down in 1891 should be followed; they prescribed that part of a species may be classed as a subspecies because it possesses diagnostic characteristics less striking than those of clearly distinguishable species. Thus subspecies are entirely the same as species except that they are less easy to recognize and distinguish. It should be a main principle in nomenclature to adopt solely practical points of view in solving problems.

Lionel Wiglesworth, assistant to A. B. Meyer, had three objections to Hartert's method. First, the naming of subspecies implies positive knowledge where there is really only a guess. Second, the procedure fosters the opinion that species are inclined to develop distinct races by a sort of leap (*per saltum*), and that intermediate links are rare and created by interbreeding. Third, there is no criterion of judgment for deciding when the new bud on the stem of a species is mature enough to be tagged as a subspecies. Trinomial nomenclature is suited only to describing distinct races within a species; the intermediate gradations between two extreme forms, to which often most of the individuals belong, must remain unconsidered. Therefore the speaker hoped that it would be possible in a short time to abolish trinomial nomenclature and to replace it by symbols or mathematical signs, so that intermediate forms could be described more or less as follows: *Haliastur* $indus_1$ $girrenera_3$; *H.* $indus_2$ $girrenera_2$, and so on.

The only person who openly rejected the theory of evolution was a student of theology, Otto Kleinschmidt (1870–1954), who had just attracted attention by a study of the palaearctic marsh tits. He spoke only briefly, because he had already published his views.[13] A purely genetic conception of the multiplicity of forms in nature has disadvantages. The natural sciences should be pursued neither historically nor philosophically but "exactly," like mathematics. Kleinschmidt could find no evidence at all for maintaining that there are gradual transitions between individual species. He considered it most probable that species are sharply defined in nature and that some of them can remain constant for a fairly long time. "From my point of view, species cannot be taken apart." "Species exist in nature, just as individuals do." Much the same sentences could have been written by J. H. Blasius or Schlegel 30 years earlier.

Only one person shared Hartert's views unreservedly; this was A. B. Meyer, who from his youth had admired Wallace and had translated his works into German. He prophesied that the time would come when no zoologist hostile to subspecies would remain, just as there was none now who rejected the theory of evolution. *Subspecies* has a scientific meaning only if it is interpreted genetically. Groups of individuals constantly diverge from other groups that are geographically separated but closely related, though if they come into contact they interbreed and create transitional forms (for example, *Rhectes* [*Pitohui*] in New Guinea, and many others). Consider, for example, the extremely specialized and isolated form *Parotia* in New Guinea; its four "species," *sexpennis, berlepschi, lawesi,* and *carolae,* are at present geographically separated from one another. On the basis of their great similarity to one another and their difference from other birds of paradise, a scientific observer could not have the slightest doubt that these forms are derived from each other or from common ancestors, and it is also certain that they are capable of fertile crosses. Thus, they should be regarded as four subspecies, not four species. Hartert developed for us the most logical system for designating such forms.

Whatever Hartert and Meyers's efforts in Dresden for a reform of nomenclature in a Darwinian spirit, they were unable to persuade their opponents. In fact, at first the opponents grew stronger. It became evident that Reichenow, the most influential German ornithologist because of his position as editor of the *Journal für Ornithologie*, had accepted Berlepsch's "practical" view. After he had already let this be understood several times, he declared categorically in 1901: "We must turn back! The species must remain what it was until now, the smallest indivisible unit in the system, the essence of an individual being. It must not become a concept for a systematic group. Therefore we must not rank a subspecies below a species, but regard it as a conspecies, an aberration, or an inferior species."[14] And in the future, as if in protest against Hartert's recommendations, he made a habit of using the concept of subspecies (he now spoke only of conspecies) precisely and almost solely on the occasions when, according to Hartert's principles, it was inadmissible (*Certhia familiaris brachydactyla, Lamprocolius chalybaeus chloropterus,* and so forth).

But it was too late to put out the fire, because even in 1901 many minds had caught up Hartert's inflammatory idea. From 1898 his views were propagated by an Austrian, Victor Ritter von Tschusi zu Schmidhoffen (1847–1923), editor of the periodical *Ornithologisches Jahrbuch.* In the same year Carlo Freiherr von Erlanger (1872–1904) adopted them when he was working on the material he had collected in Tunisia. But Hartert's most effective support came from a Viennese, Carl E. Hellmayr (1878–1944), a systematist of lively intellect and tireless energy, who from 1900 employed Hartert's ideas while grouping the allopatric forms into species. Among the Berlin ornithologists, Herman Schalow in 1900 proclaimed himself a disciple of the new school.

In the meantime, Hartert had given a more precise form to his principles of nomenclature. Stimulated by Kleinschmidt's publications, he had stressed in 1901 that geographic distribution was decisive in establishing whether two very similar forms should be treated as subspecies or as separate species. That had been demonstrated recently by the examples of the marsh and willow tits (*Parus palustris* and *salicarius*, Kleinschmidt, 1897)[15] and the North African crested and thekla larks (*Galerida cristata* and *theklae*, Erlanger, 1899).[16] In accordance with this principle Hartert gave a definition of subspecies in the introduction to his great work, *Die Vögel der paläarktischen Fauna* (1903):

> We describe as subspecies the geographically separated forms of one and the same type, which taken together make up a species. Therefore not just a small number of differences, but differences combined with geographic separation, permit us to determine a form as a subspecies, naturally when there is general agreement of the main characters.

This was indeed a pioneer book, not only because Hartert collected in it the sum total of knowledge about birds in the entire palearctic region, but also because the first part, published in November 1903, at the same time ensured the triumph of trinomialism. Hartert's group felt so strong by then that Hellmayr openly recommended the "rooting out" of the opposition.

On that principle all the contemporary British ornithologists would have had to be rooted out. No one in the British Isles dared to advocate trinomialism because the powerful P. L. Sclater, who had been accustomed for decades to set his imprimatur on English ornithological work, had a low opinion of all innovators. He was still editing the *Ibis*, as he had been doing since 1859, and dispensing his judgment of the performances of the world literature. Of Hartert's new work he said in 1904:

> But the main point of the book is that the author calls upon us virtually to give up the binomial system, which has been in universal use since its foundation by Linnaeus, for a trinomial system. Here we most decidedly decline to follow him . . . We shall, no doubt, be stigmatized by some of our friends as "fossils" and "antediluvians"; but we believe that the great majority of sober-minded ornithologists, in spite of the efforts of the new school, will stick to the binomial system.[17]

Sharpe hoped so too. On August 24, 1909, a few months before his death, he wrote:

> My views on this subject have often been stated, as for trinomials, I look upon the system as destructive. I consider that the

> burden imposed upon the Zoologists who follow this method for the naming of their specimens will become too heavy, and the system will fall by its own weight. That races or sub-species of birds exist in nature, no one can deny, but, to my mind, a binomial title answers every purpose.[18]

Therefore in his *Hand-list* he promoted all previously described subspecies to the rank of species, reaching the enormous total of 18,939 species of birds.

As a result of Sclater's and Sharpe's influence, the three zoologists of the Tring Museum, Rothschild, Hartert, and Dr. Karl Jordan (from 1893), were regarded in England as peculiar outsiders. But here too the "hereditary conservatism" about which Seebohm had already complained gradually yielded to a new point of view. In 1905 a much-traveled private collector of birds, Harry Forbes Witherby (1873–1943), decided to advocate Hartert's method in the *Ibis.* Two years later he founded the periodical *British Birds*, in which he favored the new school. Finally, in 1912, still under the glare of the elderly Sclater, the "destructive" nomenclature crept into the *Ibis* itself, and when in the same year the completely trinomial *Hand-list of British Birds* was published (by Witherby, F. C. R. Jourdain, and N. F. Ticehurst, together with Hartert), only a few people grumbled at the unfamiliar names. The stoutest bulwark of binomialism had fallen after a 20-year siege.

Trinomials might perhaps have encountered less enduring and fierce opposition in England if their use had not involved serious consequences: the rejection of the Strickland Code of Nomenclature, to which English ornithologists had adhered since 1842. They were rightly proud of it, because its well-thought-out stipulations had proved their value remarkably well. Strickland,[19] one of the brightest luminaries in ornithology, had compiled it, and in 1842 the British Association, at its Manchester meeting, had accepted it as binding.[20] At that time the lack of consistent rules had reduced the nomenclature of species and genera to chaos. Babel had broken out not because there was no critical compendium of species described, but principally because many authors willfully brushed aside the historic privilege of the first man to describe a species.

Linnaeus, the creator of binomial nomenclature, had himself laid the groundwork for this arbitrary procedure when in 1751 he set forth, in his *Philosophia botanica*, the aesthetic, philological, and other rules to be observed in the allotment of names. In ornithology, his guidelines for the limitation of the law of priority were first followed by Illiger, who in 1811 believed he had to replace all generic and specific names that were not classically formed with purer ones. Johann Wagler went even further in his *Systema avium* (1827).[21] He demanded that every scientific name be suitable and dignified ("aptus et dignus"), and inveighed against those who preferred to use the customary and familiar names, even when they were certainly ill chosen, rather than new ones.*

Swainson, in his *Natural History and Classification of Birds* (1836), carried the principle of selection to excess. He considered that the great leaders in the field of natural history (including himself) must be accorded the right to clean up nomenclature from time to time, and to choose between good and bad names without any regard for priority. Unsuitable geographic names (like *Meliphaga Novae Hollandiae*, because there are several species of *Meliphaga* in Australia) should be rejected, as should most dedicatory names, because they should be reserved to honor only really great scientists; it was never permissible that great botanists should thus be immortalized in zoology. For the rest, he followed Illiger (1811) closely, and like him threw out all names derived from mythology, the ancient gods, history, morality, pathology, and the like (such as "*Catarrhactes, Fratercula, Stercorarius, Fregatta, Cochlearius*").

Meanwhile, the reaction against such high-handedness had already set in. In 1835 Strickland energetically attacked the arbitrary alteration of established terms, and advocated that the rule of priority be strictly followed;[22] he found immediate agreement and emulation among influential ornithologists like Bonaparte (1838), George Robert Gray (1840), Keyserling and Blasius (1840), and others. Finally he worked out the firm and carefully thought-out rules that have been known since 1842 as the Strickland Code. In it he called the "Law of Priority the only effectual and just one," and he founded it on the twelfth edition of Linnaeus's *Systema naturae* (1766). Exceptions were allowed only for names that clearly contained false information (like *Picus cafer* for a Mexican woodpecker, or *Muscicapa atra* for an olive-green bird). He also prescribed a change in the species name whenever it was used as a generic name (not *Pyrrhocorax pyrrhocorax* Linn., 1766, but *Pyrrhocorax alpinus* Vieillot, 1816). He recommended that the specific name should take precedence over everything else, and that the author's name be put in parentheses whenever the specific name was joined to a new generic name instead of to the original one, *Muscicapa crinita* Linn., for example, becoming *Tyrannus crinitus* (Linn.).

By accepting these rules English ornithologists soon achieved the desired conformity in nomenclature. In other countries, too, such as Italy and France (chiefly thanks to Bonaparte's influence), the Strick-

*Necque quicquam adversus hanc rationem valet opinio istorum, qui omne novatum nomen rejicere conantur, quod illa usitata cognomina, quamvis haud apta, trita tamen et omnibus nota, ideoque magis commoda esse dicunt. [Nor is there any value at all in the opposition to this method of those people who attempt to throw out every invented name, because they say that the names in common use, even though unsuitable, are nonetheless familiar and known to everyone, and therefore more convenient.]

land Code was used as a guideline, and when the German ornithologists Hartlaub and J. H. Blasius accepted it as well, the community of scholars really seemed, in Strickland's words, to have put itself under new laws. Nobody expected that the Berlin school, headed by Lichtenstein, would react in horror against such "foreign regimentation." In 1847 Cabanis wrote, in his "Ornithologische Notizen":

> In the matter of nomenclature we, as Lichtenstein's pupils, adhere strictly to the rules established by Linnaeus and Illiger, and declare ourselves decided enemies of the nomenclatural barbarism that has recently become so rife, and that offends against the nature of all languages. With that in mind we have often cashiered generic names that broke the rules and have replaced them by other, more recent, ones, or, where these did not exist, have proposed new names . . . "The inflexible law of priority" is only one integral part of Linnaeus's rules on nomenclature, and must not be manipulated unilaterally at the expense of the others.[23]

"That parenthesis-monster, as illogical as it is unhistorical, imported from abroad and often all too well liked," also aroused Cabanis's passionate opposition.[24] For a long time he followed the procedure of his teacher, Lichtenstein: instead of indicating the describer of the species, the author's name was endowed with a peculiar historical significance by referring to the man who first assigned the combination of generic and specific names under discussion (*Cuculus sinensis* Linn. becomes *Urocissa sinensis* Cabanis). Cabanis yielded to the "parenthesis-monster" only in 1870, so that he should not be left swimming against the stream alone.

At first Finsch (*Die Papageien*, 1868) sided with those later German ornithologists who still kept Illiger's laws instead of following the Strickland Code. But a review of his book by the witty A. O. Hume in 1874 seems to have cured him of the heresy for the rest of his life: "Let us treat our author as he treats other people's species. 'Finsch!' contrary to all rules of orthography! what is that '*s*' doing there? 'Finch!' Dr. *Fringilla*, MIHI! Classich gebildetes Wort [a classically formed word] !!"[25]

But just when it seemed finally that the Strickland Code, whose text had been freshly approved by the British Association in Newcastle in 1863, would unite all ornithologists, the barely achieved stability and conformity of the nomenclature was upset by a violent blow from America.

The ornithologists there had come to the conclusion that the Strickland Code needed reform because it had proved too liberal and still allowed too much room for individual whims. The reins must be tightened and all those rules permitting any deviation from absolute obedience to the law of priority should be dropped, however annoying the immediate results of such vigorous new measures might be. Even when the orthography of a name was philologically wrong, it must be kept, unless it contained an obvious typographic error. As a logical consequence of following the rules, the law of priority should be antedated to begin not with the twelfth (1766) but with the tenth (1758) edition of the *Systema naturae.* The new rules, carefully worked out in 1883 by J. A. Allen and Elliott Coues, were accepted in 1885 by the American Ornithologists' Union and thereby became binding in the United States.

In Europe the rejection of the Strickland Code at first was violently opposed, and when the ornithologists on the Continent had begun to agree to it, Alfred Newton, one of the founders of the British Ornithologists' Union in 1858, wrote to the president of the Second International Ornithological Congress in Budapest, on May 4, 1891:

> As an ornithologist [I] especially lament the adoption by the American Ornithologists' Union of a Code, differing so entirely from that, which in principle, had hitherto been maintained. No greater blow at the much desired uniformity of nomenclature was ever delivered, and its immediate effect has been, to render a good deal of the excellent work of recent American ornithologists almost unintelligible to their European brethren.[26]

But his words fell on deaf ears. In Germany the American rules had been applauded by Berlepsch and Hartert, who influenced their adoption by the sixteenth annual meeting of the Deutsche Ornithologische Gesellschaft, held at Frankfurt am Main on May 12, 1891. A few days later the Budapest Congress also approved them, so that in future "the absolute principle of priority with all its consequences [should form] the basis for the rules of zoological nomenclature," and the law of priority should begin with the tenth edition of the *Systema naturae.*

Although this new arrangement forced many changes even among the commonest names, it was almost unanimously followed in Germany. The one exception was Hartert's friend Alexander Koenig, an explorer with a large private collection in Bonn, who declared, at the twenty-first annual meeting of the Deutsche Ornithologische Gesellschaft at Dresden in 1897, that he did not need to put on a straitjacket; every scientist was free to follow his own line. He made a temperamental attack on the "illogical" reduplicated names; to avoid using them he would choose the next oldest specific name (as the Strickland Code prescribed). Hartert, on the other hand, not merely defended before the meeting the use of the same name for genus and species, but took the next logical step and recommended that the "typical" subspecies (which until then had not been particularly identified, or had been marked by the word *typica*) should be identified in the nomenclature by the repetition of the specific name. One should not be frightened off even by such constructions as *Ciconia ciconia ciconia* (L). The members present were either amused or appalled. But in 1898 Tschusi in Europe and H. C. Oberholser in America were already following his recommendation, and soon afterward this last bitter result of the American principle, "a name is only a name, having no meaning until invested with one by being used as the handle of a fact,"[27] had established itself among all the trinomialists.

The elderly English ornithologists, who had been educated in the humanistic tradition, followed with mounting distaste the increasing mechanization and barbarization of nomenclature that the liberal rules of the Strickland Code had been intended to combat, and it is understandable that they defended themselves to the end against the threatening wave. But it was no use. After they had tried in vain to influence the decisions of the International Commission on Zoological Nomenclature, P. L. Sclater and Howard Saunders resigned from this body in 1901.

With the appearance of the last part of *Die Vögel der paläarktischen Fauna* in March 1922, Hartert had reached his goal. He had won over the ornithologists of the whole of Europe to the evolutionary view of the problem of species, and had even persuaded several American ornithologists to employ his definition of subspecies. There was no longer an American or a European, a German or an English school. Unanimity of concept and its nomenclatural expression had been established, and only a few tough eccentrics could be heard grumbling in a corner. Hartert's success was even more visible when at Copenhagen in 1926 he succeeded in bringing together as friends ornithologists from all over the world, who had been separated by the First World War. He had been unanimously elected president of this Sixth International Ornithological Congress, because all groups respected and admired him as a researcher, a man who could express in his address the quintessence of his life work in the these terms:

> It is the most important event of the last decade, that the significance of the study not only of species, but also of their geographic forms, has been recognized and has become generally accepted property, so that our knowledge of birds has grown enormously; systematic ornithology stands very high in comparison with various other branches of zoology, and is endeavoring to contribute to the solution of more general and more advanced problems.[28]

During all these years the Tring Museum had grown steadily and had become outstanding for the choiceness and the painstakingly exact arrangement of its collections. Several of those who were now leading ornithologists had been Hartert's students at Tring, and hardly any of them had failed to visit the quiet little country town in order to learn more. Hartert's life work was complete when in 1930, soon after his seventieth birthday had been celebrated by ornithologists, he retired to spend his remaining days with his wife in his native country, and to complete in the Berlin Museum the supplements to his great work. It was hard for him to leave the treasures that he had helped to accumulate over the decades: 280,000 bird skins and a library of 30,000 volumes! All this, he thought, was firmly established and would last for centuries.

But he too had to realize painfully the vanity of calculations and the mutability of things. In 1932 Ernst Hartert had a study opposite mine. I can never forget how, on a gray February morning, he came staggering in to me with an envelope in his fingertips, and sank into a chair. "My collection! My collection!" he stammered out, his chest heaving and his clear eyes swimming with tears.

The birds of the Tring Museum were even then being packed into cases, for shipment to the American Museum of Natural History in New York.

Twenty months later, on November 11, 1933, Ernst Hartert died.[29]

15

The Effect of the Theory of Mutation

By 1930, when Hartert left Tring, undescribed birds had been so thoroughly pursued that rather few species had escaped the zeal of explorers. True, two very remarkable species were among them, the Congo Peacock, *Afropavo congensis* Chapin, from the virgin forests of the Congo, and a small crow that was adapted to living in the desert and on the steppes, *Zavattariornis stresemanni* Moltoni, from southern Abyssinia, which were discovered in 1936 and 1938 respectively. Today [1949] we can venture the assertion that all but 30 species (at the most) are known.* The number of subspecies entitled to a name, according to the rules at present valid, can probably be estimated at another 2000 or more.

The following table shows the number of distinct forms recognized at different periods:

Date	Forms
c. 330 B.C. (Aristotle)	140
1555 (Gesner, Belon)	222
1758 (Linnaeus)	564
1760 (Brisson)	1,500
1790 (Latham)	2,951
1812 (Illiger)	3,779
1841 (G. R. Gray)	6,000

*[This prediction, in part based on Mayr's 1946 estimate, did not come true. In the 20 years 1950–1969 more than 60 valid species of birds were discovered, and the end is not yet in sight; about three new species are still being described annually. E. M.]

1871 (G. R. Gray)	11,162
1909 (Sharpe)	18,939
1946 (Mayr)	28,500

Both the opponents and the advocates of trinomialism had realized that its use would greatly increase the number of named forms, which was viewed either as a danger or as an advantage to general understanding and scientific progress. In Europe, Hartert was the first to urge "the recognition of even the most subtle local forms," which displeased conservative ornithologists,* and in 1899 he explained his reason: "Forms that are difficult to distinguish must be taken into account, because they occur in nature, and during our study of nature we must not allow ourselves to neglect any phenomena." But at that time Hartert could certainly have had no idea how much his advocacy would help to facilitate later research into evolution.

At first the attention of systematists was concentrated on two problems: the finest possible subdivision of species into subspecies, and the method of grouping forms with the help of trinomial nomenclature. With the knowledge of individual as well as geographic distribution and variation, the "new systematics" became more daring. Many clearly marked forms that had long been regarded as "good species" were now demoted to subspecies.

It was often impossible, however, to achieve the desirable unanimity of opinion. In cases of doubt whether a geographic form was sufficiently differentiated to be considered an independent species, many systematists were attracted by the hypothesis of Kleinschmidt, who believed that he could replace subjective and arbitrary opinion by objective criteria, and in this tended to follow the natural philosophers. The ideas he had developed in 1897 convinced him that, though the fact of evolutionary development from lower to higher forms could not be disputed, there were nevertheless no transitional forms between the species. From this he concluded that the number of species existing today had been produced by the same number of primeval germs, and to Darwin's theory of the genealogical tree he opposed his theory of the genealogical bush. According to Kleinschmidt, species vary only within certain limits; none of their geographic variations can overstep these limits and become an independent species. Thus the systematist's task should be to bring together all the races that compose a "natural" species (the "phenomenon of original creation"). We have already met with similar ideas in the

*[The argument was not only whether or not differences among populations should be described, but also whether it was advisable to recognize every distinguishable population as a separate subspecies. The modern trend is not to go as fas as was done in the 1920s and, for instance, Vaurie (*The Birds of the Palearctic Fauna*, London, 1959–1965) considers as synonyms a number of forms that Hartert had recognized as valid. E. M.]

pre-Darwinian period. Indeed, even the idea of circular arrangement, developed by MacLeay and his supporters, recurred in Kleinschmidt, when he said: "The forms of modern fauna cannot be grouped in successive series according to their resemblances. The map usually shows us a group, a circle, that does not extend in a line but everywhere returns on itself and thus creates a complete and distinct whole."[1] Therefore he recommended that the concept of species, "loaded with prejudices," be completely abandoned and that "all forms that are simply geographic representatives of one and the same creature" be called a *Formenkreis* [group of allopatric Linnaean species]. Later he replaced this term with *Realgattung* [real genus].

In arranging groups of forms, Hartert and Kleinschmidt often arrived at the same result; whenever they disagreed, it was Kleinschmidt who made the more questionable groupings, under the influence of his theories, because he included even widely differentiated forms in his *Formenkreise* whenever they merely fulfilled the two qualifications of geographic replacement and "complete representation," however large the gap in distribution might be. Soon the convenient expression *Formenkreis* established itself even in the work of those ornithologists who accepted a general phylogenetic relationship among all animals; this lasted until 1926, when Rensch clarified the situation by urging that Hartert's geographically varying species should in future be called *Rassenkreise.* These should have no theoretical boundaries; the many borderline cases could be best taken care of in the nomenclature by grouping two or more *Rassenkreise* or species in *Artenkreise*, provided they replaced each other geographically. In 1931 Ernst Mayr proposed the term *superspecies* as the international equivalent of *Artenkreis* and in 1940 Julian Huxley suggested *polytypic species* for *Rassenkreis.*

Mayr, one of the contemporary authors who have considerably expanded the concept of species, suggested in 1946 that the forms already described could be grouped together into about 8,600 species. I estimate the number as somewhat higher (10,000 at most). Trinomialism is employed in basically the same way by James L. Peters in *Check-list of Birds of the World* (1931-)* intended to record all known forms of birds.

The larger and more heterogeneous became the groups of forms that the systematists considered races (subspecies) of the same species, the more numerous became the opportunities to utilize this material for the study of geographic variation and the factors causing it. Intergradation had originally been taken as the touchstone of trinomialism, and only

*Comprising fifteen volumes, of which all but volumes 8 and 11 have been published (1931-1970), the first seven prepared by Peters, the others by various ornithologists (including Peters for a portion of volume 9) under the editorship, partly in combination and partly in succession, of Ernst Mayr, James C. Greenway, Jr., and Raymond A. Paynter, Jr.

intergrading forms that constituted an unbroken series were combined into one species. Proceeding in this way, one could be tempted to ascribe the gradual change in a species from one geographic extreme to another to the direct influence of climate on the organism (J. A. Allen, R. Ridgway, and others). This Lamarckian explanation automatically necessitated the rejection of Darwin's theory (at least for geographic variation), which saw selection as an effective factor in evolution. When ornithological systematists had been accustomed to this interpretation, they continued to be influenced by it even after they had eliminated the prerequisite of intergradation from their definition of subspecies. Hartert belonged to this group. In 1897, at the annual meeting of the Deutsche Ornithologische Gesellschaft in Dresden, he said:

> It is quite conceivable that the continued effect of local influences will increase the peculiarities they create, and that what we now know as a local aberration will in time become a subspecies or species. I am inclined to explain the creation of subspecies and new species in this way, without having to take refuge in explanations based on sexual or other kinds of selection, or in other more or less mystical theories . . . Evidence for the direct effect of local influences can be found in the fact that, in certain areas, the majority of species occurring there are marked in the same way, so that, for example, in deserts we find so-called desert forms, etc.

The parallelism of geographic variations remained for some time the favorite subject for study by avian systematists, who employed the large museum collections to theorize on evolution. They failed to notice that they were making a narrow and prejudiced selection out of the rich manifestation of geographic variation, and that they were simply pushing aside as unclassifiable anything that their thought patterns could not accommodate. K. Görnitz's article in the *Journal für Ornithologie* (1923), "Über die Wirkung klimatischer Faktoren auf die Pigmentfarbe der Vogelfedern [On the effect of climatic factors on the pigmentation of bird plumages]," belongs to this group of studies. The author maintained "that the majority of geographic races of birds differentiated by color are not the product of natural selection, but have arisen idiokinetically through the influence of climate." According to him, the germ plasm is altered by external factors, either at any time or only during a particular period of sensitivity for the germ cells.

Other ornithological systematists had already rejected the Lamarckian interpretation of the formation of geographic races: They proceeded from the established fact that most geographic variation in color markings did not follow any climatic rule. One of the first to emphasize this was Oskar Heinroth, who wrote in 1903:

> I must admit that I know of no instance in which one could be justified in assuming that substrate and climate have acted directly on the production of color or form . . . Island races in particular, which certainly arose through the immigration of only a few individuals, seem most likely to have been produced by the inherited, and thereby reinforced, number of chance variations. Because the new form could no longer mix with the original form, the fortuitous and later reinforced individual character of a stray pair of birds left its mark on the island race to which it gave rise . . . Nothing definite can be said of any advantage of a particular color for its wearer, however much the utilitarian theorists would like to have it so.[2]

Soon afterward the "new systematics" was able to rest its interpretation on the revolutionary view of inheritance that had arisen from the rediscovery of Mendel's laws and the enunciation of the mutation theory by the botanist Hugo de Vries, both in 1900. This theory led for the next 30 years to a complete split of the evolutionists into two camps. The mutationists, and indeed the majority of the Mendelians, led by William Bateson, considered major genetic changes (mutations) the crucial events in evolution. Such mutations could create new species in a single step. Natural selection, for the mutationists, was essentially a negative and therefore an unimportant factor. It could not create anything new, but could only eliminate harmful mutations.

The opponents of the mutationists adhered to Darwin's interpretation. They considered the slight and, as Darwin had called it, "fluctuating" (= gradual) variations as the true material of evolution. Natural selection could accumulate such slight variations, and gradually build them up into large differences between species. Like Darwin, the gradualists believed in a certain amount of inheritance of acquired characters and some of them even in a direct influence of the environment. The fight between these two schools carried over into ornithology.

The saltational origin of new kinds had already been presented to ornithologists as a hypothesis several decades earlier, when Dr. Gustav Jäger (1832–1916) appeared as a proponent of Darwin's ideas at the thirteenth annual meeting of the Deutsche Ornithologen-Gesellschaft at Stuttgart in September 1860, thereby promptly provoking the objections of J. H. Blasius and Bernard Altum. Jäger made a prophetic remark (which remained without consequences because it was so much in advance of the theory of mutation), reported in the account of the meeting as follows:

> Darwin seemed to him to have been somewhat one-sided in the detailed elaboration of his ideas, because he regarded change almost exclusively as an entirely sequential process, evenly distributed over

> a number of generations. The sharp distinctions separating the majority of species were evidence, not against Darwin's theory, but for it; that in many, if not most, instances change took place by saltation; that a species was the descendant of a particularly favored individual — which could be called monstrous in comparison with its parents and kindred — and which in the struggle for existence had exterminated its nearest relatives. Observations of domestic animals supported this theory. The speaker, however, emphasized that he did not thereby wish to exclude the Darwinian view of a gradual change in and splitting up of descendants, first into local races, then into climatic varieties, vicariant species, and so on.

Jäger then recommended the use of statistical method to define the extent of variability of species. It should establish, for example, "the amount of individual variation peculiar to each species, and the percentual frequency of the appearance of monstrous (not in the pathological sense) individuals."

Soon after de Vries had published his theory, various zoologists began to study mutation in domestic poultry. The American C. B. Davenport, of Cold Spring Harbor, is the most important to mention here; from 1906 on he published the results of his genetic studies, first on chickens and then, in 1907, on canaries. His work did not, in fact, influence the theories of ornithological systematists.

However, a professional ornithologist had already begun to attract attention with his evidence for the mutationist theory. In 1907 William Beebe, another American, called attention to the dichromatism known in some species of birds (the blue and white snow geese, the common and ringed murres, the Indian and black-winged peacocks), and noted that "if the intraspecific occurrence of such dichromatic phases as these adumbrates new and permanent forms, we have an interesting and significant stage of species formation by geographical variation."[3] Nevertheless, he recommended leaving open the question of whether mutation had played a part in developing any one of these examples of dichromatism, or whether they were all to be explained "as arising by the accumulation of continuous variations."[4]

The systematists ignored Beebe's stimulating work just as much as they did a publication by A. Braune, a Dresden doctor, in *Die gefiederte Welt* (1910). He had systematically crossed the two "wild" forms of Gouldian finches, red-headed and black-headed, and had thus established "an extraordinarily interesting case of Mendelism." Not until 1918 was the problem experimentally worked on and precisely analyzed by A. Adlersparre in *Fauna och Flora.*[5]

In 1918 I turned my attention to discontinuous variation and, in opposition to the ruling Lamarckian tendency in ornithology, ascribed the transitions between well-differentiated geographic races not to the influence of "climate," but (like Seebohm as early as 1887) to hybridization in the secondary contact zone. Then in 1920 (*Avifauna Macedonia*) I began to collect examples of dichromatism among bird populations and to study them in collections of skins. During the following years I found a large number of startling mutants, which until then had generally been considered separate species. Among such polymorph species the distribution of mutants within large populations as well as local and seasonal differences in their relative frequency could be fairly easily determined. One could expect that the study of this variation would make it possible to follow the spread of mutations within a population (rate of dispersion, inhibiting and promoting factors), and thus to reach some generally valid conclusions. The results accumulated by 1926 allowed me to state "that considerable importance can be ascribed to large mutations in the alteration of a species. It is only by saltation that reddish-brown can change to black, or black to white; many conspicuous markings have probably also appeared suddenly and completely (not gradually through the accumulation of many small alterations)."[6] In the meanwhile (after 1923) Frank M. Chapman of the American Museum of Natural History had also been working on the significance of mutation in the formation of races among birds. He too found (first in his study of the emberizine genus *Buarremon*) that many — especially the most conspicuous — characteristics of a race could not have evolved gradually, but must have appeared as a mutational jump.

The gradualists, however, were not convinced. Charles Otis Whitman (Chicago) had already in 1904 energetically attacked the theory of mutation and supported his views, in *The Problem of the Origin of Species*, with references to his years of extensive experiments on many species of pigeon, which demonstrated to him that not mutation but orthogenesis, "definitely directed variation," was the "method" of evolution. Whitman was sure that there is no discontinuity in the phenomena of variation, heredity, and evolution. It is therefore not surprising that Hartert wrote in a letter to Dr. A. Süssenguth in Munich on August 25, 1910: "I am no believer in de Vries's theory of mutation. It is not proved, and seems to me quite improbable, that observations on *Oenothera biennis* can be applicable to the whole range of fauna, and in particular to birds!"

In the 1920s more and more students of mammals, birds, and insects studied the phenomena of gradual geographic variation; among ornithologists there was, in particular, Bernhard Rensch, who had been working since 1924 on various research projects on the climatic parallels of characters among palaearctic and nearctic polytypic species (*Rassenkriese*), in fact on the same phenomena studied by J. A. Allen as early as 1871 and by many other ornithologists since then. Like his predecessors, he found that in many groups of races the characters show similar changes, clearly produced by climatic factors. Above all, (1) in general, the cold-climate races of a species are larger than those living in warmer climates (Bergmann's rule); (2) beaks and wings among the warm-climate races are often relatively longer (Allen's rule); (3) races from warm and humid areas are more melanic than those living in cooler and drier regions (Gloger's rule); and (4) the number of eggs in a clutch is generally smaller in warm than in cold areas (clutch-size rule).

These facts led Rensch in 1929 to the conclusion that

> This extensive, mostly finely graduated parallelism of the establishment of characters cannot be brought to agree with an origin through mutation and selection, and compels one to accept the direct influence of climate . . . The hypothesis is suggested that the persistent geographic races originally arose as phenovarieties, which in the course of many generations were continually influenced in the same way by climatic factors until they became regularly transmitted by inheritance . . . Geographic races occur by mutation only in a few instances. The fact that often extreme varieties arise on small islands suggests the effect of a kind of orthogenesis.*[7]

The geneticists now saw the need to state their views on the phenomena of geographic variation that did not seem to correspond with their hypotheses. Their explanations differed. Some of them took refuge in the assertion that these "climatically controlled" races were modifications, not genovarieties but phenovarieties, which would revert to their earlier state if the environmental conditions that had created them were to cease. This view was based on William Beebe's experiments in 1907 on doves of the genus *Scardafella;* however, it was amply refuted and soon given up even by the geneticists.

In the meanwhile, however, a book appeared that stimulated and promoted research into evolution to a degree unknown since Darwin's *Origin of Species* and de Vries's *Mutationstheorie.* This was Theodosius Dobzhansky's *Genetics and the Origin of Species* (New York, 1937). Written by a zoologist equally well trained in cytological and experimental genetics and in the study of nature, the book deserved its spe-

*[In subsequent publications on the subject during the 1930s Rensch abandoned this Lamarckian interpretation. He pointed out that by the new redefinition of mutation (by the Morgan school) even very slight variations could be classified as mutations, and that, indeed, Darwin's fluctuating variations now satisfied the definition of mutations. As such they could respond to selection. Thereby the 30-year-old conflict between mutationists and gradualists was resolved. The mutationists abandoned their saltationism and resistance to natural selection, while the gradualists abandoned Lamarckism and the inheritance of acquired characters. E. M.]

cially important position in the history of ornithology because it revealed a change of direction in genetic research, aptly dubbed, in the foreword by the editor, L. C. Dunn, "the Back-to-Nature movement." After many years of laboratory work, genetics had now reached the point of employing its acquired knowledge on the manifold problems connected with multiplex organic life and the discontinuity of organic variations. Population genetics, founded about 1930 by Sewall Wright in America and R. A. Fisher in England, began to collaborate with systematics, for which a new era had started almost contemporaneously in Germany, America, and England. It was marked by projects much more important than most of the previous work, some of which was even replaced. The intellectual élan of this renaissance reminds one very much of the period between 1870 and 1900, when systematics was reinvigorated by the theory of evolution.

Dobzhansky demonstrated that it is always and unalterably mutation that gives a fresh impetus to evolution, and that then selection, in geographic isolation, changes the genetic structure of a given population, always controlled by the prevailing environmental factors and the ecology of the species. A living being inherits not simply this or that morphological characteristic, but a definite way of reacting to environmental stimuli. Every mutation, every change in the genotype changes the reaction norm, and some of these changes can, under favorable circumstances, enable the new genotype to adapt itself harmoniously whenever the old genotype has become unsuitable. Selection affects not the genotype as such, but its dynamic qualities, its reaction norm, which alone decides its capacity in the struggle for existence.

These arguments, which gave a satisfactory genetic interpretation to the "climatic laws," immediately put an end to all Lamarckian hypotheses among ornithologic systematists. From now on, ornithologists provided the most effective support for the new research on evolution, because no other class of animals was already so completely known and so accurately worked on as the birds, from the points of view of historical relationships, geographic distribution, ecology, and the variability of species.

It was therefore an ornithologist, Ernst Mayr, who in a basic work soon reexamined from the systematic standpoint the questions concerning the formation of species that Dobzhansky had treated primarily from the genetic standpoint. Since 1932 Mayr had been the curator of the former Tring Museum collection of birds, which as the Rothschild collection had become part of the American Museum of Natural History in New York. His *Systematics and the Origin of Species*, a synthesis of taxonomic, genetic, and biological ways of viewing evolution, published in 1942, will long remain a reliable guide for systematists working in the complicated labyrinth of phenomena through which his predecessors had tried vainly to find their way during the past 150 years.* Two other books, both appearing at almost the same time, were devoted to the same task. Each was excellent of its kind and they were mutually helpful: Julian Huxley's *Evolution, the New Synthesis* (London, 1942), and G. Heberer's collection, *Die Evolution der Organismen* (Jena, 1943), in which is included a contribution on "Genetik und Evolutionsforschung bei Tieren," by H. Bauer and N. W. Timofeeff-Ressovsky.

Ornithological publications soon showed the effects of the new movement, most clearly at first in 1941 in a comprehensive monograph by Alden H. Miller, *Speciation in the Avian Genus* Junco,[8] whose author saw as his main task the pursuit of differentiation from individual variations through all intermediate stages up to the species. He paid special attention to the effects of different isolating factors, hybridization and intergradation, the recombination of racial characteristics and the spread of relatively large mutations, and the correlations of characteristics (which are less than one would have expected). The climatic correlation of color characteristics is highly variable, and even when it is very positive and well defined, it does not always correspond exactly to the geographic area in which particular climatic conditions (rainfall, humidity) occur, "probably as a result of other factors, like barriers to expansion and population pressure." Miller's extremely accurate and statistically based results were founded on an enormous quantity of material, no fewer than 11,776 skins of the genus *Junco*, of which 4,486 belonged to the Museum of Vertebrate Zoology in Berkeley. This is an indication of the increase in skin collections in some of the leading museums. At the time the American Museum of Natural History in New York and the British Museum, each with an estimated number of 800,000 skins, headed the list, followed at some distance by other American museums and the Zoological Museum in Leningrad. A hundred years earlier, when the most famous museums were those of Leiden, Berlin, and Paris, the importance of ornithological collections was rated by the number of species they contained; now research was concentrated more and more on single species and groups of species, and the systematist therefore tried to acquire the largest possible number of individual specimens.

But even when this demand for material is satisfied, the researcher is often still not yet able with certainty to carry out his classification of a group of forms according to their historical relationships. This can be seen in the treatment of the Galapagos finches of the genus *Geospiza*, discovered by Charles Darwin during the voyage of the *Beagle*

*A completely revised and greatly expanded version appeared in 1963 as *Animal Species and Evolution*, with an abridged revision in 1970 entitled *Populations, Species, and Evolution*.

and then repeatedly studied by students of evolution. Limited to a small and remote archipelago, this genus is abundantly but not clearly differentiated, and its characteristics were the chief inspiration for Darwin's idea "that species are not (it is like confessing a murder) immutable."[9] Although some thousands of *Geospiza* skins had gradually accumulated in museums, no fully satisfactory solution to the problem of their speciation had been worked out when David Lack, an English ornithologist, decided at the end of 1938 to travel to the Galapagos Islands in order to study the various questions on the spot, in the light of the new population genetics. In so doing he proved that a confusion of this sort could be cleared up only when knowledge of ecological peculiarities, means of isolation, food, population density and size, and so forth, had explained the dispersal and persistence of morphological differences, which would permit the classification, with the precision demanded today, of the forms according to their natural relationships. Lack published his results, so important for theoretical developments, in *Darwin's Finches* (Cambridge, 1947).

Ornithological systematics, in order to accomplish its most important task, is thus more and more compelled to ally itself with biologic research. At the same time the biologists have begun to pay special attention to the development of behavior characteristics, which forces them to study closely the results of the systematists. The abyss that ever since the days of Linnaeus and Buffon seemed destined to separate forever the two approaches to the material has now disappeared; today they are united in a common objective, research into evolution.

True, insofar as ornithological systematics is research into relationships, it has been able to show significant achievements only in the lower categories, as far as genera. Under the influence of post-1930 publications, especially Bernhard Rensch's book, *Neuere Probleme der Abstammungslehre. Die transspezifische Evolution* [Recent problems in the theory of descent: transspecific evolution] (Stuttgart, 1947),* more and more attention is being paid to the fact that micro- and macroevolution (or, in Rensch's terms, intraspecific and transspecific evolution) cannot be sharply distinguished; one provides a starting point for the other, and in both, only particular environmental characteristics, by means of selection, can direct the random variation into specific channels. As a logical consequence of this view, Mayr and his colleagues have been reevaluating many genera of birds that had been created and recognized on the basis of outmoded theoretical principles, and have "sunk" all those that placed exaggerated emphasis on single structural characters. This process is founded on the results provided by the study

*An English edition appeared in 1960 as *Evolution above the Species Level.* A major impetus to research on the higher categories was given by paleontologists, particularly G. G. Simpson, for example, *Tempo and Mode in Evolution* (1944).

of geographic (intraspecific) variation in characters. Because there is no doubt that existing differences among contemporary populations can be used to reconstruct phylogenetic stages, research is trying to discover what evolutionary trends can be inferred from the study of living examples in well-defined groups of forms (families). These trends, long known in comparative anatomy and paleontology, can be explained by the greater mutability of certain gene complexes and the fact that their tendency to vary is maintained as a group character, so long as the genotype remains essentially the same in all representatives of the group. The method is explained and employed by Mayr and Vaurie in their "*Evolution in the Family* Dicruridae" (1948).[10] Organic chemistry will probably soon prove helpful through its work on the analysis of plumage pigments; the latest research and publications by Otto Völker (1944, 1950) have demonstrated that the various "diffuse pigments" (yellow, red, violet) can be grouped according to their chemical, thus also according to their genetic, relationships.

However, these methods are useless when isolated species or higher groups, from families upward, have to be classified in the system of relationships; it is incontestable that, since the publication of Fürbringer's work in 1888, ornithology has made only limited progress in the systematics of the higher groups. Fossils discovered since then have merely confirmed that the division of the class of birds into orders is extremely ancient, and that even the recent orders were already sharply differentiated from each other in the early Tertiary period. No remarkable clues have been provided by Cretaceous deposits since the discovery of the "toothed" birds *Ichthyornis* and *Hesperornis* in 1871.

In fact there has been a regression. Some structural similarities in higher groups, which were thought to be useful in reconstructing a genealogical tree, have proved to be convergences, after they had been functionally explained. Thus the types of skull formation, on which in 1867 Thomas Huxley had based his highly regarded ornithological systematics, were in 1947 shown by Helmut Hofer of Vienna to be due to the kinetics of the skull. Blasyk proved in 1935 that differences in size and form of the scales covering birds' legs are functional in origin, and in the same year Stolpe showed that the structural similarities among *Colymbus*, *Hesperornis*, and *Podiceps* are simply the necessary products of the same kind of movement in swimming. To the frustration of the phylogeneticists, much of their material for major reconstructions evaporates into thin air as soon as it is accurately tested by the "biological anatomists."

Therefore arbitrariness has not as yet been successfully excluded from the classification of birds. Even contemporaneous systems (Erwin Stresemann's in Kükenthal's *Handbuch der Zoologie*, 1933–34, and

Alexander Wetmore's, Washington, 1934) do not agree completely either in the grouping of the large units or in their sequence. Where they differ, each author has followed his own opinion on the "best" arrangement, and there is still no sign that the considerable remains of the "practical" systematics conceived by Ray, Brisson, and Linnaeus will ever be completely scrapped and replaced by a demonstrably "natural," or phyletic, classification.

PART THREE

THE DEVELOPMENT OF BIOLOGY

16

Early Theories and Collections of Material

In the early days of ornithology, systematics and biology were inseparable. For 2000 years ornithologists had looked for distinctive biological characteristics in order to classify birds according to their natural degrees of relationship; though in the process they had discovered many peculiarities of ecology and behavior, they had not achieved their goal. Finally, they revised their attempts and took structure as the basis of systematics. As shown in Chapter 3, this decisive change was introduced in 1676 by Willughby and Ray's *Ornithologia.* Then systematics and biology took opposite directions, and only after 250 years were they successfully and finally reunited. Even as late as 1900 the present complete union seemed inconceivable to many people, and in that year one of America's leading ornithologists, Robert Ridgway, wrote: "Popular ornithology is the more entertaining, with its flavor of woods and green fields, of riverbanks and seashores, of birdsong and the many absorbing things that exist outdoors. But systematic ornithology, as a structural part of biology, is the more instructive and therefore the more important." In this chapter we are going to trace the development of "popular" ornithology.

The English physiologist William Harvey (1578–1658) followed Aristotle's lead when working on "the bird" as a structural and functional type. He chose the domestic chicken as his subject in his famous *Exercitationes de generatione animalium* (London, 1651), in order to penetrate, with this easily accessible object, as far as possible into the little-known world of physiology and behavior. Nevertheless he did not fail to observe other species of birds. Like Aristotle, Harvey paid close attention to the processes of courtship and mating. He noted that solely by means of psychological stimuli to the libido (stroking, gazing), an unmated female can be induced to lay an egg. Like Pliny, he made observations on the biological significance of the number of eggs. The militant and long-lived species, "because every form of breeding is developed by Nature only to preserve the species," needed fewer offspring than the defenceless and short-lived kinds. The doves, like *Turtur* and *Palumbus*, brood only two eggs at a time, "but frequency of laying compensates for the lack of numbers." Like the Stoics, he found in instinctive behavior evidence that a small quantity of divine afflatus resides in the bird.

> Indeed, it is admirable that the little creatures choose their nests with such forethought, and create, construct, strengthen, and conceal them with such inimitable skill and ingenuity, that it must be acknowledged that there is in them a particle of divine breath, and one must admit that the skill and wisdom inculcated in them are rather to be admired than comprehended.

He studied with reverence nature's provisions in the impulses of the broody hen, who defends even artificial eggs "with equal courage."

Harvey's older contemporary, Caspar Schwenckfeld (1563–1609), aroused interest in another biological field. He also was following Aristotelian suggestions when he contributed factual observations to the comparative biology of birds.

Even as a child Schwenkfeld, who was born at Greiffenberg in Silesia, was devoted to catching and studying local animals, and his enthusiasm did not cool when he took up the practice of medicine, first in his native town and then in Hirschberg. What most attracted him to animals was their behavior, because it revealed the wisdom of the Creator.

He became famous as the author of the first regional avifauna known to world literature, *Theriotropheum Silesiae* (Liegnitz, 1603), because he listed and described (mostly very well) the birds that he knew in his district — about 150 species. But it was chiefly his biological data that affected the development of ornithology. As far as his experience went, he described very aptly ecology, song, nest, number and color of eggs, and food; unfortunately, for the sake of "completeness," he credulously took over statements by Gesner and Aldrovandi, many of them false. As a student of Aristotle and the Renaissance zoologists, Schwenckfeld imitated them in trying to come to some general conclusions and to classify birds according to different criteria. He considered that they could be grouped:

(1) according to their "residence" (habitat; he divides them into 17 ecological groups);

(2) according to their degree of mobility (resident, transient, migrant, winter visitor);
(3) according to their foot structure (fissipeds with two subgroups, palmipeds with four);
(4) according to their food (nine groups);
(5) according to their color.

But because none of these classifications seemed to him to facilitate a view of the whole, he followed Gesner in using the alphabetic arrangement of species.

There was no progress worth noting in this field during the seventeenth century, except in England. It was no longer the philosopher's zeal for knowledge that drove men to seek out birds in the open, but only love of money or a passion for hunting. Thus the reading public demanded books on bird catching and hunting, like the one with the descriptive title, *Schertzendes Lufft-Waidwerk. Oder: Kurzweil in der Waidmannschaft mit Vogelstellen* [A merry book on aerial hunting. Or: diversion in hunting with the trapping of birds]. But the *Uccelliera* (Rome, 1622) of Giovanni Pietro Olina, a doctor of laws in Novara, is a famous exception among these books, because, as it says on the title page, it is a "vero discorso della natura, e proprietà di diversi uccelli, e in particolare di que' che cantano, con il modo di prendergli, conoscergli, alleuargli, e mantenergli" [true discourse on the nature and characteristics of various birds, and in particular those that sing, with the way to catch them, recognize them, breed them, and keep them]. The text (most of which, on the behavior of the birds treated, is lifted from older writings — Belon, Gesner, Aldrovandi, Schwenkfeld, and especially *Il canto de gl' augelli*, by Antonio Valli da Todi, Rome, 1601) is much less praiseworthy than the plates. The gifted Florentine engraver Antonio Tempesta (1555–1630) took the opportunity to demonstrate, in some 40 plates of birds, the superiority of engravings over the woodcuts previously used to illustrate bird books, and thus helped the work to a wide distribution.

After Schwenckfeld there was a full century's wait for the man who laid the true foundations for a study of comparative behavior, the ducal privy councilor Freiherr Johann Ferdinand Adam von Pernau, Lord of Rosenau.[1]

He was born on November 7, 1660, at Steinach (Lower Austria). Ten years later the family, being Lutheran, was forced to leave Austria, and settled at Sulzbach in Franconia. When he was only 16, Pernau entered the Bavarian university of Altdorf, where he took his doctorate. Fairly lengthy journeys through Italy, France, and Holland and a career at the Saxe-Coburg court followed, till shortly after 1690 the young baron and court councilor bought the Rosenau estate, near Coburg. There, retired from the bustle and clamor of the world, he devoted himself to studying God's creatures, until his death on October 14, 1731.

Birds proved to have the strongest attraction of all. Not satisfied with carefully observing them in their natural environment, or keeping many of them in cages in order to learn more about their peculiarities, he induced them, with indescribable patience and skill, to regard the cages as their "territory," and to come back after each flight into the surrounding country. In 1702 Pernau published the first record of his rich experiences, not in order "to chase after renown with this book," but because he was impelled to show other people the way of salvation from insipid dependence on pleasure and shameful gluttony. He had "learned from much experience, that the delights of the mind, which one seeks *in rebus naturalibus*, are not subject to any such mutability as underlies those diversions, like gambling and similar matters, which merely depend entirely on human invention." He published his little book anonymously (enlarged editions followed in 1707 and 1716) under the title *Unterricht, Was mit dem lieblichen Geschöpff, denen Vögeln, auch ausser dem Fang, nur durch Ergründung deren Eigenschafften und Zahmmachung oder anderer Abrichtung man sich vor Lust und Zeitvertreib machen könne* [Lesson on what can be done, for pleasure and amusement, with those delightful creatures, birds, in addition to catching them, only by inquiring into their qualities and taming or otherwise training them]. Encouraged by the demand, he published a second volume in 1720, as *Angenehme Landlust* [Agreeable country pleasures], "that may be innocently enjoyed in town and country, without exceptional expense."

Schwenckfeld had recognized clearly only a few biological problems; Pernau was the first to see a large number of others. According to him, there are important differences among the species of birds:

(1) in the way food is taken (it can be crushed in the beak, swallowed whole, or licked up);
(2) in their "residence" (ecology);
(3) in their degree of mobility (resident, transient, migrant);
(4) in the choice of nesting spot and the number of broods;
(5) in their sociableness after the breeding period;
(6) in their sociableness during the breeding period;
(7) in their periodic change of color, owing to the wearing out of plumage and alteration of beak color;
(8) in the length of the song period;
(9) in the ways of bathing (water, sand);
(10) in the ways of feeding the young (from the crop or the beak).

As a conscientious student, Pernau refused to fill the gaps in his knowledge by taking over other people's ideas, because then "things that one knows, and things that one does not know, would be mixed together, and stated as true, so that one would make the other suspect." Therefore he treated only a limited number of birds, and made up for this by the accuracy and intelligence of his interpretations.

Pernau discovered "territory" in the bird world, and its biological significance; he established that complicated bird songs are not inherited but must be acquired by listening to another singer; he correctly interpreted a series of forms of behavior (for instance, that of the young blackbirds who scatter at once after leaving the nest) as "instinct, given by Nature to young birds"; he was also the first to recognize clearly that the migrating bird is not started on its way by hunger and cold, but "is actuated at the right time by a hidden impulse and obeys it, whereas Man on the contrary often resists such." He even studied intensively the meaning of different bird calls.

Pernau's books indicated a method and an aim for German ornithology, and before much time had elapsed a researcher appeared who, while following the principles of the anonymous author of those fundamental books, nevertheless quickly outstripped him in experience. He was a chaplain to Count Pappenheim, Johann Heinrich Zorn (1698–1748),[2] the author of a book to lead men to God through the observation of birds.

Since the beginning of the eighteenth century, ecclesiastical dogma had been threatened more and more by the growing pressure of skepticism, buttressed by the theories of French and Italian scientists, whose efforts to explain biological problems by causation seemed to their opponents to substitute the arrogance of finite human reason for reverence for God's wise dispensations. Until that epoch philosophy had always been dominated by the Aristotelian concept of the world, which saw in divinely ordered purpose a specific kind of cause. This teleology deduced the assumed purpose from the means, or worked from the assumed purpose to the means. The mechanists, however, denied the effect of purpose, and postulated instead that every change resulted from a physical or mechanical cause. In principle (but rarely in practice) they worked by induction, seeking the causes of certain effects and then following up the effects of the hypothetical causes. They understood nature as a mechanical system, whose basic explanation lay in the general laws of motion, and soon they ventured with their attempted interpretations into fields where they lacked even the most elementary knowledge, so that here their opponents had an easy time.

To close observers of animal behavior the causal explanations proffered by Descartes, Borelli (1679), and especially Claude Perrault (in his *Essais de la Physique*, Paris, 1680), must have seemed simply ludicrous. For Zorn (1742), "all those who come marching in with mere senseless and lifeless materialism and mechanism are not the wise men they imagine themselves to be, but sophomores and fools." The Freiherr von Pernau thought the same (1716):

> That the owl has such a silent flight that it is impossible to hear it fly, serves for the astonishment and reproach of atheists . . . God did this so that it could catch other birds asleep at night . . . What do the so-called *esprits forts*, the messieurs Atheists, say to that? Are some of their atoms or little bits of dust the cause, have they just been mixed up together at random, so that the owl, which must prey in silence, will not be heard?

When the Protestant pastor Zorn wrote his edifying book on ornithology, the reaction against the impertinent causal researchers had been in full flood for some time. Not only theologians but natural scientists were busy shoring up the edifice of Aristotle's interpretation of the world, which Descartes had undermined. One of the first was the great John Ray, in his widely read book, *The Wisdom of God Manifested in the Works of Creation* (London, 1691). In many passages he refers to the structure and behavior of birds, and his friend and pupil, the Reverend William Derham, does the same in his *Phisico-Theology, or a Demonstration of the Being and Attributes of God from the Works of Nature* (London, 1713), which was reprinted many times and translated into several languages. The imagination of the teleologists was stimulated by innumerable natural phenomena, and Johann Kanold even explained (in Neickel's *Museographia*, 1727) that "the rarities in our museums ought to be spectacles through which the eyes of our minds should be led to the Almighty GOD and Eternal Being."

Zorn's aim, in his two-volume book on birds (1742, 1743), is described in the title, *Petino-Theologie, oder Versuch, die Menschen durch nähere Betrachtung der Vögel zur Bewunderung Liebe und Verehrung ihres mächtigsten, weissest- und gütigsten Schöpffers aufzumuntern* [Ornithotheology, or an attempt, to encourage men, through closer observation of birds, to the admiration, love, and reverence of their most powerful, wise, and good Creator]. In order to achieve it, he ranged through all the areas of ornithology in search of evidence for the purpose of Creation. "The more one directs one's observations to the different forms and shapes, plumage, songs, colors, ways of procreation, and so on, the more Divine Wisdom will reveal itself," because "God made the world, so that his invisible Being, in particular his wisdom, power, and goodness, should be recognized," and birds are "particularly made for the service of men, who are specially in a position to profit by them." Thus, for example, "the intention of the great Master with the various magnificent colors of birds is specially the revelation of his power, wisdom, and goodness." They serve mankind in distinguishing many species of birds and make it easier for him to recognize their sex. "But the various colors are also of great advantage to the birds themselves," as an indication either of sex (to help in mating) or of species (to help in distinguishing dangerous from harmless species).

In his search for purposive arrangements Zorn was the first to draw attention to many facts, which he interpreted as the relations between cause and effect. Certainly fools would say:

> It is false to conclude that certain groups of birds have this kind of limb and structure and no other, because, following the Master's purpose, they must live, for example, in and near water; whereas on land they must survive in another manner. One should rather conclude that, because they are constituted in this way and in no other, they can live in no other way than they do. Or: they are not created for a particular way of life, but must follow that which is practical for them, because they are created in this way and no other. But it seems to me, that this is the same sort of conclusion as if I were to say, when looking at a roof-gutter which I notice just as I write these lines, this was not made in order to collect water and carry it off, but the water must run off because the gutter is there and nowhere else . . . But the craftsman's purpose was that water should collect in it and run off where it would cause the inhabitants least trouble, therefore he placed it in this spot and nowhere else, to fulfil this purpose . . . The various groups of birds not only have various arrangements of limb for a particular purpose . . . but they also have an instinct to behave according to this purpose, although they could equally well survive in another way. Our wild and domestic ducks can simply live on the land and will breed when they are given bran, draff, and grain to eat, but when they see water they let everything lie and go in, even when they are still very young, just hatched, and have not been led by any older bird.

Zorn thought that all species of birds that usually live, breed, and brood on the ground, such as partridges, quail, and larks, whose feathers are earth-colored, are so colored in order that "the bird will not be recognized."

> The most surprising thing is that these birds seem to know that their feathers protect them, so when a bird of prey stoops at them, they do not rise and try to save themselves by flying and running, but huddle down, or stay lying on the ground . . . The white color of the birds in the cold and snowy lands of the midnight sun must . . . especially on account of the many birds of prey living there, add greatly

> to their safety whenever they huddle into the snow, or fly away just above it, since they cannot so soon be discovered. Experience teaches that our hawks chiefly catch the white, checkered, or white-tailed pigeons, so that these, and especially the pure white ones, are not willingly kept in dovecotes.

"The cuckoo gives its eggs to small birds to hatch, and therefore they cannot be large, so that they can be covered and hatched." The color of the "many-hued" eggs of birds breeding in the open protects them, "so that if often happens that human eyes overlook the nests of these birds that breed on the ground, and they remain undisturbed." On the other hand, birds that brood in holes usually have white eggs, "so that they can see their eggs in the dark, take charge of them, and not easily lose sight of one during the brooding." "God and Nature do nothing in vain, therefore the long claw or spur (of the lark) does not exist in vain. I think that it is not merely a distinguishing mark (whereby this group of birds is distinguished from others), but is present for another purpose"; for it "serves well against sinking too deeply into the ground, and because the feet therefore cover a greater surface, they are able to run better across the rough fields."

In building nests and looking after the brood, the stereotyped behavior of birds does not come "from imitating older birds, or others of their own kind, or from premeditation, but neither can it be called a mechanical arrangement, and such incomprehensible things must be looked for in the order and prevision of the eternally wise Master." "The Creator, who gave birds the miraculous instinct to migrate into another part of the world for their survival, also gave them strength and intelligence enough to start the journey at the right time."

What Zorn had to say in his first volume about the birds' choice of mates, choice of nesting site, nest building, incubation, and rearing of the young still meets modern standards of accuracy. He deals much more profoundly than his predecessor Pernau with the cause, process, and direction of passage and migration, and discusses special problems, like migration by day and by night, influence of weather, and influence of wind. Of the arrival and departure of migrants (which Pernau had already made into a calendar in 1702), Zorn kept an exact record, in combination with meteorological notes.

In the first chapter of the second volume (1743), he makes an uncommonly stimulating comparative study of German birds with regard to their external parts, flight, song, voice, food, breeding, feeding of young, number of broods, number of young and their degree of development at the time of hatching, nest building, eggs (size, shape, color), migrating instinct, bathing, parasites, intelligence level, and so on. In the second chapter he deals with "the sexes and species" of the birds known to him and to be met with in the domain of the counts of Pappenheim. "There is not one of the species mentioned, which total 127 without counting our domestic poultry, that I have not myself seen either several or many times." As a result Zorn's descriptions are usually very accurate.

From these quotations it should be clear that Zorn asked many new questions and answered part of them correctly, not in spite of, but because of, his strict logic in basing his observations on the teleological principle. During the whole of the eighteenth century this, rather than the causal, principle was bound to lead to more accurate conclusions, because chemistry and the physiology based on it were still rudimentary, and therefore it was impossible to check causal hypotheses by induction. Nevertheless, natural scientists during the Enlightenment began more and more to discredit the search for a purpose; they enjoyed replacing metaphysical interpretations with rationalist "explanations." Buffon did not hesitate to maintain that food and climate were the most important causes of variation and mutation of species, a statement believed for almost 100 years. In 1770 he solved the riddle of the sense of direction in migrant birds by saying, "The sensorium of the bird is chiefly filled with images produced by its sense of vision . . . it carries in its brain a map of the places it has seen."

Aristotle's harmonious image of the world could not withstand the attack for long, even though efforts to replace it with another failed. The fantastic hypotheses proposed by the materialists as causal explanations could not satisfy any serious biologist, and the excellent animal psychologist J. A. H. Reimarus must have spoken for many when he complained in 1798 about the spirit of the age:

> People with artificial preconceptions attempt to shake our conclusions, deduced from everything that we ascribe to the purposes of a wise Creator, and seek most industriously to dissuade us from reflection, efforts which, so far as I can perceive, serve neither to promote our rational inquiry, nor to establish our morality, nor to calm our outlook.[3]

In ornithology the immediate result of this development was that a prudent specialist no longer concerned himself at all with interpreting the facts he had established, because he distrusted the causal scientists just as much as the teleologists. An example representative of many is *Gemeinnützige Naturgeschichte Deutschlands nach allen drey Reichen* [General natural history of Germany covering all three kingdoms], by the forester Johann Matthaeus Bechstein, whose three volumes on birds (Leipzig, 1791–1795) are nothing but a rigorously classified collection of material, though unprecedentedly rich in detail. For him, "natural history is a science that teaches us to know natural objects in a certain definite order." The author does not dare to draw any sort of conclusions from his facts. His restraint corresponds exactly to the theory of a famous member of the Paris Royal Academy of Sciences, who (according to Blumenbach in 1790) "maintained that in our day it would be as ridiculous to believe that the eye is designed for seeing, as to assert that stones are designed to crack one's skull."

However, this period of resignation did not last long. As early as the first decade of the new century, the religious needs of the enthusiastic younger generation in Germany found a surrogate in the natural philosophy of Schelling and Oken, whose influence on the development of ornithological systematics is described in Chapter 10.

Just at the turn of the century, when people were still concerned only with the increase of single areas of knowledge, the *Traité élémentaire et complet d'ornithologie, ou histoire naturelle des oiseaux*, was published in Paris in 1800. This complete handbook of ornithology was compiled with thoroughness and scientific method by a young Parisian, F. M. Daudin (1774–1804), who had been crippled in both legs since his childhood, and it gives a good idea of the high degree of knowledge that had already been attained at the end of the eighteenth century. Vicq d'Azyr, Cuvier, and Blumenbach had been particularly active in the fields of anatomy and physiology during the last two decades, and the biological data obtained by Zorn had been considerably enlarged. In his "Explanation of the chief points to be studied in order to complete the account of each bird," Daudin listed 35 questions, including these:

> How often, and at what season, does the molt take place? What changes occur in male and female during the breeding period, and at other seasons, and how do they behave toward the young when they are growing? How long does incubation last? Does only the female brood, or does she take turns with the male? How does the sound of the call vary, according to the different emotions, in the male, the female, and the young bird?

Daudin found answers to many of these questions in the existing literature. He was able to enter 40 species in the chart of the length of brooding period, and 27 in that of the longest life span.

Not long after this French handbook a German one was published, *Anatomie und Naturgeschichte der Vögel* (volume I, 1810; volume II, 1814), by Friedrich Tiedemann (1781–1861). He was an anatomist and physiologist, whose unrivaled thoroughness in studying the literature proves that he had spent years in preparation. He reveals himself here not simply as a master of the subjects that he taught as a college professor in Landshut (the first volume treats of sense organs, locomotion, food, circulation, respiration, and reproduction), but also as an expert on the literature about the behavior and distribution of birds, which even then was very difficult to survey. When he wrote the second volume, on the "history of reproduction and growth," "metamorphosis," "location and distribution," and "bird migration," scarcely 14 years had elapsed since Daudin's *Traité*. But much had happened in this short period. Ornithology no longer interested only collectors, systematists, anatomists, and occasional naturalists; it had suddenly become popular.

This change took place earliest in Germany. The rapid increase in enthusiasm had been prepared for by Gesner, Pernau, Zorn, and the 252 colored plates of birds in Frisch's *Vorstellung der Vögel in Deutschland* (Berlin, 1733–1763). Philosophers, poets, and painters had contributed to advertising the charms of unspoiled nature. Knowledge of the nests and eggs of indigenous birds was widespread in Germany earlier than in neighboring countries, thanks largely to the color-illustrated *Sammlung von Nestern und Eiern verschiedener Vögel* (Nuremberg, 1772), by the city physician of Kahla, Friedrich Christian Günther (1726–1774), who was widely experienced. After 1790 several excellent bird specialists began to make their presence known. Bechstein's *Gemeinnützige Naturgeschichte Deutschlands* (bird volumes 1791–1795) was closely followed by *Naturgeschicte der Land- und Wasser-Vögel des nördlichen Deutschlands und angränzender Länder* [Natural history of the land and water birds of northern Germany and adjacent countries] (Köthen, 1795–1803), written by Johann Andreas Naumann, a farmer from the district of Anhalt. The growing desire for the information provided by colored plates of German birds was amply satisfied. The beautiful plates by the brothers J. C. and J. T. Susemihl inspired three men from Darmstadt, Borkhausen, Lichthammer, and Bekker, to produce an expensive folio, *Teutsche Ornithologie oder Naturgeschichte aller Vögel Teutschlands* (1800–1812). Only a few parts had been issued when two ornithologists, Johann Wolf from Nuremberg and Bernhard Meyer from Offenbach, appeared on the scene with a still more splendid work in folio entitled *Naturgeschichte der Vögel Deutschlands* (Nuremberg, from 1805, with plates by A. Gabler and J. M. Hergenröder).* These two collaborated again in a *Taschenbuch der*

*Johann Wolf (1765–1824), a schoolteacher in Nuremberg, had published from 1799 three issues (of six plates each) of a folio, *Abbildungen und Beschreibungen der in Franken brütenden wilden und zahmen Vögel* [Wild and domestic birds of Franconia, illustrated and described], for which Ambrosius Gabler had painted and engraved the illustrations. From 1805 onward, with the collaboration of Dr. Bernhard Meyer (1767–1836), it was continued as *Naturgeschichte der Vögel* *deutschen Vögelkunde* (two volumes, octavo, Frankfurt am Main, 1810). It sold quickly because of its useful information and practical illustrations, although Bechstein had already written a similar short book, *Ornithologisches Taschenbuch von und für Deutschland* (in three parts, Leipzig, 1803 and 1812), and his *Gemeinnützige Naturgeschichte* had been reissued between 1805 and 1809 in an improved and much expanded form. In short, by 1810, when Tiedemann's *Naturgeschichte* began to appear, a great many Germans interested in nature were studying birds.

Matters were very different in France. Native birds had been almost forgotten while people admired the exotics appearing in many luxurious illustrated volumes — an interest first stimulated and developed by Brisson and Buffon. Levaillant was right to complain in 1799, "Not only do we have no natural history of our own country but . . . we have none for the environs of Paris, and, worse than that, we do not even possess the history of one, yes one, species!" There was not a single handbook on French birds. Anyone who wished to learn about them had to look through Buffon or track down the key word in one of the great dictionaries. In the huge *Encyclopaedie méthodique* published in Paris by Panckoucke, he could find French birds treated with great scholarship and from small experience, at first in 1782 and 1784 by Mauduyt de la Varenne, and then in fragmentary fashion by Bonnaterre, from 1790 to 1792. Later on he could refer to the 24 volumes of the *Dictionnaire des sciences naturelles appliqué aux arts* (1803-04), to which the ornithologists Sonnini and Vieillot contributed.

England also, with its seafaring tradition, had been more interested in exotic than in native birds, although there was no lack of attractive picture books, like Pennant's *British Zoology* (first edition 1766, fourth 1776-77), William Lewin's *Birds of Great Britain* (1789-1791), Edward Donovan's *Natural History of British Birds* (from 1794), and Thomas Bewick's *History of British Birds* (1797-1804). What Gilbert White wrote about "Faunists" to Barrington on August 1, 1771, applies to all these authors.

> Faunists, as you observe, are too apt to acquiesce in bare descriptions, and a few synonyms: the reason is plain; because all that may

Deutschlands by the same publisher (J. F. Frauenholz in Nuremberg), who had the outstanding plates of the Wolf book newly engraved for a slightly different format. This splendid work has always been a great rarity, because the *Teutsche Ornithologie* by the Darmstadt ornithologists reduced its distribution. Finally the publication (from 1820) of Johann Friedrich Naumann's much more complete *Naturgeschichte der Vögel Deutschlands* brought the Nuremberg project to an untimely end. The thirtieth issue, with the last of the 180 plates, appeared in 1821 (see Johann Wolf in Oken's *Isis*, 9 (1821), Beylage No. 4; Claus Nissen, *Die illustrierten Vogelbücher* (Stuttgart, 1953), No. 1007, and p. 53, where a total of only 16 copies is indicated.

be done at home in a man's study, but the investigation of the life and conversation of animals, is a concern of much more trouble and difficulty, and is not to be attained but by the active and inquisitive, and by those that reside much in the country.

But the active and inquisitive field ornithologist still needed by British ornithology was already at work — Gilbert White himself. This perceptive observer of nature spent almost his entire life (1720–1793) as a cleric in his native village of Selborne, in Hampshire, and enjoyed himself by studying the life and conversation of animals, especially birds.[4] He communicated his findings in letters to Thomas Pennant (from 1767) and Daines Barrington, F. R. S. (from 1769), including biological details of the most diverse kinds, which were not published in any existing English book. His criterion was that "good ornithologists should be able to distinguish birds by their air as well as by their colours and shape, on the ground as well as on the wing, and in the bush as well as in the hand."[5] This outstanding man would have taken the great wealth of his knowledge into the grave with him, if he had not been encouraged by Barrington to publish a selection of the letters (110 from the years 1767–1787) written to him and Pennant. So *The Natural History of Selborne* came into existence; it has been reissued over a hundred times since its first appearance in London in 1789, and is numbered among the English classics because, in the words of Alfred Newton, White was "a prince among observers, nearly always observing the right thing in the right way, and placing before us in a few words the living being he observed," in a style marked by "its unaffected grace, its charming simplicity, and its natural humor."[6] His careful attention to all forms of life made him the first to differentiate the chiffchaff (*Phylloscopus collybita* Vieillot) from the willow warbler (*Phylloscopus trochilus* Linnaeus), and in so doing to demonstrate that in doubtful cases the systematist must rely on the decision of the biologist.

No two birds can differ more in their notes, and that constantly, than those two that I am acquainted with; for the one has a joyous, easy, laughing note; the other a harsh loud chirp. The former is in every way larger, and three-quarters of an inch longer, and weighs two drams and a half; while the latter weighs but two: so the songster is one-fifth heavier than the chirper. The chirper (being the first summer-bird of passage that is heard, the wryneck sometimes excepted) begins his two notes in the middle of March . . . The legs of the larger of these two are flesh-coloured; of the less, black.[7]

Many years elapsed, however, before the same characters were rediscovered by Bechstein (1793) and a few systematists got around to describing the specific distinctions between these two warblers.

It is certainly true that anyone wishing to learn something in particular from White's collection of letters would have difficulty in finding his way among the chaos of treasures. The first author to organize biological knowledge under key words (following the example of the French dictionaries) and to enlarge upon them from his own experience (like White) was a retired colonel, George Montagu (1751–1815), a close observer and intelligent interpreter of behavior. His *Ornithological Dictionary; or, Alphabetical Synopsis of British Birds* (two volumes, London, 1802; supplement, 1813) became the fundamental book on British field ornithology.

During the same period the investigation of North American birds was almost better carried out. This was largely the work of one man, Alexander Wilson (1766–1813), who had started out in his native Scotland as a journeyman weaver. After eking out a meager living by writing verses and peddling, he decided in 1794 to emigrate to Pennsylvania. It so happened that after a few years in Philadelphia he made the acquaintance of the naturalist William Bartram, who interested him in ornithology in 1803. With fanatic intensity Wilson steadily observed, collected, and drew birds, until in 1808 he had enough material to begin to publish his book, *American Ornithology; or the Natural History of the Birds of the United States*, illustrated by himself. He could hardly have imagined that he had thereby made his name. Before he died in 1813 he had already published 7 volumes, with 63 colored plates.

New light had also been shed on the behavior and nesting of South American birds, and it was provided, startlingly enough, by a Spaniard. In 1781 Don Felix de Azara (1746–1811), a Spanish army officer, was sent to Paraguay to take part in the cartographic work of a commission that was to settle the disputed boundary between Portuguese Brazil and Spanish Paraguay. At first as a pastime, and then, after he acquired the duodecimo edition of Buffon's *Histoire naturelle*, with scientific seriousness, he devoted himself for a good 20 years to studying the behavior of birds and mammals. As soon as he reached home he published his famous book, *Apuntiamentos para la historia natural de los paxaros del Paraguay y Rio de la Plata* (three volumes, Madrid, 1802–1805). It is a mine of reliable information, not simply on the appearance, but also on the biology of the indigenous birds. Sonnini, who had at one time taught the Comte de Buffon about bird life in Cayenne, translated Azara's book into French (four volumes, Paris, 1809), and then, as we saw in Chapter 6, Illiger got to work on it.

In 1777 Sonnini himself, shortly after his stay in the tropics, was sent on a journey by Louis XVI. Sonnini had formed the fantastic plan of crossing Africa from north to south, but he got only as far as Upper Egypt, and in 1780 went home by way of the Aegean islands and the Peloponnesus. For a long time he kept his experiences to himself, publishing his *Voyage dans la Haute et Basse Égypte* only in 1799, a work frequently consulted by ornithologists for its important information about the migration and wintering of European birds in Egypt and on the Aegean islands.

Most of these latest sources, as well as all the older ones, starting with Aristotle, were used and often quoted at length by Tiedemann in his *Naturgeschichte der Vögel*, because he intended to reach important conclusions by collating everything then known about birds. In spite of all his efforts, the yield was slight; he looked for causal explanations and could not find them. For example, why do the smaller birds in South America, as Azara stated, lay far fewer eggs than those in Europe — usually only two, and at the most four? Wherever he could, Tiedemann gave physiological explanations, in order not to concede anything to the teleologists. Polygyny among some birds resulted from their having small lungs and not very extensive air sacs: "they also live in the lower strata of air, which contain little oxygen, so that in consequence their blood is not very much oxidized and is very rich in inflammable elements, carbon and nitrogen, and therefore perhaps it is better adapted to emitting semen, which consists principally of the latter." To a person like Tiedemann the egg chart did not mean a greater or lesser need for the species to multiply (Harvey, 1651); he thought "that in general birds living on animal substances are the least fertile, and on the other hand those living principally on vegetable substances are the most fertile. The reason for this seems to be, that the birds living on animal substances find food less abundant." He paid no attention to the fact that the color of the plumage may often be related to the bird's need for protection (Zorn, 1742). As a physiologist, he considered that food and temperature chiefly determined plumage color:

> If it is principally carbon, in its many degrees of oxidation, that forms the chief coloring agent in the animal and vegetable kingdom, as is very probable, then this phenomenon would be explicable in that the food of birds living on vegetables and insects contains a larger quantity of the coloring agent, carbon, than does the food of birds living on animal substances. The influence of warmth on the color of the plumage is revealed by the general increase in its variety and vividness from the polar circles toward the equator . . . Perhaps warmth is effective only in producing a larger quantity of carbon in the animal juices . . . Thus the plumage of a bird would be the more vividly colored the more it feeds on vegetables, the warmer the air temperature is, and the more extensively light falls on its plumage.

Tiedemann had no sympathy with such terms as "instinct," which are not subject to analysis. "It is very remarkable that birds usually return to the same place where they have bred" (according to the determinations of Linnaeus and Spallanzani). "If one tries to explain this phenomenon also in terms of instinct, then it is not properly explained, because instinct is still not explained." The stimulus for migration lies, according to Tiedemann, "very probably in the peculiar composition of the atmosphere, perhaps in a different electric tension in the air, which is created by the different position of the earth in relation to the sun or by the different seasons." "The return of birds in spring in a direction away from the equator and toward the poles seems to be effected partly by lack of food, partly by the rise in air temperature and the altered electric tension, and finally by the urge to breed." And from his laborious tables on the geographic distribution of birds (220 pages!), the author could conclude nothing more than that the number, variety, and singularity of birds depend on the number, variety, and singularity of the animal and vegetable products of their biotope, from which it can be deduced that "every part of the earth, in its creation and formation, by its peculiar climatic and physical conditions has produced its own peculiar vegetable and animal forms." Birds therefore have not spread from one place around the earth, as Linnaeus had taught on the authority of the Books of Moses.

That was basically all that causal research had been able to learn about the natural history of birds up to 1814. Did this meager and, moreover, still uncertain result justify the centuries of effort?

Dedicated biologists were not bothered by this doubt. They did not ask where their zeal was driving them, and followed with enthusiasm a man who, warning against the current fashion in science, declared:

> The fruits of such an irreligious study of nature are easy to see. The head is filled with knowledge that produces nothing except swelling; such men keep babbling names in a manner neither useful nor agreeable, and make natural history odious to sensitive persons; the unorganized mass of knowledge finally brings its possessor to the point where he can no longer see the wood for the trees. This method of pursuing the study of natural history is the lowest and meanest, and can at most serve to provide material that at some future time may become useful. It does nothing to elevate the mind and ennoble the heart.

The words are those of Christian Ludwig Brehm.

17

The Naumann Period

After 1790, budding ornithologists in Germany made more rapid progress than Bechstein's contemporaries, who had with difficulty filled in the outlines of the study of indigenous birds, whereas their sons were familiar with the subject even before leaving the classroom.

By 1820 there were already many devotees of this *scientia amabilis* in all regions of Germany. The increase could be attributed not only to the good handbooks by Bechstein, J. A. Naumann, and Bernhard Meyer, assisted after 1815 by Temminck's *Manuel d'ornithologie* and Meisner and Schinz's *Die Vögel der Schweiz*, but also to the cultivation of ornithology in the universities, where it took an important place in the zoological lectures of Blumenbach in Göttingen, Merrem in Marburg, Nitzsch in Halle, Lichtenstein in Berlin, and Tiedemann in Heidelberg.

However, it would not have advanced so rapidly if it had not been led by three gifted men who swept the young along with them. Brehm, Faber, and J. F. Naumann differed in habits and preconceptions, but were united in their aim.

Pastor Christian Ludwig Brehm (1787–1864) opened the eyes of his contemporaries, who had forgotten to look for the relations between birds and their environment, between form and function, to the "view of the whole."[1] As a young man he could already draw on unrivaled experience, and a profound insight into relationships. He looked for them because, like his fellow pastor Zorn a hundred years earlier, he was a teleologist. For him "the study of nature can and should be a true service of God." "The more our minds are inclined to faith," he wrote in an essay, "Die Naturwissenschaft aus religiösem Gesichtspunkt betrachtet [The natural sciences viewed from a religious standpoint]" (1827),[2]

> the more religious our hearts are, so much the more the veil is lifted that conceals God's working from our dim eyes, so much the more clearly our newly sharpened vision sees in even the insignificant and apparently chaotic the endless wisdom of the Most High . . . Just as little should the naturalist forget the work of the Creator so far as to succumb to the idea of reconstructing nature, that is, shaping and developing it according to his own intentions; he must rather, if I may be permitted the expression, follow the Creator and attempt everywhere to recognize his footsteps . . . "Every creature is adapted, in its entire structure, to the place in which it lives, and to the meals that it eats . . .
>
> Plumage is denser in cold countries than in warm. Northern birds are very thickly plumaged, indeed the Iceland and Greenland ptarmigan have such a dense and impermeable coat that their toes are buried deep in feathers. Birds in warm countries not only have thinner plumage but sometimes have completely bald spots. Even color is adapted to circumstances, so that only at a particular season is it different from what it is at other times. The ptarmigan that live in the north and in the Alps are yellow, gray, and black in the summer, like the rocks on which they live; in winter they are white.

Through his thoughtful explanations Brehm, like Zorn, became a precursor of functional or, as it is called today, biological anatomy. In the third volume of his *Beiträge zur Vögelkunde* (1822; see below) he compared the climbing equipment of the nuthatch and the woodpeckers:

> The form of the woodpeckers can be regarded as the basic shape for climbing birds. Their short, strong feet, with toes in pairs and equipped with large, outward-curved claws, their wedge-shaped tail consisting of hard, curved-back feathers, and their usually slim and low-slung body, enable them to hop up trees with the greatest speed and sureness. The whole equipment is so well adapted to the purpose that it seems impossible to change anything without interfering with their ease of climbing. But the nuthatch is very different; its feet are longer, with three forward-pointing toes; its body is rather short; and its tail has such weak and pliable feathers that it can give no support during a climb. And yet the nuthatch not only climbs up trees quite as nimbly as the woodpeckers, but also — what they cannot do — climbs down, often hanging head downward and clutching the trunk so tightly that in this position it can crack a beech or hazel nut. How did the Creator manage this? Entirely through the shape of

toes and claws. The toes are much longer than in the woodpeckers, and therefore cover a much larger area; the ends of the claws on the middle and rear toes are 22½ lines [about 5 centimeters or 2 inches] apart when the toes are spread out (that is, almost as far apart as the body is long); they have very large, needle-pointed claws curved in a semicircle, and a number of pads below. Thanks to this arrangement they cover in climbing a relatively large area, which naturally offers more uneven places and more holds than a smaller one. The warts on the sole evidently help in ensuring a tight grip, and the binding at the base of the toes prevents them from spreading too far apart while increasing their strength. Since the arrangement of climbing equipment in the nuthatch is so different from that of the woodpeckers, its way of climbing is also very different.

Such expert knowledge could not be acquired by anyone content only to watch live birds, no matter how closely; he must also hunt birds with trap and gun, in order to study their bodies carefully and take them apart. The pastor of Renthendorf was both an ingenious bird-catcher and a passionate bird collector.

Christian Ludwig Brehm, born on January 24, 1787, in the parsonage at Schönau (not far from Gotha), was intended to be a theologian, but he found the nearby Thuringian Forest more attractive than school. His father was a friend of Bechstein, who taught the natural sciences at the Salzmann Institute in the neighboring town of Schnepfenthal, and before the boy was 10 his thirst for knowledge had been partly satisfied by the present of Bechstein's recently published *Gemeinnützige Naturgeschichte.* It may have given a direction to his interest, because while he was still at the Gymnasium in Gotha he started a bird collection. In 1810, after his theological studies at Jena, he spent two years as a tutor with a landed family living near Neustadt on the Orla. He enjoyed nothing more than combing the neighboring woods, carrying his fowling piece and, according to his own words, observing the creatures and products of nature with a devout mind. Then in 1813 he was appointed to the nearby parish of Renthendorf, where he was active for more than 50 years, and his way of honoring God in nature made the name of his little woodland village world famous.

When Brehm published his first book in 1820, he was already thoroughly familiar with Thuringian birds. As he knew by then, there was still much in natural history to be corrected, completed, and added to. Therefore he examined "each bird with exactness, for its different regular changes of color, its internal structure (about which so extremely little has been published), its entire existence, its food and procreation," and he limited his reports to those species about which he had something new to say, something to correct, or a complete description to make. In this way he created his valuable book, *Beiträge zur Vögelkunde in vollständigen Beschreibungen mehrerer neu entdeckter, und vieler seltener, oder nicht gehörig beobachteter deutscher Vögel* [Contributions to ornithology by means of complete descriptions of several newly discovered, and many rare, or insufficiently observed German birds] (three volumes, Neustadt a. d. Orla, 1820, 1822, 1822). Few works had such a lasting effect on German ornithology. Even J. A. Naumann's much admired *Naturgeschichte der Land- und Wasser-Vögel* was outclassed, and the pastor of Renthendorf became the center of an admiring circle of experienced ornithologists. He was able to kindle his own burning enthusiasm in others, and through his letters alone he exercised an irresistible force. E. F. von Homeyer, who received his first letter from Brehm in 1833, still felt the effect when he was an old man. As he wrote in 1881, "Brehm, more than anyone else, understood how to attract a young and energetic person to the study of nature, and to demonstrate what endless pleasure it can provide, and what a loyal friend in joy and sorrow science, and above all natural science, can be."[3]

As early as 1824 Brehm saw such a group of specialist friends around him that he risked founding an ornithological journal — the first in the world. He called it *Ornis, oder der Neueste und Wichtigste der Vögelkunde* [the newest and most important in bird lore], and promised the subscribers an intermittent series; the issues would appear "only" every three to six months, and each issue would be six to twelve pages long.

The golden age of European ornithology had begun. Side by side with the enthusiastic Brehm appeared his friend and critic, the cooly observant and clear-thinking Frederik Faber. Faber was born in 1796 in a village near Henneberg, not far from Fredericia in Denmark. His father, a lawyer, gave him a good schooling in Copenhagen and then had him study law at the university there. "Fritz" must already have been an ardent zoologist, because as soon as he had passed his final law examination, in October 1818, he obtained a state grant for research in Iceland. With really fervent energy he quartered the island from May 1819 to September 1821, observing and collecting chiefly birds and fish, and bringing home astonishingly precise notes. A professional zoologist would have been generally admired but he, as a lawyer, could hope for nothing but the acclaim of a small group of like-minded men, who were nearly all active in Germany. They were the audience he had in mind when (though already embarked on an official career as regimental quartermaster and auditor of the Schleswig Curassiers, based on Horsens) he published in 1822 his *Prodromus der isländischen Ornithologie* (Copenhagen). This was the promising forerunner of a significant monograph, "Beiträge zur arktischen Zoologie," which appeared in Oken's *Isis* in the years 1824, 1826, and 1827, unfortunately split into numerous sections. A critic called it "diffuse." Only later was its true value recognized, and today it is considered a model of detailed description, equaled only rarely at any time, although many authors, from J. F. Naumann onward, attempted to imitate it. Faber had all the hallmarks of the true expert: complete mastery of the literature, anatomical training, a keen eye for the details of a bird skin, and a real genius for what the field biologist must note when he wishes not only to describe, but also to understand. More critical than the enthusiastic Brehm, and more intellectually developed than the quiet Naumann, Faber unquestionably takes precedence of the others in this illustrious group.

He attracted more attention with his two-volume book *Über das Leben der hochnordischen Vögel* [Bird life in the far north] (Leipzig, 1825, 1826). In fact this publication, appearing so soon after Brehm's *Beiträge*, was yet another milestone in the history of ornithology. As Faber said in the foreword, "The time is past when natural history consisted of the names and descriptions of a few natural objects, and when its admirers had no higher aim than to know and enumerate the names of as many natural products as possible." What was earlier regarded as the goal of science was only the means for penetrating its true essence. Now it should be important to discover the inner continuity of nature by means of its laws, because according to Kant the whole of nature is nothing else than a linking together of phenomena according to laws. Therefore in this book Faber took pains to organize the majority of biological phenomena under headings: geographic distribution, migration, the many subdivisions of reproduction, the kinds of locomotion, the taking of food, and so on, and then he tried to discover the effective causes.

Faber looked for regulating forces partly in the animate and inanimate environment (thereby becoming a pioneer in ecological method), partly in the bird itself (to which he ascribed a large number of instinctive urges, such as rearing of the young, feeding, protection, keeping warm, warning, and defense). Tireless in his efforts to decipher nature's legal code, this trained lawyer believed that he had discovered a series of natural laws that are followed according to rules; "when law and rule collide, the latter must yield." According to Faber, the effect of the "homesickness urge" on migrating birds is meaningfully regulated, like that of other urges, by external factors, in this case by air temperature and the availability of food. He recommended this hypothesis for further study, and provided the first basic material for it in a long chart of observations arranged according to dates, which he made in the spring of 1824 near Horsens, "in relation to the arrival of migrants and the simultaneous flowering of plants," as well as scattered notes on the arrival of fish and the emergence of animals from hibernation. "The motives that prompt birds to visit different places in their own zone" — that is, the ecological factors on which the distribution of each species depends — could arise from: *(a)* their being accustomed to certain kinds of nesting material, to particular locations for nests, or to certain sleeping places; *(b)* the climate, especially the air temperature; *(c)* food; *(d)* the "desire for security," the urge for safety from enemies. In innumerable places he risked perceptive explanations, and nearly always found the right ones. Thus Faber's book today, when his method of observation is highly regarded, can be read from beginning to end with interest and admiration.[4]

Ornithologists at the time were happier with Johann Friedrich Naumann's book on birds, which made steady progress from 1820 on, reaching its fifth volume six years later, because its author was governed less by analytic intellect than by lively powers of observation, and above all his attention was centered on bird life in Germany.

The Naumanns,[5] an active family of farmers from the little village of Ziebigk near Köthen, had been zealous birdcatchers for generations. For Johann Andreas Nauman (1744–1826), who took over the farm in 1760, the pursuit had become a passion. As he wrote of himself in 1797, "The love of the beautiful inhabitants of the air seemed to be so strongly rooted in me that I could not watch birds unmoved, and in my boyhood this became a complete passion."[6] When Johann Friedrich (1780–1857), the eldest of his three sons, showed great pleasure in and aptitude for drawing and painting, Johann Andreas had him take lessons, and whenever the father caught or shot a rare bird his son painted it for him. "We at last began to make a collection of all birds that pass through our district, to our own satisfaction, because at last some good friends advised me to describe them scientifically and have engravings made of them."[7] The result was finally printed at the author's expense, with the young artist's engravings, and entitled *Ausführliche Beschreibung aller Wald-, Feld- und Wasser-Vögel, welche sich in den Anhaltischen Fürstenthümern und einigen umliegenden Gegenden aufhalten und durchziehen* [Detailed description of all birds of wood, field, and stream that reside in or pass through the principality of Anhalt and some of its neighboring districts] (four volumes, 1795–1803), by Johann Andreas Naumann. In 1801 it was renamed *Naturgeschichte der Land- und Wasser-Vögel des nördlichen Deutschlands und angränzender Länder.* This excellent work was hardly finished before the Naumanns made new and remarkable discoveries, which they did not wish to keep to themselves and therefore published, in eight supplements (1804–1817), although their idealism cost them a great deal; in spite of all their efforts, only 60 of the 500 copies printed were sold.

During this period, in the summer of 1816, Johann Friedrich Naumann received a letter from Amsterdam. It came from Temminck, who was working on a second edition of his *Manuel d'ornithologie* and was looking about for an artist. He asked Naumann to paint him a series of plates of birds, which Naumann agreed to do only under certain conditions. In his reply from Ziebigk on August 28 he wrote:

> I own a small farm, which supports me and my family, but neither allows us to live in great style, nor leaves anything over beyond the bare necessities of life. I must attend to anything that happens, and unfortunately often lend a hand myself; the income does not permit me to keep regular servants. My garden, in which I have a nursery and cultivate 700 species of exotic plants (I am also a botanist), I must look after largely myself, and therefore I am also a mechanic and make all the smaller tools for farming and gardening, and also my guns and other items of wood, bone, and metal. You therefore find me always occupied; now I am a carpenter, now a locksmith and gunsmith or turner, or I work as a gardener in my garden or I oversee my farmhands in the fields. To recuperate I go for a while into my little collection room and cheer myself up among my pets, or I hunt or catch birds. I am a good shot, and hunting is my greatest pleasure, because then the practical side of ornithology and in addition botany and entomology can be employed. That with so many occupations I have no time left to work for anyone else, you will readily understand, because though many jobs do in fact cease as soon as the bird migrations end in the fall, they are replaced by new ones more suitable to the season, which last until the spring migrations; for now I begin to organize the notes that I collected in the summer and to do my literary work in the evening; during the day I make paintings and copperplate engravings of my birds. My elderly father cleans and polishes the copperplates, and he also prints from them during the winter, having also built the presses himself. So therefore, when in winter the farm and garden work is over, the jobs for our book, such as drawing, painting, engraving, printing, [and] describing the birds are our welcome occupations. Only the printing of the text and coloring of the plates we do not do ourselves; the latter is copied from sample sheets that I have made. Then the winter is also the time when I can study books on natural history [and] travel, and the latest periodicals, at my leisure.[8]

This was the sort of life led by the man who had undertaken to write a long natural history of German birds. He made the first preparations for it in 1817. To enlarge his experience he followed the advice of his eager correspondents Friedrich Boie and Captain von Wöldicke, and from May to June 1819 he visited the seabird breeding grounds on the islands off the west coast of Schleswig-Pellworm, Amrum, and Sylt.

A year later readers interested in ornithology were most pleasantly surprised by the first volume (on birds of prey) of a book whose title, *Johann Andreas Naumanns Naturgeschichte der Vögel Deutschlands nach eigenen Erfahrungen entworfen . . . aufs neue herausgegeben von dessen Sohne Johann Friedrich Naumann* [Johann Andreas Naumann's natural history of German birds, drawn up in accordance with his experiences . . . newly edited by his son Johann Friedrich Naumann], gave no sign that it resembled the "first edition" only faintly. In fact, the son had begun on a much wider scale and had incorporated only fragments of the original. Brehm, who was hard to please, called it "splendid" in his long review of the second and third volumes for his *Ornis* in 1824:

> The younger Mr. Naumann is, like his father, a born naturalist, for he possesses the eagerness that fears no obstacles and does not cool during difficult and protracted investigations; the gift of observation that catches what thousands have overlooked, and that knows how to discover the essential; the zeal that never flags while collecting, and that by rugged endurance succeeds in capturing what is hard to come by; and finally the skill that understands how to classify and expound what has been discovered.[9]

To the qualities of the author were added those of his collaborator. Thoroughly convinced that the anatomical characteristics of the different groups and a general introduction to bird structure should not be omitted from a scientific book on ornithology, Naumann had made the most fortunate choice in his search for a specialist. Christian Ludwig Nitzsch (1782–1837), his collaborator, was one of the most accurate, cautious, and imaginative morphologists who ever concerned themselves with avian anatomy. He had begun his studies in Wittenberg in 1800, taken his doctorate in medicine in 1808, and, after the unification of the universities of Wittenberg and Halle in 1816, had moved to Halle as full professor of zoology and director of the Academic Zoological Museum. His contributions on anatomy and physiology to Naumann's book helped not a little to earn its immortality.

Assisted in all these various ways, the progress of German ornithology from 1820 to 1826 was perhaps unparalleled in the history of the science. Each of the triumvirs Naumann, Brehm, and Faber made his own particular contribution: Naumann collected all known facts with unflagging energy and described them vividly and attractively; Brehm illuminated them from his teleological standpoint; and Faber, the critical scientist, dissected them to find out their causes. The far journeys made by German ornithologists provided a copious quantity of material for study and comprehension: Heinrich Boie investigated Java with an idealist's enthusiasm, while the ambitious Rüppell was in Abyssinia, the Freiherr von Kittlitz in Kamchatka, and Eversmann in Tartary; and, when the great Pallas's *Zoographia Rosso-Asiatica* finally appeared in 1827, it revealed the first glimpses of bird life in the enormous Russian Empire.

This smooth sailing did not last long, and difficulties began when Frederik Faber was no longer at the helm — he died unexpectedly on March 9, 1828, at the age of 31. German ornithology suffered from his loss for over 30 years, while other countries, especially England, were almost able to catch up.

Brehm took over and steered into a fog of metaphysics. When lucky discoveries about goldcrests and tree creepers had attracted his attention to the lowest units in systematics, he became convinced after much pondering "that creatures are arranged in groups or series, because different shapes seem to have developed from one basic shape; these, like works on a single theme, are in the same key and show great similarities, but also significant differences."[10] Now he began fanatically to search for these "variations on a theme," which were for him the unalterable creations of God's hand. He thought there were three species (or, as he later called them, genera — meaning subspecies) of snow buntings, three of pied flycatchers, the garden, the wood, and the true chaffinches, the deciduous-tree, the rush, and the grass-tree pipits, and so forth. To the skeptics he explained, "Anyone without an inborn scientific eye will fail, even after repeated observation, to see what the gifted man notices at once."[11] After that it was not surprising that a rapidly increasing number of people parroted him. The cautious Naumann was certainly not among them, but he was too little of a collector and a theoretician to be able to oppose the prophet from Renthendorf effectively. Faber was different. As he wrote in 1826:

> Brehm shares a failing of many investigators and scholars generally, who easily accept a pet theory and very often base their proofs on it, or at least are guided by it, though the proofs of a theory should not have to rely on the theory itself . . . Dead skins are only a small part of birds . . . Observation of the bird and its behavior is the clue used by Nature herself to guide the observer back along the right path, even though her meanderings may seem to have misled him.[12]

But this gifted naturalist was no longer there to direct and admonish. Brehm, who had readily listened to him, drew further and further away from biology, a field in which he had once performed outstanding service and still occasionally did so, and lost himself, along with his fascinated admirers, in the maze of his "research into subspecies." Naumann shook his head and let him go. He himself continued unswervingly on the path he had chosen from the beginning, and did not rest until he had brought out (1844) the twelfth and last volume of his natural history — a monograph on the birds of northern and central Europe that no German work ever equaled for detail. There is much to be said in favor of it, but also a little against it. Generations of ornithologists grew up on it, and more than any other it helped to spread genuine knowledge throughout Germany and the neighboring countries. However, the neatness and polish of its descriptions were more instructive than stimulating. "The reader . . . feels as if now everything were known about our native birds, and there were nothing left to discover."[13]

In fact, German ornithology now entered a dubious stage. The old verve was gone. The flock of Brehm's disciples had thinned out, and those who were inclined to metaphysics now listened to the revelations of the natural philosophers Kaup, F. A. L. Thienemann, and Reichenbach. The best minds, like those of Cabanis, Hartlaub, Heuglin, and Burmeister, were no longer attracted by the prospect of gleaning a few poor ears from the already well-harvested fields of European ornithology. They switched to the inexhaustible riches of exotic ornithology, and left amateurs and egg collectors to look for details that had been ignored or differently described by Naumann. It was true that his masterly work had rapidly increased the number of field ornithologists in Germany, but an intellectual leader of Faber's caliber was needed to direct all this lively interest toward a rewarding goal. Instead of a leader, an organizer appeared. One of Naumann's admirers, an ardent oologist called Eduard Baldamus (1812–1893), arranged a meeting of German ornithologists in Köthen, where he had been teaching religion since 1838. From September 27 to 29, 1845, thirty-two friends of bird study met to found an ornithological section of the Gesellschaft deutscher Naturforscher und Ärzte [Society of German Naturalists and Physicians].[14] They listened to lectures, looked at eggs and skins, and finally were invited to Ziebigk by Naumann. The business manager of the new club was Baldamus, and the secretary Dr. Friedrich August Ludwig Thienemann (1793–1858), who had earned great respect among oologists through his book, *Systematische Darstellung der Fortpflanzung der Vögel Europas mit Abbildung der Eier* [A systematic account of the breeding of Europen birds with illustrations of their eggs] (Leipzig, 1825-1838), but was now a disciple of the natural philosophers. In order to publish the minutes of the meetings and suitable articles, he founded *Rhea* in 1846, as the club organ, and intended to publish it anually. But he miscalculated as badly as Brehm, whose *Ornis* had quietly died soon after its third issue in 1827. *Rhea* did not survive even thus long, for no sooner had the second volume been published in

1848 (with the minutes of the 1846 meeting in Dresden and the 1847 one in Halle) than Baldamus started a rival publication, called *Naumannia.* This also contained the germ of its own demise, and the club was not in much better condition though it had given itself new bylaws as the *Deutsche Ornithologen-Gesellschaft* [German Society of Ornithologists] at Leipzig in 1850 and had elected the 70-year-old Naumann as chairman. *Naumannia* died of intellectual desiccation in 1858, a year after Naumann's own death (August 15, 1857); he had gloomily written to the helpless Baldamus in 1854: "*Naumannia* can go the way of all earthly things whenever it wants to. I can't preserve it."[15]

This course of events had long been foreseen by Cabanis and Hartlaub, who in 1853 had founded the *Journal für Ornithologie* as a central German periodical for all branches of ornithology. They had realized that the problems taken over by German ornithologists from the Naumann period were exhausted, and that only beginners and trivial minds would still find any attraction in them. The crisis of European ornithology stood revealed.

18

From Fixed to Evolving Species

"Everyone is taking a hand," as the exclusive Temminck was supposed to have sighed in comic despair when the "democratization" of ornithology was making great strides during the 1840s, and the flood of publications was steadily rising. This was true not only in Germany but in neighboring countries, and it was helped along by good, instructive handbooks. In Italy the highly skilled Paolo Savi (1798–1871) published three small volumes of *Ornithologia Toscana* (1827–1831); in Sweden, Sven Nilsson brought out a two-volume *Skandinavisk Fauna* (1824–1828), supplemented by 178 lithographed plates of Swedish birds, published in oblong folio at Stockholm (1828–1837) by the very gifted artist Magnus von Wright. France at least achieved a few books on birds of the provinces, beginning with Polydore Roux's *Ornithologie Provençale* (1825–1830). England's contribution came quite late, when in 1837 two works entitled *History of British Birds* began to appear simultaneously. One was in five volumes, written by the imaginative and temperamental MacGillivray; the other was in three volumes, by the conventional William Yarrell. From 1832 to 1837 John Gould put out five folio volumes of *The Birds of Europe*, whose 449 plates portrayed in lively fashion most species of European birds. During the same decade North America produced two works full of biological detail, Thomas Nuttall's *Manual of the Ornithology of the United States and Canada* (two volumes, Cambridge, Massachusetts, 1832–1834), and John James Audubon's *Ornithological Biography; or, an Account of the Habits of the Birds of the United States of America* (five volumes, Edinburgh, 1831–1839).

The successful adventurer John James Audubon, who in many ways resembled François Levaillant, was born (according to the most recent evidence) on April 26, 1785, on the island of Santo Domingo, as the illegitimate son of a French Creole, Mlle. Rabin, and the sailor (later Lieutenant) Jean Audubon. In 1789, when he was 4, his father, a radical revolutionary with a legal wife in Nantes, took him back to France and adopted him there. He was allowed to grow up without any regular schooling, however, so that he was able to wander freely outdoors and draw whatever caught his eye. In the legal registration of his baptism, which took place only on October 23, 1800, he was entered as Jean-Jacques-Fougère Audubon. In 1801 the family moved from Nantes to nearby Couëron. From here his father seems to have sent him in 1802 to study for a while under the famous Parisian painter Jacques-Louis David, but the next year he was sent to America to avoid conscription into Napoleon's army. There he followed his early inclination, wandering through the woods and beginning a collection of bird pictures painted by himself. In the winter of 1805–06 he appeared again in Couëron, where he received some instruction in the natural sciences from Dr. Charles d'Orbigny, father of the famous traveler Alcide d'Orbigny, before he left again for America. There he worked at widely varied professions, getting to know the ornithologist Alexander Wilson briefly in 1810, until he finally made a name as a painter of birds.

At first there were many disappointments, because he was a good observer of bird life but no great expert; his pen, guided not only by experience but also by an imagination inclined to embellishment, was more successful than his brush, on which he had first set his hopes. Since he was unable to remember form and movement and render them artistically, he frequently attached dead birds to a board in theatrically distorted positions and so copied them "after nature." In 1824 he failed to convert his many paintings into cash in Philadelphia, New York, and Boston; as we have seen (Chapter 9), he was unfortunate also with Charles Lucien Bonaparte. So he tried his luck in Europe, where he arrived in 1826. His imposing appearance, winning manners, and mercurial temperament quickly gained him an entrée into influential circles in Scotland; the pictures exhibited in Edinburgh (all life size) were much admired, and the famous engraver and publisher William Lizars was prepared to risk the issue of a costly plate work of enormous dimensions, in double elephant folio. In July 1827, the first plates of *The Birds of America* appeared (the last not until 1838, 435 in all), and Audobon, who had traveled through England and France as his own drummer, succeeded in collecting a surprisingly large number of subscriptions.

Soon after his arrival in England he made the acquaintance of such devoted ornithologists as Selby, Vigors, and Swainson, and their example inspired him with the idea of adding a text to his plates. He therefore returned to the United States in April 1829, to collect material and observations, and a year later went back to Edinburgh, this time with his wife. Swainson, who earned his living by writing books, and was just about to finish the volume on birds in *Fauna Boreali-Americana* (sponsored by Sir John Richardson, the polar explorer), immediately tried to sign a contract with Audubon that would lead to fame and fortune, but luckily for Audubon and for science the plan fell through, because he could not meet Swainson's demands. The task originally intended for the "Quinarist" went to a much more qualified ornithologist, William MacGillivray, who was at that time (1830) a poor private scholar and was only too ready to edit Audubon's manuscript for a suitable sum and amplify it with accurate morphological (and partly anatomical) details. Audubon's *Ornithological Biography* was thus really a collaboration between Audubon and MacGillivray. As a sequel he republished text and plates together in seven octavo volumes, as *The Birds of America*, which came out in orderly sequence from 1840 to 1844.

After spending the 1830s partly in America and partly in England, Audubon returned in 1839 to live permanently in America, where he died in New York on January 27, 1851, with the reputation of being America's greatest ornithologist. A yearning for romance has always found satisfaction in his dramatic pictures and his dynamic personality (even wishing to see in Mlle. Rabin's child the son of Queen Marie Antoinette), and finally has canonized the passionate hunter as a patron saint of bird protection.[1]

At the same time knowledge of tropical birds greatly increased; a mass of new facts was acquired not only by the systematists but also by those who wished to learn more about world distribution.

However, at that time ornithologists tried only rarely to make constructive use of the mountain of zoogeographic details. The doctrine of the fixity of species provided the theoretical background to such efforts: birds were still regarded always as the realization of a "type," a basic idea of the Creator, a realization which, as for Aristotle, was the work of a purposive and effective vital force, supposed to be the "nisus formativus" (Blumenbach) or the "organic force" (Johannes Müller). To it was ascribed the power of shaping the variously organized genetic material, in equally varied but purposively adapted ways, into specific forms. The center of creation was understood to be the actual area of distribution of each species. According to Louis Agassiz, there was only one way to explain the observed distribution of animals, namely, to suppose that they are autochthonous, born like plants on the land where they are found.[2] Faber thought much the same:

> So that the global distribution of birds could take place according to uniform laws, and so that every habitable place could be evenly

> populated by them, the first individuals of each bird species were settled at the very beginning, by a special natural law, in locations of various size, within whose bounds they were to have their true home. The nature of the species of bird was created to suit the natural constitution of its zone.[3]

Lesson added to these concepts the idea that the species of birds could hardly have been created all at once, because in the beginning the globe was covered with water, out of which the land gradually rose, in order to become green with plants, and finally to blossom with flowers. The first created birds must have been aquatic, followed by the birds of prey (who lived on the refuse of the sea), then by the grallatores, and so forth, and finally by the graminivores.[4]

According to K. A. Rudolphi in 1812,[5] whenever creatures resembling one another occur in several places, this is a result of the fact that the same external effects (that is, the same climatic and edaphic conditions, the humidity, and the geographic situation) must work in the same way on the same "type." The environment in which the individual natural bodies find themselves at their moment of creation affects them, in Hermann Burmeister's words, "like a stamp, whose particular effects imprint a particular character on an idea that was until then typical only in a general way."[6] Schmarda concluded from all this that "the middle region of the distribution area [must be] regarded as the original focus and center of distribution, the center of creation. For widely distributed animals, separated by large land masses, one must necessarily assume several centers of creation."[7] The physiological adaptation of each species to its effective area is so exact that, according to Agassiz in 1850, no improvement at all is possible. Therefore at no time have animals been able to withdraw from the influence of their natural environment and abandon their home.[8]

The history of animals at that period thus had nothing to do with the history of the distribution of species, but was a "history of creation." The attempts to divide the book of creation into chapters and, guided by the distribution of birds, to distinguish a large number of regions (=centers of creation) began with Lesson in 1838 and were carried on by several other authors, notably Louis Agassiz in 1850, until the closing publication of this epoch, P. L. Sclater's "On the Geographic Distribution of the Members of the Class Aves" (1858). Here the organization of the earth's surface into regions was based on the following statement:

> It is a well-known and universally acknowledged fact that we can choose two portions of the globe of which the respective Faunae and Florae shall be so different, that we should not be far wrong in

> supposing them to have been the result of distinct creations. Assuming then that there are, or may be, more areas of creation than one, the question naturally arises, how many of them are there, and what are their respective extents and boundaries, or in other words, what are the most natural primary ontological divisions of the earth's surface?[9]

The basic opinions of the systematists and zoogeographers were therefore still not very far removed from those of the Aristotelian school. At a time when the naturalists propounding theories of animal behavior drew some conclusions from the mass of their detailed observations that differed radically from Aristotle's, the zoogeographers still thought that a bird was attached to its "original" habitat as if by a divine decree, which warned it at the border, "Thus far and no further." But the psychologists attributed free will, judgment, and reason to the bird, and derided the Aristotelian concept (sanctioned by ecclesiastical dogma) that interpreted all stereotyped animal behavior as an expression of instinct (an "analogy to intelligent ingenuity"), and allowed only to man a "thinking" (and immortal) soul. At least in Germany, around 1858, the "freethinkers" considered that they had won their battle against "the Church's psychology of animals," so long revered.

The rebellion had started during the Age of Reason, at first in France (La Mettrie, 1845), but initially encountered energetic resistance. The intellectual leader of the reactionary movement, who to some extent reopened the old war of the Stoics against the Epicureans, was a philosopher at Hamburg, Hermann Samuel Reimarus (1694–1768). In his book, *Allgemeine Betrachtungen über die Triebe der Thiere, hauptsächliche die Kunsttriebe* [General observations on the instincts of animals, chiefly the mechanical instincts] (Hamburg, 1860), he opposed the ideas of the Abbé de Condillac in the latter's *Traité des animaux* (Amsterdam, 1855). Condillac, the founder of sensationalism, attempted to demonstrate that the so-called instinctive behavior of animals consists of an acquired ability, developed from experience, reflection, comparison of objects, connection of ideas, judgments, invention, consideration, and practice, but, when mature, requires no particular further reflection. Reimarus observed:

> If M. Condillac had become better acquainted with the natural history of animals, and had surveyed the multiplicity and real character of their mechanical instincts, he could not have submitted such a hypothesis (which could seem valid only in his study) as the clue to the mystery. Anyone observing Nature herself must immediately acknowledge that the mechanical instincts of animals are not capabilities discovered by themselves through their use of reason, and acquired by practice (paragraph 118).

According to Reimarus,

> the predetermined powers of body and soul [the innate instincts] save animals from being hindered in their own preservation and welfare and those of their kind by their lack of understanding and experience, because they do not have to acquire the necessary skill, but bring with them into the world inherited capabilities of the most skilled and unerring kind, for all the needs of their sort of life (paragraph 152).

The blind but determined effort that accompanies the course of instinctive behavior is related to an inner sensibility and vague conception of itself, and every conception of such natural effort is associated with pleasure and creates pleasure in the appropriate behavior. Therefore the actions themselves, in which Nature's lure is followed, remain constantly agreeable (paragraph 41).

> Many actions of animals seem so constituted that everything in the body is arranged and prepared in advance for the movement, which needs only the first impulse from physical sensitivity and blind appetite, just as a firework on the stage is so constructed that it starts with a single spark and impulse, and provides the desired spectacle (paragraph 132).

Animals act without ideas, judgments, or conclusions, and therefore cannot think, in the true meaning of the word. Still less do they possess reason, because that is the ability to think clearly. They can, however, through their vague conception of the present and the past, and in expectation of similar events, be inventive after their own fashion.

> A tame seed-eating bird, tied by a small chain and unable to reach water that is too low for him, invents something that he never would have done had be been free, and that never would have been necessary. With his beak he pulls up a little bucket on its cord, holds the cord tight with his feet, pulls again, holds tight again, and pulls again, until his beak can reach the water; then he drinks, and when he has slaked his thirst he lets the bucket fall back into the water trough. Does he invent this by means of ideas and deductions? It is certainly easy for a reasoning man to transform the conceptions of the bird into ideas,which I do not think necessary to go into here. However, we must not attribute to animals our own kind of thought, and still less must we insist on it; they manage with their vague conceptions. If one wishes to train the bird, one pulls up the little bucket by its cord in front of him, lets him drink, and then lets the little bucket fall back into the trough. Then the bird's skill is clearly

> nothing but imitation, and the expectation of a similar effect. But if the bird makes the discovery by himself, it proceeds in the same way. There is a connection between water, bucket, and cord, and they arouse a sense of connection also in the bird. Because he is used to fetching other things that he wants with his beak, and holding them with his feet, he invents, in expectation of similar results, the means of pulling his drink toward him in its vessel and holding it tight (paragraph 25).

Reimarus had no more doubt than Pernau and Zorn that birds are compelled to start their migration by internal and not external stimuli (paragraph 25 of Anhang, 1762 edition). Observation shows "that one cannot connect with their departure any noticeable sensitivity to the external world. They must therefore become inwardly alarmed at a particular time, so that they develop a longing to migrate and at the same time the place grows too hot for them." Evidently birds feel it necessary to migrate to a particular part of the world. "Just as dogs and other animals track game with their delicate sense of smell . . . so one may suspect that birds too in their passage are compelled by a sensitivity to a certain route (that is, something that draws, excites, and attracts) to take a particular direction in their flight."

The eighteenth century greatly valued this accurate observer and acute thinker. His excellent book was republished in 1798 by his son, Johann Albert Hinrich Reimarus, a Hamburg doctor (1729–1814), a friend and intellectual counterpart of Erasmus Darwin. To weaken the current of opposition, which was swelling threateningly, he furnished the book with many intelligent additions, including this: "One should not deduce that everything in the animal occurs merely through organic instincts, unconsciously, or that animals should be considered simply dull machines. It is true that they act with a certain conception and will, but these reach only as far as the nearest object; they know nothing of the ulterior ends."[10]

It was a vain effort. The scientific theoreticians of the Age of Reason scented in the doctrine of instinct the odor of outlawed teleology, and therefore jettisoned it. The French again set the fashion. The "best authority" in the *Encyclopédie méthodique* was the master of the hunt for the park of Versailles, Charles George Le Roy (who died in 1789), an intellectual disciple of Condillac. He gave a blanket explanation of all skilled capabilities in animals as the product of reason. Under the influence of natural philosophy these opinions began to prevail after the turn of the century in Germany too, and there they were systematized. Oken and his disciples followed their own whims entirely in constructing a graduated scale of intellectual achievements, which corresponded to their *a priori* gradation of physical achievements. This development culminated in the *Tierseelenkunde* [Science of animal souls] (two volumes, Stuttgart, 1840) by P. Scheitlin, an emotional professor from St. Gall, Switzerland. As an admirer of "the mighty Oken, for depth and breadth of view unexcelled by a contemporary German naturalist," he demonstrated that the "psychological" characteristics of birds make them a link between fish and mammals, whereby the lowest birds are closest to the fish and the highest to the mammals. Water birds are fish birds and, because they are neighbors of the fish, are partly related to them and are less skilled at making distinctions than the land birds. On the water they cannot learn much about plants, animals, men, land, or the world in general. The stork is at the highest bird level. "He already has a wide world around him, much knowledge of the senses and of conditions, considerable power of thought, emotion, and will. He is like a human child, a masterly study for the still more complete animals." Sympathy, compassion, love and hate, gratitude, vanity, reverence, conceit, pride are manifest in animal behavior.

However superficial and inept Scheitlin's rhetorical explanations may seem, they were swallowed enthusiastically by the members of a romantic generation who saw themselves in the new role of protectors of animals. Christian Ludwig Brehm's exuberant son, Alfred Edmund Brehm, was among the book's most devoted readers. Scheitlin's science of animal souls is reechoed in old and new combinations in Brehm's successful *Leben der Vögel* (Glogau, 1861) and later in his popularizing *Thierleben* (*Vögel*, 1866–67). For Brehm (1861) the bird is "highly developed intellectually." Birds have "character." There are gay birds as well as sad, honest as well as thievish, noble as well as vulgar, frank as well as sly. "We have evidence for the abundant emotions of birds: love of mates and offspring, compassion for weaker members of the species or class, gratitude for favors received." In Brehm's opinion, "every unprejudiced observation demonstrates the existence of reason in birds, indeed of a highly developed reason." Like his intellectual predecessor, Le Roy, he explained, "Relying on my experience, I consider it possible that the bird learns how to build a nest partly when it is only a nestling and partly through later observation of the work of older birds of the same species."

The biological sciences had arrived in a blind alley. Gustav Jäger wrote in 1864, "I could adduce a number of well-known zoologists who would in private reveal a certain weariness with life." Many people who thoughtfully reviewed the whole area of ornithology at the end of the 1850s and compared the facts with the attempts at their interpretation must have felt the same. They turned into scoffers at comprehensive theory and worked patiently on, piling brick on brick, without any constructive plan.

Then in 1859 appeared Charles Darwin's book, *On the Origin of*

Species by Means of Natural Selection, which upset all existing conceptions. He tried not merely to demonstrate the general process of development of all organisms from their simplest beginning (which other authors, including his grandfather Erasmus Darwin, had already undertaken), but above all to discover the mechanism that produced the development and differentiation. From the facts that, first, all organisms vary to a noticeable degree and a large number of these variations are perpetuated by inheritance, and second, always only a very small proportion of the next generation reaches the age of procreation, he decided that the mechanism he sought was the principle of selection of suitable variations, in the course of the struggle for existence.

Thomas Huxley commented on the effect of this intellectual lightning bolt on his contemporaries:

> My reflection, when I first made myself master of the central idea of the "Origin" was, "How extremely stupid not to have thought of that!" . . . The facts of variability, of the struggle for existence, of adaptation to conditions, were notorious enough; but none of us had suspected that the road to the heart of the species problem lay through them, until Darwin and Wallace dispelled the darkness, and the beacon-fire of the "Origin" guided the benighted.[11]

To Aristotelian teleology, which had until then been able to maintain its position in biology despite all attacks, Darwin opposed a convincingly logical, mechanistic theory — or rather, in his own view, he succeeded in reconciling mechanism and teleology. By ascribing the origin of variations to chance, and the selection of the most suitable form to the hazards of the struggle for existence, he resembled Democritus and the "Atomists" in replacing the purpose of providence by blindly operating chance. But, as he wrote on May 22, 1860, to the botanist Asa Gray, "the effect of what we may call chance" applied only to single details of the exposition on which he based his theory. Chance itself could serve a higher design. Darwin was not in the least convinced that "this wonderful universe . . . is the result of brute force." "I feel most deeply, that the whole subject is too profound for the human intellect. A dog might as well speculate on the mind of Newton."[12]

This theory produced a change of direction in all fields of ornithological research. The logical system of relationships had to yield to a genealogical system. Zoogeography added a historic-dynamic point of view to the static one; from now on it had to explain the actual distribution of organisms as the transitory result of changes undergone, and still to be undergone, not only by the organisms themselves, but also by the earth's crust and the climate, during the course of geologic ages. Further, according to Darwin (1859), "the range of the inhabitants of any country by no means exclusively depends on insensibly changing physical conditions, but in large part on the presence of other species, on which it depends, or by which it is destroyed, or with which it comes into competition."[13]

As a result of this new theory, animal behavior also became an item in the process of historical change, subject to the same laws as form. Instincts provide the permanent scaffolding of behavior, according to Darwin, who returned to classical animal psychology; they are just as important as physical structure and can be traced back to a hereditary modification of the brain. If the conditions of existence alter for a species, it is at least possible that small changes in insincts would be useful to it. In fact, observation teaches that instincts vary from one individual to another, and so provide an opportunity for selection.

> No complex instinct can possibly be produced through natural selection, except by the slow and gradual accumulation of numerous slight, yet profitable, variations. Hence, as in the case of corporeal structures, we ought to find in nature, not the actual transitional gradations by which each complex instinct has been acquired — for these could be found only in the lineal ancestors of each species — but we ought to find in the collateral lines of descent some evidence of such gradations; or we ought at least to be able to show that gradations of some kind are possible; and this we certainly can do.[14]

In many cases the instincts may first change as a result of the effect of selection, while the extracerebral structures follow; in other cases the procedure may be reversed.

Darwin was convinced that the cooperation of the two factors in development, heritable variation and selection, created a natural explanation of adaptation, which until then had been merely the object of astonished admiration. Thus research into adaptation, so long avoided by strict scientists, became respectable once again, and the proponents of the new theory gladly availed themselves of the previous work performed by the teleologists, because they likewise traced the means from the end, the cause from the effect. The younger Reimarus, a teleologist, had already foreseen this development when he wrote in 1798: "The presupposition, founded on experience, that everything serves a certain use, far from hindering research into the means whereby it is carried on, spurs it on and gives it a direction."[15]

Long before Darwin, ornithologists had chosen the coloration of birds and their eggs as a subject for observations on adaptation. C. L. Brehm (1827) and C. W. L. Gloger (1829) had already pointed out that the females of many species that breed in the open, especially ducks and gallinaceous birds, are inconspicuous in color, because nature gives special protection to the more endangered parent, the one more important for the preservation of the species. In a more extended study[16] Gloger explained how useless even the most adaptive instincts in nest building would be if the eggs betrayed the presence of the nest. Therefore nature provides conspicuous colors only when they cannot be harmful under the circumstances (white eggs are to be found especially among pugnacious species, which keep a constant watch on the nest, or among species that cover the clutch before leaving, such as ducks and grebes, or among birds that breed in hollows or in covered nests. In many other instances the coloring of the clutch is more or less clearly adapted to its surroundings, especially among those birds that breed on the ground, like nightingales and skylarks. The egg color may even match the color of the nesting material (green eggs in the moss nests of hedge-sparrows).

According to Gloger, the cuckoo differs from all its relatives, the Scansores, in having nonwhite eggs. If they were white, they would clearly

> not only be too conspicuous to the birds that are supposed to accept them as theirs and look after them like parents, inciting them often to reject them as too unlike their own, abandoning the whole clutch, but also they would attract the attention of an enemy from far away. Thus, variable as they in fact are, they always preserve a mean between the extremes of all those eggs with which they will be incubated.

When 20 years later German ornithologists believed that they had discovered that "the cuckoo's egg shows the color and markings of the eggs of the bird in whose nest the egg is to be laid" (G. H. Kunz, *Naumannia*, 1850), it was no longer possible to be content with merely observing the adaptation; the means used by nature to achieve it must be sought. At that time Kunz thought "that, because egg coloration is organic, the sight of the eggs lying in front of her in the nest has such an effect on the parturient cuckoo that the egg about to be laid takes on their color and markings," whereas Opel (*Journal für Ornithologie*, 1858) suspected that it was the food received by the nestling cuckoo that created the shape and color of the host's egg, and forced her to look for the same host species later. Altum, on the other hand, returned to Kunz's theory in a speech delivered to the thirteenth annual meeting of the Deutsche Ornithologen-Gesellschaft at Stuttgart in 1860:

> Nature transmits to the products mental impressions received by the sense organs; she transfers from the maternal organism to the

developing offspring what has been very strongly and lastingly felt. The eggs of many marsh- and ground-breeding species resemble their surroundings, and the same eggs in verdant grass steppes are more intensely green, where in dry landscapes they are duller and browner, etc. In the well-known "error" of the cuckoo, in its ability to imitate in its eggs the colors of all the different nest companions, this kind of influence on color reaches its height.[17]

After the publication of the *Origin of Species*, these and similar phenomena were favorite demonstration materials for the effect of natural selection. Gustav Seidlitz explained as selection the adaptation of the cuckoo's egg in its color, size, and thickness of shell,[18] though without revealing what factor, in his opinion, produced a selection leading to steadily improved adaptation.* As early as 1859, Darwin himself had drawn attention to the protective coloration of many birds, and explained it as the result of selection by birds of prey. The conspicuous coloring or voice that mark the male of many birds is, according to him, the result of "sexual selection"; individual males in succeeding generations have certain small advantages over other males in weapons (spurs), or means of defense or attraction (color, build, song), and, because these advantages impress the females, have transmitted them to their male offspring.

Alfred Russel Wallace, Darwin's fellow-discoverer, then demonstrated the effect of selection with many examples, at first in an article published in 1867 on "Mimicry and Other Protective Resemblances among Animals," in which he maintained

> that none of the definite facts of organic nature, no special organ, no characteristic form or marking, no peculiarities of instinct or of habit, no relations between species or between groups of species — can exist, but which must now be or once have been useful to the individuals or the races which possess them.[19]

The key to understanding the sexual color difference in birds lies in their breeding role. The paleness of the desert birds, the snowiness of the polar birds, the bark color of nightjars and owls, and the green of the woodland birds, are the work of selection. Wallace even believed that he had discovered defenseless birds that imitate pugnacious ones to deceive their enemies: for instance, the Moluccan orioles (*Mimeta*), which have grown very like the aggressive honeyeaters of the genus *Philemon*. "There are, no doubt, some special enemies by which many small birds are attacked, but which are afraid of the Tropidorhynchus

*Alfred Newton (*Zoologist*, 1877) was the first to suggest that the protective coloration of the cuckoo's eggs is a result of the selective effect of the host bird.

[*Philemon*] (probably some of the hawks) and thus it becomes advantageous for the weak Mimeta to resemble the strong, pugnacious, noisy and very abundant Tropidorhynchus."[20]

Though in this article Wallace was simply expanding Darwin's ideas (to the point of exaggeration), he left them far behind as soon as he dealt with psychological, not morphological, developments. Wallace believed no more than his contemporary Alfred Brehm in the existence of unchanging and effective instincts unmixed with reason. He considered the instinct theory to be the invocation of a mysterious ability, and thought that all a bird's capabilities were produced by rational acts that had become routine, "which are well known to be inherited."

In his article, "The Philosophy of Birds' Nests" (1867), Wallace maintained that the materials, position, and manner of building the nest could be explained by the habits of the species, the kind of tools they used, and the materials easiest for them to obtain — by the simplest adaptation of means to a particular purpose, completely within the intellectual capacity of birds. In his view, the art of nest building is handed down by tradition.

> Very young birds can both hear and remember, and it would be very extraordinary if they could live for days and weeks in a nest and know nothing of its materials and the manner of its construction. During the time they are learning to fly and return often to the nest, and as their daily search for food invariably leads them among the materials of which it is constructed, and among places similar to that in which it is placed, is it so very wonderful that when they want one themselves they should make one like it?

One may suppose that often only one member of a pair is a bird hatched the previous summer, so that it could have been taught by its mate.

> That excessive uniformity in the architecture of each species of bird which has been supposed to prove a nest-building instinct, we may, therefore, fairly impute to the uniformity of the conditions under which each species lives . . . When, however, new conditions do occur, they take advantage of them just as freely and wisely as man could do . . . I cannot find a particle of evidence to show the existence of anything beyond those lower reasoning powers which animals are universally admitted to possess.[21]

Wallace tried to solve the problem of migration in the same way.[22] According to him, the "instinct" to travel arises from the habit of cruising to find food (a theory already proposed by Faber in 1825). The restlessness of a caged bird at the time of migration means nothing more than "social excitement." To find their way to their winter quarters birds fly at a great height, thereby probably picking up guidelines from the physical characteristics of the land beneath them. True, young birds are reported often to leave after the older ones, but one may suppose that they are following the last stragglers of the experienced group.

Palmén amplified these opinions very painstakingly in his book, *Über die Zugstrassen der Vögel* [On the migration routes of birds].[23] He, too, attacked the concept of instinct. The "so-called migratory instinct" can be attributed "partly to physical heredity, partly to a traditional heredity, which means a transmission or inheritance of habits, one may say of experiences, from older to younger individuals." Flocks of birds usually have older and stronger birds as leaders. These note the route and therefore take certain paths, clearly to be observed from the landscape, especially along lengthy watercourses, and they follow them in all their most varied bends. According to him, the species that he had studied (especially ducks and waders) do not deviate from these migration routes. He emphatically rejected an "innate sense of direction." It is not true that northern European birds maintain a general northeast to southwest direction. Individuals keep to the same routes in spring as in fall, because they have learned them, and they are shown unknown areas by other birds to whom they attach themselves. "The habit of migrating arose and grew, like all other characteristics of birds, by gradual development from quite insignificant beginnings; it is a habit transmitted by an inherited mechanism."

August Weismann, in his popular book *Über das Wandern der Vögel* (Leipzig, 1878), relied completely on Palmén. "We now know that birds, during their migration, pay no attention to compass direction but only to landmarks. They follow well-defined migration routes." These are nothing but the primeval paths used during their gradual spread northward.

The premise that such "transmission of habits" could occur only with the help of selection was left so vague by Palmén and Weismann that it was hardly noticed; even entrenched opponents of Darwinism, like E. F. von Homeyer, could completely subscribe to these opinions. In the *Journal für Ornithologie* (1878), he noted with satisfaction that the power of "inherited habit solves many riddles," not only migration but many other peculiarities of behavior, such as nidification, choice of nest site, and so forth. "The famous word *instinct* should finally be relegated to the attic," because it is made up of "irreconcilable contradictions," on the one hand supposedly representing an intellectual activity that in some mysterious fashion always hits the mark, yet, on the other hand, standing in no sort of relation to human reason.[24]

Neither Darwin nor his followers excluded an inheritance of acquired characters. The question simply was to what extent evolutionary change was due to selection of inheritable variants and to what extent due to "use and disuse" and other Lamarckian forces. For some characters, Wallace was the more consistent selectionist; for others, Darwin. In the case of sexual dimorphism, Darwin thought that the differences were strictly the result of "selection" by the female. Wallace, however, ascribed the evolution of showy plumage in the male not to "sexual selection" (the choice of chance variations by the female), but to the "increased vital force" of the male, "which simultaneously influences the color." He said in so many words:

> Those individuals who were most pugnacious and defiant, and who brought these erectile plumes most frequently and most powerfully into action, would tend to increase them by use, and to leave them further developed in some of their descendants. If, in the course of this development, colour appeared — and we have already shown that such developments of plumage are a very probable cause of colour — we have every reason to believe it would be most vivid in these most pugnacious and energetic individuals; and as these would always have the advantage in the rivalry for mates (to which advantage the excess of colour and plumage might sometimes conduce), there seems nothing to prevent the progressive development of these ornaments in *all dominant races;* that is, wherever there was such a surplus of vitality.[25]

The idea was at once hungrily seized upon by Wilhelm von Reichenau, in 1880.[26] According to him, not sexual selection alone, but the vital energy of the male, appearing in definite forms, leads (in combination with selection) to the higher development of secondary sexual characteristics in male birds. The male's sexual excitement is therefore a formative factor — "a confirmation of direct adaptation in the Lamarckian sense."

But that was not enough. Even the more profound study of geographic variation, whose nature Darwin had explained in 1859 as selection finding ample scope for its improvement in altered conditions of existence, seemed more and more to confirm Lamarck's views. The leading American ornithologists, like Baird, Ridgway, and J. A. Allen, had all become convinced Lamarckians. In such a state of affairs even Darwin himself became uncertain:

> In my opinion the greatest error which I have committed has been not allowing sufficient weight to the direct action of the environment, *i.e.*, food, climate, etc., independently of natural selection. Modifications thus caused, which are neither of advantage nor disadvantage to the modified organism, would be especially favoured,

> as I can now see chiefly through your observations, by isolation in a small area, where only a few individuals lived under nearly uniform conditions.
>
> When I wrote the "Origin," and for some years afterwards, I could find little good evidence of the direct action of the environment; now there is a large body of evidence.[27]

Thus the working area of chance, "spontaneous variation," which according to Darwin produced the material for selection, was steadily narrowed down. To his Lamarckian opponents were now added the vitalists, who reintroduced teleology in a form suitable to the idea of evolution ("dynamic teleology"). One of them was the ornithologist Henry Seebohm, who flatly declared: "There is no such thing as Fortuitous Variation; and . . . Spontaneous Variation, like Spontaneous Generation, is a myth."[28] In Seebohm's view variations either are "definite," directed from the beginning to a goal, or else arise from the influence of use and disuse (an example is the long, pointed "traveling wing" of migratory birds). Therefore the evolution of species can progress without the intervention of selection, whose only effect lies in hastening it. The "law of necessary completion," whose validity Darwin had attacked, was respectable once more.

No wonder many ornithologists thought classical Darwinism could be added to the pile of disproved theories.

19

The Return to Empiricism

Before Darwin came forward with his theory in 1858, he had spent almost 15 years collecting evidence for its validity. Many of his supporters were not so careful; in their employment of Darwinian theory they far exceeded the limits of certifiable experience and collided with other theories that they could not convincingly disprove, thereby enormously increasing once more the tasks of empirical research. Serious scientists now abandoned the noisy debates and withdrew to continue accurate research on individual problems; although variation, heredity, hybridization, selection, behavior, and instinct had been studied for years by Darwin, under new aspects they remained important subjects for investigation.

During the first period of Darwinism, attention was focused on the dynamics of the evolutionary process: the history of organisms, the history of their distribution, and the history of their physical and psychological development.

At such a time, Bernard Altum's *Der Vogel und sein Leben* [The bird and its life] (Münster, 1868), a book that ignored all the results of causal research and tried simply to demonstrate the balanced harmony of creation, must have seemed a horrifying anachronism. Almost no book on birds was so fiercely attacked on publication. Only after 1885, when the intoxication with Darwinian optimism had worn off, did it gradually begin to have an effect.

Bernard Altum was born on December 31, 1824, in Münster, Westphalia.[1] His father, an artisan, was himself an enthusiastic naturalist and collector, who gladly encouraged the same traits in his son and sent him to the Gymnasium. Birds soon became Bernard's great passion; while he was still at school he learned taxidermy, and went out shooting in search of anything missing from his bird collection. When he had taken his final examinations in 1845 he went to the Academy in Münster and, "after long hesitation," decided to enter the priesthood. But his love for animals did not diminish. Chancing upon the first issue of *Naumannia* in 1850, he fell completely under the spell of ornithology, and already a curate, moved to Berlin in the fall of 1853 in order to learn more about it. There he continued his philological and zoological studies, and soon a little ornithological club gathered around him, because he was an experienced woodsman; Bolle, Kutter, Hansmann, and Albert Günther became his good friends. In 1854 he delivered an address to the eighth annual meeting of the Deutsche Ornithologen-Gesellschaft in Gotha (attended also by Baldamus, C. L. Brehm and his son Alfred, Cabanis, Hartlaub, Kaup, J. F. Naumann, and Ludwig Reichenbach) on "The Structure of Feathers According to Their Coloration," in which he laid the groundwork for all future research into this problem and was the first to explain iridescence as interference colors ("perhaps the colors of thin plates"). A philological dissertation earned him his doctorate at Berlin in 1855, but he immediately returned to his studies and attended lectures by Johannes Müller on comparative anatomy and physiology. Before his return to Münster in 1856, he took the opportunity, as Lichtenstein's assistant, to enlarge his acquaintance with bird forms in the Zoological Museum.

In 1857 he became curate of the Münster Cathedral, and in 1859 also a *Privatdozent* (lecturer) of descriptive natural history at the Royal Academy. There he gave a lecture on oology and another on ornithology, among other subjects, and beginning in 1862 combined his courses with ornithological field trips. A series of important publications testifies to his activity at that time; among them is an essay on the food of German owls,[2] which laid the foundations for systematic research on pellets "for the identification of the population of small mammals in a given district." At the same time he demonstrated the usefulness of owls. He grew steadily more intimate with the last important supporters of the Deutsche Ornithologen-Gesellschaft, particularly with their intellectual leader, Professor J. H. Blasius of Brunswick. Blasius saved the Gesellschaft from threatened dissolution by taking over the post of general secretary (which meant sole executive) in 1867 from the now apathetic Baldamus. But in the next year, because he was overburdened with offices, he had to surrender it to his pupil, Baron Ferdinand von Droste auf Hülshoff.* In 1869 Altum was awarded the professorship of zoology at the Forestry Academy in Eberswalde — an

*Baron Droste's unskillful tactics wrecked the Gesellschaft in a few years, and in 1875, a year after his death, its remains were incorporated in the youthful and vigorous Deutsche Ornithologische Gesellschaft, which had been founded in Berlin by Cabanis in 1868; see Chapter 17, reference 14.

agreeable surprise — when J. T. Ratzeburg retired. He began an active and rewarding study of the zoology of forests, without forgetting ornithology, and when Friedrich Kutter died suddenly in 1891, the Deutsche Ornithologische Gesellschaft elected Altum president in his stead. He died on February 1, 1900, active to the last.

At the time of his nomination to Eberswalde, Altum was reputed to be one of the stoutest opponents of the theory of evolution. The fourth impression of his little book, *Der Vogel und sein Leben*, had just appeared, only seven months after the first. It was a polemic against Darwinsim which, as he had rightly expected and had written to Baldamus (January 6, 1868), procured him "much pleasure and many enemies." He fought fiercely on two fronts, because he attacked both the materialistic theory of Darwinism and the anthropomorphic conception of animals' souls, which was then so popular in Germany. The church's teleological idea of nature could have had no more skilled defender than Altum, who had far more practical experience and logical training than most of his opponents.

For him, as a pupil of Johannes Müller, nature stood poised as a preestablished, unchangeable harmony. "All individual beings are harmoniously interrelated and thus form a total unified reflection of Nature." The peculiarities of individual beings are calculated and arranged to fit the harmonious portrayal of the whole. In accordance with this, the world of animals is systematically a well-ordered unit, split into groups of forms which are in turn split into smaller ones, yet in such a way that star-shaped rays from each group extend toward and into other groups. Thus skuas link gulls to predators. "The animal fits into its surroundings just as a shape cut from a single piece of paper fits back into it. And the bird, too, by means of its color and markings fits into these surroundings." Even the particular kind of bird song fits the singer's surroundings. Every bird is created only for certain local conditions, every species has its particular sphere of activity and particular tasks within it. The graminivorous birds flock together in the autumn not chiefly for their own sake, but to establish and maintain a balanced plant life. The cuckoo must be released from the whole business of incubation and rearing so that it may be free to visit and remain in any place that has begun to suffer damage from caterpillars. Migratory birds are sent to specific winter quarters "so that there also they may play a sure and effective role in the process of natural development." On the other hand, birds are also useful in the preservation and suitable distribution of many organisms; in fact, berries are eaten by many birds not simply to spread the seeds but also to enhance the probability of their germination. An animal does not act for itself, but under a higher commission: *animal non agit, sed agitur* [an animal does not act, but is acted upon].

This view of animal behavior bears no resemblance to Alfred Brehm's descriptions and explanations of it in his widely distributed and greatly admired books, *Das Leben der Vögel* (second edition, 1867) and *Ilustriertes Thierleben: Vögel* (1866, 1867). "Not so much an insufficient love of truth as insufficient specialized knowledge, inadequate observation, prejudice, and quite often also unconscious exaggeration or easily misleading embellishment in reporting what has been observed, in many instances obscure the fact under consideration." Altum gave plenty of examples to prove it, and thus, by opposing Alfred Brehm and Wallace, deserves the greatest credit for reintroducing to ornithology the concept of instinct, and also for discovering a large number of significant relations between the individual and his environment.

"The animal does not think, does not reflect, does not establish aims for itself, and if it nevertheless behaves purposively, then someone else must have thought for it," argued Altum, exactly like Johannes Müller, and in agreement with the practical and rational conclusions of Ray and Zorn. Birds whose coloration matches the ground huddle down when in danger instead of flying away, but they know nothing of the advantage of their coloration, because albino partridges behave in just the same way as those with normal color — one of countless proofs that birds, without knowing what they do or why, make directly and unerringly toward their goal. Many forms of instinctive behavior are rigid, but others through the possibility of adaptation show variability, whereby many an observer has been misled. "A possibility of adaptation exists and must exist in the endless multiplicity of external circumstances in which birds are forced to live." This is expressed in nest building, for instance. "The environment of birds is not a complex of mathematical figures but nature in all its variety of form and color, and the suitability of a nesting site depends on a combination of many factors, which never follow a blueprint." Nevertheless a nest is "a purely natural product and not a work of art; it is simply the product of the procreative urge. The stronger the urge in the life of a particular bird, the more complete the nest; the weaker the urge, the poorer and weaker the construction." An improvement in "artistic technique," provided by the experience and practice of the individual, is out of the question. "The so-called love for offspring is the compulsion to feed birds that have a certain appearance, a certain cry, and a certain way of fluttering their wings and opening their beaks; it is not love." Birds also do not think when they are incubating; they are compelled to brood, and for a specific time.

Every pair of birds has to have its own breeding territory, whose size is determined by the behavior and specific food of each species. Every song is primarily a mating call; it is the overture to the process of breeding and attracts the female. Second, it notifies other males of the species of the existence of the singer's territory; third, like coloration, it serves to identify the species. The song's degree of perfection stands in inverse ratio to the sociability of the singer. "Only birds that have a breeding territory sing well." Birds know nothing of the meaning of their call for other individuals.

> Adult birds emit a certain cry when danger appears, just as we exclaim when we are startled or surprised, and the young understand it in a general way, that is, passively. They are alarmed by it and react to it in the manner of their species, not as individuals, without any kind of previous instruction from the adults or any experience of their own, because they behave just as appropriately the first time as they do later. The voice of the young birds stimulates the adults to definite actions, like feeding; that of the adults stimulates the young to come or to hide; the voice of the prey excites the raptor, and that of one sex excites the other. There is absolutely no sign of teaching, or agreeing together. The actions of migratory birds, too, can be compared to behavior without roots in experience, teaching, or the knowledge provided by reflection.

All this contradicted the "vulgar psychology" (in Wundt's phrase) that Alfred Brehm had introduced into ornithology, where it had become strongly entrenched. But the brothers Adolf and Karl Müller promptly appeared to defend their friend Brehm and his pets that had been deprived of their souls. In the *Journal für Ornithologie* (1868) they published a passionate reply,[3] maintaining that Altum's writings, full of superficial, irrelevant, or even totally false observations, were not likely to exert any corrective influence on the healthy tendency of nature study at the present time (meaning the Scheitlin-Brehm "science of animal souls"). The authors were in fact right, because it took German ornithologists over 25 years to submit to a revision of their ideas in the light of Altum's theories.

This reconsideration began with more accurate observation of bird migrations. The elderly field ornithologist E. F. von Homeyer, who had approved Palmén's hypotheses in 1878, became with succeeding tests more and more doubtful about the theory of migration routes. After thorough preparation he decided to produce a rebuttal, *Die Wanderungen der Vögel* [The migrations of birds] (Leipzig, 1881).[4] From the many facts he had collected he concluded that "everywhere that migrating birds have been seen and carefully observed, they have all been following a specific direction." Members of the same species travel through Europe in a flock whose front is as wide as the width of their breeding area; in the autumn they fly mostly from northeast to southwest, and in the spring the other way, unless geographic peculiarities crowd them together in certain places as if they were in a defile. Homeyer considered that everything Palmén, Wallace, and Weismann had written about migration smelt of the lamp. "It must be admitted that in recent times natural history in general, and this field in particular, has suffered from abandoning the accurate method of observation and replacing it by imagination." There was still no adequate explanation of bird migration. The hypothesis that migrating birds are guided by tradition was completely misleading; as early as 1825 Frederik Faber had clearly proved it untenable. It was just as false to propose lack of food as the cause of the autumn exodus. "The naturalist must set himself the task of finding the motive force that causes birds to start traveling at a particular time . . . Almost more amazing are the senses of place and of direction that guide them on their journey. At the moment it is better not to attempt any kind of explanation."

Although many students, untroubled by these weighty objections, might have preferred to cling to the respectable account of "traditional migration routes," their faith in Palmén's theory was finally extinguished by Heinrich Gätke's book, *Die Vogelwarte Helgoland* (Brunswick, 1891).[5] For decades Gätke, a painter, had been regarded by English and German ornithologists as one of the best authorities on research into migration. He was born in 1814 at Pritzwalk (Pomerania) and had chosen to live permanently in Helgoland in 1837. From the 1850s on he attracted the attention of specialists to this way station for migratory birds by publishing articles first in German, and then also in English, periodicals. From 50 years of experience, Gätke concluded that the naturalist studying the wonderful ability of migrants to choose the right route, and the immediate cause of migration, is confronted by a riddle that up to now has resisted every attempt to solve it. "What at present has been ascertained in reference to the migration of birds furnishes us with no clue, by the aid of which we are enabled to penetrate the depths of this wondrous mystery."

Gätke, confirming Homeyer's assertion that the migration takes place usually on a broad front, considered that the spring return flight is often made by a shorter route than the autumn one. "If we represent the line of their autumn migration by the base and perpendicular of a right-angled triangle, the path of their return passage in spring will be represented by the hypotenuse of such a triangle." He did not believe the Palmén-Wallace hypothesis that the direction of a migration is the result of inherited experience, asking, "Can experience be something of which the subject is altogether unconscious? and, further, can experience the result of which is positive knowledge, be actually inherited?" Even the supposition of a tradition immediately acted upon is indisputably refuted by observed facts. One must therefore conclude that "migrants, with regard to the time and direction of their movements, act with a means to an end, but unconsciously, or, in other words, by instinct." So everything was right back where the much abused teleologists, from Zorn to Altum, had begun and ended.

Even the many cooperative attempts to discover the secrets of bird migration could not make the slightest headway. These cooperative groups had been fashionable for some time, following the establishment, in May 1875, on the urging of Anton Reichenow at the constituent assembly of the Allgemeine Deutsche Ornithologische Gesellschaft, Brunswick, of a committee (Alexander Bau, Reichenow, Schalow) on observation stations in Germany, with an "appeal to all German ornithologists" to help fill the existing gaps in the knowledge of German birds. The gaps were indicated as particularly disturbing "in the section on distribution of individual species, but especially on the map of the migration routes of our northern migrants." From 1877 on, the results were published in the *Journal für Ornithologie*, and this procedure was soon imitated. In England, John Cordeaux was the prime mover in establishing a "Committee for the Migration of Birds," whose first "Migration Report" appeared in 1879. The Ornithologische Verein in Vienna, which had been publishing its own periodical, *Die Schwalbe*, since 1877, also wanted to employ its members usefully, and founded in 1882 a "Committee for Ornithological Observation Stations in Austria-Hungary." The newly formed American Ornighologists' Union followed in 1883 with a "Special Committee for the Migration of Birds." In order to extract the full value from this development, which spread almost at the same time to still other countries, two industrious organizers, Dr. Rudolf Blasius in Brunswick (a son of the eminent J. H. Blasius) and Dr. Gustav von Hayek in Vienna, worked out a fantastic plan for international cooperation. They attempted nothing less than a network of ornithological observation stations around the whole habitable globe. The plan was approved by Crown Prince Rudolf and provided the motive for calling a First International Congress of Ornithologists at Vienna in April 1884. After long debate a resolution presented by Rudolf Blasius was accepted, and a large international committee (under his chairmanship) was nominated. Now the two sponsors of the idea were able to use Congress funds to found a periodical, *Ornis*, which started appearing in 1885 and printed the innumerable data sent in. In Blasius's mixed metaphor, a noble scientific contest would now blaze up among the nations. Many hundreds of "stations" were in fact registered, and thousands of printed pages in *Ornis* and the *Journal für Ornithologie* were filled with "annual reports." In 1885 A. B. Meyer even spirited away the reports of the stations in the kingdom of Saxony and published them in Leipzig under his own name and that of his assistant, Helm. But the dreary mass of material remained, reliable and unreliable mixed, and because no one ventured to evaluate it, the German committee ceased functioning in 1886, the English in 1889, and finally the International Committee and the Saxons in the 1890s.

This European failure contrasted with a contemporary American success, which employed the same methods but was in fact better administered and set up in more favorable territory. C. Hart Merriam rightly called a compilation by W. W. Cooke, *Report on Bird Migration in the Mississippi Valley in the Years 1884 and 1885* (Washington, 1888), "the most valuable contribution ever made to the subject of Bird Migration."[6] It was an extensive collection of notes on observations combined with meteorological data, whose theoretical results were summarized by J. A. Allen in the *Auk* (1889) as follows. As spring approaches, the rising temperature in the winter quarters awakens the periodic activity of the bird's reproductive system. The result is physiological restlessness, and the release of the irresistible urge to leave at this particular season for the breeding area. The birds are guided thither by the unerring instinct of inherited memory; an individual's ultimate choice of a particular district may be determined by "a true home love." It is that which doubtless guides birds for years together to the same fields and nesting trees, and possibly continues through generations.

These results and deductions made an impression in Europe, too, and led to the periodic meetings of the international ornithologists becoming a permanent fixture, because enthusiasm for the study of bird migration was the stimulus also for the Second International Congress. In 1891 Otto Herman, a Hungarian, summoned it to Budapest so that he could campaign for the study of migration in his own country. He was completely successful, for shortly after the Congress he founded the Hungarian Ornithological Center and a periodical, *Aquila*, chiefly devoted to migration.

Only certain limited problems could be studied with the statistical methods then in use, because avian phenology, working only with arithmetic on an unanalyzed mixture of data, could not provide reliable results on the speed of migration of individuals, the direction and distance of their journeys, or on their "true home love."

We are naturally surprised that none of the existing organizations happened on the obvious idea of improving the method by marking individuals. For a long time it had been customary to equip carrier pigeons with inscribed rings on their legs, and even in the seventeenth and eighteenth centuries falconers had occasionally put similar markers on their hunting birds to indicate ownership. The 150-year-old question whether the buzzard's coloration changes with age, or whether its conspicuous variations last all through its life, had tormented Johann Andreas Naumann, the "philosophical farmer":

> In these last collecting seasons I have been trying to follow up this difficult problem more closely, by fastening a copper ring on the shin of all buzzards that I caught; engraved on it were my name, address, time of capture, and a letter to indicate what variety it was. Then I let them fly in the hope of catching them again, perhaps with different coloration.[7]

In the course of the nineteenth century occasional amateurs or hunters repeatedly hit upon the same or similar markers for large birds, without finding any imitators among the ornithologists. The avian phenologists of the 1880s were so unprepared to try an unusual system that nobody responded when, at the Viennese International Congress in 1884, the Prussian forester Borggreve recommended the introduction of an "experiment" in the study of migration, by cutting off the middle toe of individual birds in certain parts of the country, so that there would be reliable evidence for establishing their migration routes.

Thus the credit for permanently introducing the technique of banding into ornithology goes to an inconspicuous Danish teacher, Hans Christian Cornelius Mortensen (1856–1921). While he was reading Kjaerbölling's *Skandinaviens Fugle* [Scandinavian birds], his eye lighted on a quotation from Thienemann's *Rhea* (1849), according to which a Dutchman, Baron van der Heyden, tried for several years around 1830 to tame wild ducks and geese. He equipped their offspring with a brass collar, on which the marker's address was engraved. One of these birds was reported from the neighborhood of Danzig. This gave Mortensen the idea of trying the same thing with starlings. In 1890 he caught a few of them and put thin strips of zinc around their legs, with "Viborg 1890" written on them in zinc ink. But when he noticed that the birds disliked the metal strips, he gave up the attempt. A few years later aluminum came on the market at a reasonable price and began to be used for poultry rings. Mortensen fastened one of these rings to a live red-breasted merganser that had been brought to him and, as luck would have it, the bird was shot soon afterward and Mortensen learned of it. That encouraged him in 1899 to make the old experiment again on a large scale. In the autumn he caught 162 adult starlings in his nesting boxes, and let them fly again after equipping them with aluminum rings, on which he had with great difficulty stamped "Viborg" and a serial number, or only "C.M." or "S.v." But he waited in vain for a single report. Repeating his efforts in 1900, he stamped the rings "M.

Danmark." This time one of his starlings was shot in Holland and another in Norway. The method had stood the test.[8]

Circumstances favored its rapid spread. In 1900 the Deutsche Ornithologische Gesellschaft, whose president (after Altum's death) was Rudolf Blasius, supported the founding of a permanent ornithological observation station on the Kurische Nehrung, in East Prussia. It was called the Rossitten Bird Station of the Deutsche Ornithologische Gesellschaft, and was headed by Johannes Thienemann (1863–1938), a great-nephew of the oologist F. A. L. Thienemann. Its chief object was the observation of migrations; after the failure of the cooperative endeavors, hope was now placed on a single permanent station in a favorable location. Two years after Rossitten was founded, Thienemann began to use Mortensen's method. In October 1903 he banded with aluminum 151 hooded crows caught on their migratory flight. Only a serial number and the year were stamped on the rings. That same winter he got news of five birds that had been shot in Pomerania. In 1904 "Vogelwarte Rossitten" was added to the inscription, and the system was extended to small migrating birds, including 62 robins. It proved to be a dangerous undertaking. No sooner had the daily press spread the news than the animal protectionists raised a storm of protest against the "den of murder" at Rossitten. They demanded that the government forbid banding, because it encouraged the people living inland to catch and kill large numbers of birds in order to find one from Rossitten among hundreds of others.

Even a few ornithological writers failed to understand Thienemann's scientific program. Curt Graeser, in his book *Der Zug der Vögel* [The migration of birds] (Berlin, 1905), considered it useless to band robins, because their breeding grounds were scattered all over Europe and were already known. He thought the same about the other marked birds; their breeding grounds and winter quarters were also known in general, and the distribution among them of individual birds was without scientific interest or any significance for the general questions of bird migration. In the same year Hans Duncker doubted "that this Thienemann enterprise would yield much new for us about bird migration" (*Der Wanderzug der Vögel* [The migration routes of birds], Jena, 1905). The avian phenologists still had great difficulty in understanding the importance of studying individual birds, even after Thienemann had explained it in detail. Nevertheless, the victory of the new technique could no longer be postponed. In 1908 Jacob Schenk persuaded the Hungarian Ornithological Center to order aluminum rings on the Rossitten model. Collaborators helped to band 1064 birds, including many storks, one of which was shot on January 20, 1909, in Natal. That was the most impressive success to date; others followed so rapidly that as early as 1910, on the basis of 35 banded birds found, Thienemann was able to make a map of the route followed by East Prussian storks across southeast Europe to Asia Minor, Palestine, and then along the Nile Valley to South Africa — a "serpentine" migration route that fully corresponded to Palmén's theory.[9] In the meantime other institutions had started banding: the University of Aberdeen (A. Landsborough Thomson), the periodical *British Birds* (H. F. Witherby), and the American Bird-Banding Association in New Haven (Leon J. Cole), all in 1909. Now "a noble contest" really did "blaze up among the nations."

At first the metal ring seemed merely to be a new aid to migration research. But then the marking of individuals proved to have a much more comprehensive significance, because it helped the study of behavior on the breeding grounds and of the buildup of small populations — two subjects that were steadily attracting more precise attention from ornithologists. All this was not simply the result of technical innovations, which included not only the development of the aluminum ring but also improvement in optical instruments; it was produced primarily by the revolutionary theories that had once again radically transformed the study of animal psychology.

20

The Reform of the Theory of Behavior

The decisive impulse for the new animal psychology, which successfully vanquished the psycho-Lamarckism of the post-Darwinian period, came from August Weismann. From his discovery that in multi-cellular organisms there is a rigid division of labor between somatic cells and germ cells, and that the content of the germ cells remains constant from one generation to another, he deduced, in his lecture *Über die Vererbung* [On heredity] (Jena, 1883), that all changes important for the development of the organic world proceed solely from changes in the primary germ plasm, or, more accurately, are set up by changes in the molecular structure of the germ cell. Characteristics acquired in individual lives alter only the somatic cells, and therefore cannot possibly be inherited. Also, "all instincts are produced purely by selection; they are rooted not in the experiences provided by the individual life, but in variations in the germ plasm."

This thesis, heavily attacked at first, prompted C. Lloyd Morgan, professor of zoology at University College, Bristol, to undertake detailed research into the relations between instinct and experience. The result was his classic book, *Habit and Instinct*, published in 1896. The examples on which he relied to reach important conclusions are mostly taken from avian behavior (partly from his own observations, partly from the literature or his informants); however, as he frequently noted himself, they are by no means sufficient to provide a sound basis for his conclusions. At one stroke ornithology was confronted with a large number of new and important tasks; but at that time the ornithologists themselves took so little interest in psychological questions that not a single one of their reference works mentioned Morgan's book.

After an exact description of the instincts and habits of young birds, certain habits and instincts at mating time, nidification, rearing of the young, and the urge to migrate, the author found himself forced, for better or worse, to answer the vexatious question of whether habits can become instincts (as the famous philosopher Herbert Spencer and many of his school maintained), or whether they are to be fundamentally distinguished (as Weismann taught). Morgan tried to find a middle way: "We feel that though the evidence for the transmission of acquired habit is insufficient, yet some connection between variation [meaning inheritable alteration, now called mutation] and modification is suggested by the facts."[1] "Yet these modifications afford the conditions under which variations of like nature are afforded an opportunity of occurring and of making themselves felt in race progress."[2] One can imagine that variations proceeding in the same direction as modifications are not suppressed like other variations, but find opportunity to develop fully. In other words, "plastic modification leads, and germinal variation follows: the one paves the way for the other."[3]

Morgan doubted as little as Weismann that true instinctive behavior is inherited and cannot be acquired by individual experience. It shows inheritable variation, which can be established by natural selection in the course of generations and stereotyped as changes in instinct. (Here he acknowledged Darwin's original concept.) Instincts are manifested by "a certain regularity in their course" and are closely related to reflexes; in fact, "many forms of instinctive behaviour include reflex actions." As a result, "completely inherited instinctive behaviour" is chiefly a motor reaction or a chain of motor reactions, though the expression of instincts can be modified by conscious interference. A chicken's first pecking for grains and small objects, and young birds' scratching and cleaning, are ready-made and inborn instinctive ways of behaving, but in repeating instinctive behavior this raw material can be changed by the influence of consciousness into what the author calls instinctive habit. Consciousness enables the animal to bring the light of past experience to bear on its own behavior and adjust it accordingly. Morgan (who appropriated William James's theory, published in 1884) believed that instinctive behavior becomes conscious in the following manner: external stimuli excite the lower brain centers and produce an effect that spreads through centrifugal channels, and creates motor as well as visceral reactions. These in turn set up a reverse stimulus that excites the cortical area and produces a conscious emotion created by the action of instinct. The association of the two is called into being by the direct connection between two different cortical excitations. The capacity to create associations is innate, but to become effective it needs the stimulus of individual experience. Even a single experience can create an association. To peck at flying insects is at first purely instinctive in young birds, but this behavior is modified as soon as an animal has an unpleasant experience, such as a bee sting. In this way, on a basis of inherited instincts, the individual creates an acquired disposition; its behavior becomes "instinctive habit" and departs further and further from its primitive basis as experience grows. The cortices of the cerebral hemispheres work through the pyramidal cells and their equivalents, so to speak, on the subcortical brain centers.* Each individual must forge the links of the chain of association for itself. The principal task of consciousness, therefore, is to increase individual adaptability. Thus with birds we have: (1) innate stability — the riches of instinct (inherited automatism); (2) youthful plasticity — the ability to form associations; (3) acquired stability — a product of the associations formed by experience (acquired automatism).

Wallace was therefore mistaken when he believed that knowledge of nest building was transmitted by tradition from one generation to another, because since his day it had been experimentally demonstrated that hand-reared young birds build a nest typical for their species; however, it seems that though the building is based on instinct it can be varied according to individual experience. The innate faculties employed in nest building are so extraordinarily complicated that the author found it rather difficult to imagine the selective factors through which a particular kind of nest structure becomes inherited in a particular species. The same difficulty arose when he had to interpret the apparently intelligent behavior of decoying, which simulates that of an injured or otherwise handicapped bird for the purpose of leading an enemy away from the brood. If one accepts decoying as having been created entirely by breeding working with variations, "its development, in complete independence of intelligent guidance . . . puts no inconsiderable strain on the theory of the all-sufficiency of natural selection."[4]

The emotions involved in the activity of instinct are important not only subjectively, by arousing feelings of pleasure, but also as signifying to other members of the species or to enemies the emotional state of the individual affected. "The action and attitude, then, to which we apply the term emotional expression is, biologically, of suggestive value. It has been developed and organized as an expression suggestive to others."[5] For example, calls and cries are expressions of emotional states and have a definite, often precisely circumscribed, suggestive significance (as Altum had already explained). Morgan therefore agreed

*This theory fully agreed with Max Schrader's researches ("Zur Physiologie des Vogelgehirns" [On the physiology of the brain of birds], *Pflügers Archiv,* 44 (1889), which succeeded in proving that a bird whose cerebrum has been removed can still see quite well but can form no idea of objects, and behaves toward a cat as it would toward an inanimate object. "A male pigeon without a cerebrum follows its instinct in the breeding season, going about with enthusiastic coos and all the movements of courtship, but does not recognize the female for what she is and is indifferent to her" (H. E. Ziegler, *Tierpsychologie,* Berlin, 1921, p. 103).

with Darwin and opposed Wallace in being convinced of the effectiveness of sexual selection among birds. The preferred partner is the one that by its postures, calls, or other similar behavior most strongly evokes the mating instinct.

Directly or indirectly stimulated by this outline of a new theory of behavior, various biologists soon devoted themselves to the study of birds with an accuracy that earlier would have been considered pedantic. They were Francis Hobart Herrick in America, Edmund Selous and H. Eliot Howard in England, and finally Oskar Heinroth in Germany.

Herrick's book, *The Home Life of Wild Birds: A New Method of the Study and Photography of Birds* (New York, 1901), fulfills the promise of his subtitle because it marks a turning point in the research methods of field ornithologists. Even in this his first book the author succeeded in observing all forms of nesting behavior — partly nidification and partly rearing of the young — from very close quarters and with intense concentration, documenting the results with photographs as far as possible. Most of the behavior seemed to him purely instinctive. Birds seem to have no choice in these matters, "but act because they must."

In England, where Wallace's work was highly respected by biologists, belief in his theory was so strongly rooted that any progress was difficult at first. In 1902 Charles Dixon, a well-known compiler, devoted an entire book to nest building — *Birds' Nests: An Introduction to the Science of Caliology* (London) — in which he tried to reduce the theory of instinct to absurdity. It was supposed to be amply refuted by facts. No bird could build a nest typical of its species "without the aid of imitation and experience," as Wallace had already established in 1867, "memory, reason, and hereditary habit playing the minor parts."

But in England, too, a new day was opening for field ornithology. Although noticed by few at first, Edmund Selous (1858–1934) had declared war on all previous ornithological writing in his book, *Bird Watching* (London, 1901). His language became crisper in his next two books, *Bird Life Glimpses* and *The Bird Watcher in the Shetlands* (both London, 1905). "The zoologist of the future should be a different kind of man altogether: the present one is not worthy of the name. He should go out with glasses and notebook, prepared to see and to think."

> Some men have strange ambitions. I have one:
> To make a naturalist without a gun.[6]

By turning to a natural history of British birds, one can always find how many eggs are laid by any species, their coloration — often illustrated by costly plates — and when and where the laying takes place; but in regard to the matters above-mentioned [the relative parts played by the male and female bird in nidification, incubation, and rearing of the young] — or, indeed, most other matters — little or no information is forthcoming. One might think that such works were written for the assistance of bird-nesters only, and whether they are or not, that is the end which they, principally, fulfil. I believe, myself, that if the habits — especially the breeding habits — of but one species in every group or genus had been thoroughly studied, so that we knew, not only what it did, but how it did it, the result would make an infinitely more valuable work, even in regard to British birds only, than any now in existence, though all the other species were left out of it, and little or nothing was said about the number of eggs, their coloration, and the time at which they were laid.[7]

What is the use of knowing that some bird or other goes through "very extraordinary antics in the season of love"? This is not nearly enough. One requires to know what, exactly, these antics are, the exact movements of which they consist — the minutest details, in fact, gathered from a number of observations. When one knows this one may be able to speculate a little, and what interest is there, either in natural history or anything else, if one cannot do that? *Mere* facts are for children only. As they begin to point towards conclusions they become food for men.[8]

Such a thoughtful observer was more fascinated by avian emotions and their related behavior than by anything else. Anyone who "has Darwin in his soul" was compelled to ask himself about their genesis. The part played by elementary reflex movements in instinctive behavior seemed to him to have been insufficiently considered. All "corporeal activity [produces] mental excitement, which again reacts upon the former" and can increase it.[9] "What delight can be greater than that satisfying an imperious and deep-seated instinct?" The superfluous nests built by many birds are therefore not "extra labour" but "extra delight."[10]

According to Selous, the groups of instinctive behavior patterns are not motivated by intelligence but are blind, unemotional movements that only through selection are related to an object. Courtship movements ("antics") he considers to have arisen chiefly from the continued existence of energy that was formerly used for a particular task (such as bathing), but now is used, at least sometimes, for a different purpose. In this way bathing movements have become a ceremony among water birds. Certain kinds of automatism seem to exist whose original object has entirely disappeared and, because their activity is accompanied by emotion, can be evoked only by various sorts of emotion; since no excitement is as strong as the sexual, they easily become a characteristic of the mating season. Sexual selection must play an important part, because instincts vary with the individual, and birds do not select their partners blindly under the influence of the all-powerful sexual drive. They seem, rather, to "mate by preference and elective affinity."[11]

Two years later H. Eliot Howard (1873–1940) began his highly regarded monograph in two volumes on British Sylviidae, *The British Warblers: A History with Problems of Their Lives* (London, 1907–1914). In studying the many relevant species he made a particularly thorough investigation of the details of mating behavior and incubation. Like Selous, he sought to reconstruct the development of instinct, for which a comparison of the behavior in closely related species offered him various points of reference. The individual, seasonal, and specific differences in emotional intensity, hence differences in "temperament" and emotional tension, set him a good many puzzles. In 1910 he pointed out that the males of many avian species possess a "territory" which they defend against rivals — a discovery that had been made considerably earlier by Pernau (1702) and by Altum (1868).

Although Howard was a less powerful and acute thinker than the incomparable Selous, he provided many important stimuli which were soon taken up in English ornithological literature. Enthusiastically fostered by H. F. Witherby, the editor of the periodical *British Birds*, which he founded in 1907, the new spirit in English field ornithology created effective propaganda for itself in the four volumes of *The British Bird Book*, edited by F. B. Kirkman (London, 1910–1913). A number of distinguished ornithologists collaborated with Kirkman, including Ernst Hartert, F. C. R. Jourdain, W. P. Pycraft, A. L. Thomson, and especially Edmund Selous, whose ability again shone at its brightest. His program provided the basis for the work because, as stated in the foreword, the principal aim was to treat the habits of birds in detail and to collect all important data on the subject from every available source, foreign (especially German) as well as British. Knowledge of the habits of native birds still showed serious gaps compared to what was known about their structure and geographic distribution. Ornithologists should therefore regard the filling of these gaps as their most important task. The immediate response to this appeal led to the tremendous increase in field ornithology in England.

In Germany, on the other hand, the invigorating influence of research into behavior was still barely noticeable. The country that had been so incomparably creative and productive in ornithology during the first half of the nineteenth century had long surrendered her leadership in new developments to her Anglo-Saxon neighbors. Since Altum had tried in vain to attract his countrymen's attention to new problems, no penetrating commentator on bird life had appeared. Even in the first decade of this century there were only clumsy attempts at analyzing problems of animal behavior. Almost all activity was confined to the study of local fauna, avian phenology, and old-fashioned biology. Such was the state of affairs that an enterprising small publisher, F. E. Köhler in Gera, could offer German ornithologists a new edition of J. F. Naumann's *Naturgeschichte der Vögel Deutschlands*, first published from 1810 to 1844, with an additional volume in 1860, as an up-to-date work of reference! He chose as editor a Gera oculist, Dr. C. R. Hennicke, who edited the periodical of the German Society for the Protection of Birds. Completely inadequate for the task, Hennicke surrounded himself with a large staff of unevenly qualified collaborators, and got them to make insertions in Naumann's text — almost without exception limited to details about local fauna. *Naumann, Naturgeschichte der Vögel Mitteleuropas*, appeared from 1897 to 1905 in twelve volumes, not like its classic original in a practical large octavo format, but in a showy folio, with tasteless "full-page illustrations" instead of J. F. Naumann's graceful colored engravings. This discordant production did immense damage to ornithology in Germany because it held up progress for a long time.

But one man finally opened a breach in the walls erected by the Naumann cult. He was Oskar Heinroth (1871–1945), born on March 1, 1871, at Kastel, near Mainz, into a family of musicians and scholars.[12] His special gifts showed themselves early. He learned to walk in a hen house, and when he was only 3 or 4 years old he could recognize all the individual hens by their voices. During his preschool years he kept lightweight domestic geese as well as chickens, which he allowed to fly loose, and he studied their behavior — to the astonishment of his parents, who had no understanding for such ideas. Later he built a small aviary in his room and populated it chiefly with tame young native birds. While studying them he established many facts that influenced his thoughts and feelings during his lifelong concern with animals.

After graduating from school in 1890, Heinroth studied medicine, was licensed to practice, and finally took his doctorate at the Physiological Institute at Kiel in 1895. In the autumn of 1896 he went to Berlin to study zoology. There he worked half days as a volunteer in the Zoological Garden, and for the second half day studied avian systematics under Anton Reichenow in the bird department of the Zoological Museum. He used bird skins for research on molting, and in 1898 published a pioneering paper, "Über den Verlauf der Schwingen- und Schwanzmauser der Vögel" [On the course of wing and tail molt in birds].[13]

In 1900 a rich private citizen, Bruno Mencke, asked Heinroth to accompany his "First German South Sea Expedition" as zoologist. The expedition left Hamburg in July 1900, in its own steam yacht, the *Eberhard*, and went by way of Colombo and Batavia (Djakarta) to German New Guinea, and then to the Bismarck Archipelago. There, on March 31, 1901, natives attacked the expedition's camp, pitched on the shore of St. Matthias Island. Heinroth, seriously ill with malaria in his tent, was hit in the leg by a spear, but he succeeded in pulling it out and then rescuing by boat the mortally wounded Mencke (who died soon afterward). The dismal residue of the expedition reached Rabaul, in New Britain, a week later. In October 1901, Heinroth arrived back in Berlin with his collections and a fairly large number of live animals, and for the time being continued his apprenticeship at the Zoological Garden.

In working over the birds he had brought back (the results were published in the *Journal für Ornithologie*, 1902–1903), he broke new ground:

> I have introduced a novelty, figures for weight. A little skill with them will provide a much clearer idea of the real size and bulk of a bird than the measurements of length . . . Further, I have considered the conditions of molt everywhere, and not only that, but from the degree to which the plumage is worn down I have drawn my conclusions about the incubation period and so on . . . In my notes I have tried to describe as exactly as possible the development of the gonads[14]

— simple techniques that have since become routine, but at that time caused astonishment. However, the new "tricks" do not account for the continuing significance of Heinroth's treatment; much more important were the reflections and questions lavishly scattered through the text. The great biologist and skilled anatomist, the author of *Vögel Mitteleuropas*, is visible in embryo, the man who in 1910 wrote of himself: "I had from the beginning aimed not at what is usually called biology — the sort of things that we all know from Naumann or Brehm — but at the observation of the finer details of habits, customs, and uses, what students call fraternity regulations."[15]

Heinroth soon found the best possible opportunity for this. In 1904 he was appointed assistant to Ludwig Heck, director of the Berlin Zoological Garden (which at that time was second in importance only to London). Shortly afterward he married Magdalena Wiebe, a tireless collaborator as long as she lived, who played a decisive part in the success of rearing birds, which he began to do at home. The description of his first great triumph, rearing nightjars in the living room (*Journal für Ornithologie*, 1909), began his long series of classic contributions to the biology of birds, native as well as exotic. He became known as "the father of comparative behavior study" after his great paper on the behavior of the Anatidae, which he delivered in 1910 at the Fifth International Ornithological Congress, meeting in Berlin with Reichenow in the chair. "I believe," he said, "that in voice, social behavior, and similar matters we often possess very good points of reference for the degree of relationship among species, genera, and subfamilies; these things, not vitally important to the continued existence of the forms under consideration, often survive better unchanged than the external and internal characteristics important for systematic identification, which are in a state of perpetual flux because of the struggle with the surrounding world."[16] Heinroth's reasoning about characteristics of coloration followed the same lines. In recent times coloration had remained simply a study for systematists, whereas Heinroth stressed that our backwardness in explaining the purpose of nuptial plumage and similar details comes from our ignorance of the habits of the individual species. For example, he had discovered from his observations that the speculum in the wing of the Anatidae serves as a signal to members of the same species, and that in certain species it is replaced by a sound as means of attraction. In future, the qualities animals value most highly in their opinion of a member of the same species should also be noted.

This was the first move toward bridging the gulf between biology and systematics; the two great branches of ornithological research, separated for centuries by "fashion," began to approach each other and to touch — a most important process, whose effects are demonstrated ever more clearly in our present period of research into evolution. That modern systematics is no longer satisfied with the morphological identification of bird groups, and tests the accuracy of its genealogical conclusions against biological species or group characteristics, is to a large degree the work of Heinroth, whose connection with the richly populated Berlin Zoological Garden and with animal dealers enabled him to study down to the smallest details the behavior of representatives of almost all families of birds, their movements, vocal utterances, length of incubation, postembryonic development, and so forth. In the introduction to the famous book that crowned the patient life work of the two Heinroths, *Die Vögel Mitteleuropas in allen Lebens- und Entwicklungsstufen photographisch aufgenommen und in ihrem Seelenleben bei der Aufzucht vom Ei ab beobachtet* [The birds of Central Europe, photographed in every stage of life and development, and observed in their behavior from the egg onward] (four volumes quarto, Berlin, 1924–1933), occur these characteristic sentences:

> How is it that this species is able to cope in nature with the struggle for existence? Why does it look like this, or behave like that? To what

> extent are activities and special qualities valid for the tracing of genealogy, and so on? Every bird presents countless puzzles to one who knows how to put the right questions . . . We are concerned here not with the study of skins, with geographic forms and subspecies, but rather with subtleties of behavior, with growth and development, with molting, with instinctive actions and mental abilities — in short, with matters that up to now have been scarcely considered.

The effect of this work, which had been 20 years in the making, was soon noticeable; acting on Heinroth's suggestions, the new generation of Dutch and German ornithologists soon became leading investigators of behavior. This development was encouraged by a periodical, *Beiträge zur Fortpflanzungsbiologie der Vögel* [Contributions to the reproductive biology of birds], founded in 1924 by Leo von Boxberger, and continued after 1926 by Ludwig Schuster. Expert reviews of all foreign literature kept its readers informed about every development, and helped to train a quick grasp of the most important details. It transformed many an egg collector into a biologist.*

The new biological questions posed by the Morgan school, penetrating into all areas of ornithology, had begun after the turn of the century to affect anatomy. The dissector of old had also been a physiologist, and even the pre-Darwinian zootomists took for granted the inner relation between form and function; in their view it was the Idea, the *nisus formativus* or organic force, that purposively formed the organism as a whole, down to its smallest parts. This view was not difficult to combine with a theory of changes in form. "As soon as the principal functions of flight atrophy or are lost, the entire body form changes more or less, as in the penguin and the ostrich," succinctly wrote J. J. Prechtl (1778–1854), a famous Viennese investigator, in *Untersuchungen über den Flug der Vögel* [Investigations on the flight of birds] (Vienna, 1846). The great difficulties in explaining theoretically the facts of the harmonious adaptation of organisms first began with Darwinism, which did not regard purpose as the principal cause. Darwin himself took refuge in the idea that this condition was produced not only by selection but also directly by use or disuse. In the *Origin of Species* (1859) he gave an example: "We may imagine that the early progenitor of the ostrich had habits like those of a bustard, and that as natural selection increased in successive generations the size and weight of its body, its legs were used more and its wings less, until they became incapable of flight."[17] Among Darwin's followers, Carl Semper tried to relate the

*Subsequent notable periodicals in the general field of ethology include *Behaviour* (1947), *Zeitschrift für Tierpsychologie* (1953), and *Animal Behaviour* (1958).

theory of evolution to physiology (1879) by adding behavior to the principal conditions needed to produce new species or variations:

> But in order to decide how far this selective (or even transforming) influence of such conditions of life may here have preserved or encouraged a certain tendency in development, or there lessened another or even cut it off, we need a profound knowledge of the effect of these conditions of existence and their variations on the different animal species. Knowledge of these physiological relations of the animals to the environment (taken in its widest meaning) is also necessary, in order to acquire a more complete understanding of those processes that took place during the transformation of one species into another, because only the physiological necessity of such changes makes them comprehensible to us.[18]

But such difficult investigations into the factors of evolution did not much concern the enthusiastic disciples whom Darwin soon acquired among English and German anatomists. They were attracted by the historical point of view recommended by Darwin in studying their material, the search for homologous formations, and the reconstruction of their genetic development. The functional aspects were relegated more and more to the background, until they were dropped entirely by F. E. Beddard, prosector to the Zoological Society, from his book, *The Structure and Classification of Birds* (London, 1898).

The reaction was introduced by *The History of Birds* (London, 1910). The author, W. P. Pycraft (1868–1942), was a close friend of Lloyd Morgan and Eliot Howard, with much experience in avian anatomy. In this book he was concerned to reestablish the relation that had been broken off between biology and anatomy. He went so far as to maintain, like Aristotle, that "habits precede structure," and counted as an evolutionary factor, besides variation and selection, the immediate effect of environmental influences. Therefore he particularly concentrated, not only on "adaptive modifications," but also on analogies, "convergent evolution," to which he dedicated an entire chapter.

A kindred spirit was Hans Böker (1886–1939), a German anatomist and ornithologist, who regarded it as his historic mission to call attention to the direct relation in animals between environment, behavior, and anatomical structure, and who started immediately after the First World War to organize the Freiburg anatomy collection according to behavioral instead of organic systems. He regarded research into analogies, not homologies, as the duty of modern comparative anatomists. The observation of living animals in their natural habitat must lead to anatomical research, and not the reverse. In his study, "Die biologische Anatomie der Flugarten der Vögel and ihre Phylogenie" [The biological anatomy of types of bird flight and their phylogeny] (*Journal für Ornithologie*, 1927), he attempted to demonstrate that characteristic proportions of the arm skeleton belong to each type of flight. These anatomical structures are far too complicated to have arisen from aimless mutations, and are much more the result of "a causal relation between function and form." The interest stimulated by Böker's writings contributed greatly to the increased enthusiasm for morphological research on birds after 1925, and to its combination with studies in locomotion (flight, swimming, running, climbing), feeding, food intake, courting behavior, and so on. Such studies were not without effect on systematics, so that a second link was forged between biology and systematics.

21

Ramification and Interconnection

At the beginning of the 1920s, ornithology changed fundamentally. Up to that time, anyone could be regarded as an expert who was well acquainted with systematics, distribution, and "habits." Very few young disciples of the "scientia amabilis" were interested in the achievements of anatomists, physiologists, geneticists, and psychologists in adding to the knowledge of birds. In other disciplines the situation was much the same. Most representatives of "scientific zoology" viewed ornithology as the province of amateurs, whose findings could not mean much to researchers into causation. The developments between 1920 and 1950 altered all that. When the highly organized physical and psychological structure of birds was recognized as offering better opportunities for causal research than many other subjects, the barriers that protected our special field of knowledge were demolished on all sides. Ornithology has progressed with such breathtaking speed that nothing important can be achieved in it nowadays except by keeping up with the pace, without losing sight of the whole.

The modern period is characterized by ramification and interconnection: ramification through particular lines of research that have proliferated enormously from the fruitful soil of old and new problems; interconnection as the result of persistent investigation of the organic complex, leading everywhere to contacts with neighboring disciplines.

Annual publications on ornithological topics increase steadily; the *Zoological Record* for 1971 lists over 3000 titles. In America, Germany, England, and elsewhere ornithology has been introduced into a number of universities as an optional subject, and almost all civilized countries have their own ornithological societies and periodicals, often several.

Among the problems taken up only recently by ornithologists is research into bird flight. Though Emperor Frederick II laid the foundations for it, 700 years passed before a competent ornithologist worked on it again successfully. The subject had long been the province of physicists, technologists, and physiologists, and hardly counted as part of ornithology. It is therefore not surprising that as late as 1913 a German handbook on ornithology could say:

> The mechanics of flight are the same in all birds. The wings have a screwlike motion, as in rowing. Turns in the horizontal plane are produced by the hind part of the wing. The tail controls vertical direction, in rising and falling. It is clear that the gliding of petrels over the sea is maintained by rising air currents produced by the movement of the waves.

One could hardly suspect from this collection of errors that a general explanation of the aerodynamic foundations of the different kinds of flight had already been worked out. The Italian physicist G. A. Borelli (1608–1679), in his famous book *De motu animalium* (Rome, 1680), was the first to study the powerful air currents created by a wingbeat. According to him, the elastic air particles resist the compression forced on them by the downbeat of the wing, which cuts through the air like a wedge, and thereby is driven upward and forward. Like an oar blade moving in water, it rests on a movable support. Though research into "rowing flight" made steady progress from the eighteenth century on, gliding remained a riddle to the physicists until more exact research into atmospheric currents began in the 1880s, and enabled a distinction to be made between static and dynamic gliding.* During the same period the French physiologist E. J. Marey, who had aimed his "photographic gun" at middle-sized birds in flight (pigeons, gulls, and others), and had made series of 10 to 12 pictures per second at exposures of 1/720 second, established the shape of the moving wing at takeoff and in flight, and treated the whole subject thoroughly in his classic book, *Le vol des oiseaux* (Paris, 1890).

The development of the airplane after the turn of the century diverted attention temporarily from rowing flight to gliding, but a new stage in research into rowing flight began in 1930 with R. Demoll's treatise on "Die Flugbewegung bei grossen und bei kleinen Vögeln [Flight movements in large and in small birds]."[1] As a result of the rapid development of cinematographic technique, Demoll had been

*Basté (1881) and Hubert Airy (1883), internal wind effects; Lord Rayleigh (1883), increasing wind velocity with increase in altitude; Otto Lilienthal (1889), vertically rising air currents from heated surfaces; S. P. Langley (Washington 1891), experiments in aerodynamics.

able to make the first slow-motion photographs (150 pictures per second) of the rowing flight of small songbirds. They showed that a scarcely suspected difference exists between the wingbeat of small birds and that of pigeons, gulls, and crows. The discovery touched off further research in this new special field, in which ornithologists too played an active part. Only a few years later series cameras were developed that permitted up to 1500 exposures per second. With their help Max Stolpe and Karl Zimmer were able to photograph the whirring flight of hummingbirds in clear sequences.[2] The same method led to complete data on the wingbeat of a song bird, proving that it differed from that of hummingbirds.[3] In 1949–50 Eric Hosking published in *British Birds* many illuminating pictures of the flight of small and middle-sized birds, for which he used exposures of 1/10,000 second, facilitated by a powerful electronic flash.

The knowledge of such flight movements, which the human eye could only now begin to see accurately, immediately stimulated a functional investigation into the structural peculiarities of the entire apparatus of flight, which hitherto had simply been recorded as facts. In the period of man's dynamic flight we view these detailed analyses differently from Johann Esaias Silberschlag, who wrote in 1781:

> I wished at first to calculate each position of the wing and each spreading and folding of the same, and the additional effect of every position of the tail, but to what purpose? It is quite as unnecessary to teach a bird to fly as a fish to swim, and men are not meant to fly. Suffice it, then, simply to indicate the forces that are at work in its flight.[4]

By 1930 significant successes had been achieved by the English pilot and airplane builder R. R. Graham.[5] He explained the aerodynamic significance of the transformation of the wingtips into narrowed-down single feathers during a downbeat, as well as the alula and other structures, and urgently called the attention of "biological anatomists" to a field that soon attracted further researchers.

Study and analysis of single shots or series of photographs provide only a patchy impression of a bird's flight technique. For more profound knowledge there still remained the necessity for close observation of the free-flying bird, something the Hohenstaufen emperor had devoted himself to. Konrad Lorenz, in his article "Beobachtetes über das Fliegen der Vögel und über die Beziehungen der Flügel- und Steuerform zur Art des Fluges [Observations on bird flight and on the relation of wing and tail shape to the form of flight]," [6] demonstrated that this approach could still furnish new and important facts. He studied the details of flight in his free tame birds, with which he became so familiar that he sometimes knew "what the bird wanted and where it wanted to go." In this way he also succeeded in discovering what tasks were beyond the capacity of the species under observation, because "the flying apparatus is always a compromise between different demands," and thus he developed points of reference for a classification of particular qualities of flight.

Because these problems are certainly very complicated, they have not been studied by many researchers. On the other hand, bird migration has steadily attracted more attention. Thanks to the active participation of large groups of people, only a short time was needed to realize the hope of Mortensen, who had prophesied that the results of the banding projects would produce a clear picture of the paths of migrating birds. Interest grew so much after the First World War that in 1927 there were banding stations in seventeen European countries. Today there are far more, and India, Japan, and other Eastern countries have joined in. By 1937, 763,803 birds had been equipped with Rossitten rings, and tens of thousands of these birds had been found again after months or years. It now became steadily clearer that none of the theories proposed in the period of avian phenology was generally valid, but that each of them applied to individual cases. Many species migrate on a broad front, in which the individuals pursue a general migratory direction (Homeyer, 1881), or they spread out fanwise in the autumn; other species make a "loop" migration, by flying over different territory in the spring from that traversed in the autumn (Gätke, 1891); many follow for certain stretches the former route of their distribution (Weismann, 1878), and yet others, led by experienced individuals, follow traditional narrow paths (Palmén, 1874).

These results were used in a search for a causal explanation of the differences, which in turn produced studies on resistance to hunger, speed of flight, the ecology of the species under consideration, the meteorology of the areas traversed, and so on.

But the old fundamental questions, what stimuli guide the migrating bird, and what provokes it to undertake the migration, remained unanswered until the method was radically changed. It had long been recognized that only tests would help, but because ornithologists, in Heinroth's words, usually had "an almost medieval horror of experiment," the fame to be won by the discoverer of new techniques eluded them.

Two American zoologists, J. B. Watson and K. S. Lashley, were the first systematically to employ an obvious method of narrowing down the problem of orientation; in 1907, and again in 1910 and 1913, they took terns from the nest and marked them individually, and then transported them by ship to far distant places. The results were as startling as they were encouraging. Nevertheless, many years passed before any professional ornithologists were tempted to try the experiment. The Stimmelmayr brothers were the first, in 1930. Werner Rüppell (1908–1945), who had been prompted by Ernst Schüz, continued it on a large scale; since then the technique has been used almost annually with intelligently varied questions and methods, chiefly by the tireless and imaginative Rüppell, but soon after him by English and American ornithologists (R. M. Lockley, Donald R. Griffin, William Rowan). It became steadily clearer that birds can use two completely different means of orientation when they try to return to the nest from a far-distant place where they have been left. Unless they perceive familiar landmarks, they are led back home by a stimulus that indicates the right direction. The search for this unknown stimulus led to a number of large-scale experiments. Again Rüppell took research one step further and in 1944, after several years of working on the homing of hooded crows that had been transported southwest from Rossitten, wrote:

> The attempt to take [a northeasterly] direction when seeking home is inborn. What external stimulus serves to orient the birds is still unknown. The crow flies mostly in company and only by daylight. When migrating it usually starts early in the morning. It is therefore conceivable that it orients itself by the position of the sunrise.[7]

Such a method of orientation could not be adequate for small songbirds, which migrate to distant points and, for physiological reasons related to their feeding, mostly at night. The results of many displacement tests permit us to suppose that small birds (and certain sea birds like the storm petrel and the shearwaters) reach home by means of "goal awareness." But neither the stimulus "which directs to the goal," postulated by Reimarus as early as 1760, nor its receptive organ has been discovered. With the help of further experiments and suitably adapted methods it should be possible to progress beyond speculation; indeed, Gustav Kramer's recent work[8] suggests that this expectation will soon be gratified. But up to now there is no experimental evidence that the irregular distribution of polarized light in the sky, which changes with the position of the sun, is used by birds, as it is by bees and ants, to help in orienting themselves.[9]

The second basic problem, that of the nature of the impulse to migrate, had a long time ago been handed over to the psychologists to solve. The situation changed only after 1920, when information about the effect of hormones rapidly mounted. At that point the physiologists joined in, beginning with William Rowan, a Canadian, whose experiments with the effect of light on *Junco hyemalis*, the slate-colored junco, started the ball rolling in 1924. At that period, because physiology was concerned in detail only with the products of the gonads, it was at first assumed that spring migration was part of sexual behavior and depended on the condition of the gonads. Rowan's experiments seemed to confirm this. But when it was demonstrated that the restlessness preceding migration persisted even when the testes had been surgically removed, this hypothesis collapsed. Today [1951*] it is known that the hypophysis, among all the endocrine galnds, represents the most important switching apparatus. It receives signals from the lower brain centers (the hypothalamus in the midbrain?), which excite activity in its anterior lobes. Therefore other anterior-lobe hormones, not the gonadotrope, start the migrating mood in birds. Among them is one that produces a change in the function of the thyroid gland controlling the metabolic rate. Therefore the swelling of the testes begins contemporaneously with the deposit of fat reserves, preceding spring migration. This deposit, however, precedes not only the spring but also the autumn migration, and is therefore a reliable indication of readiness to migrate. By artificially lengthening the day (light stimulus), premature activation of the hypophysis and thereby of the testes can be induced, but neither this nor any other artificial interference can create the migrating mood in members of a permanently settled breeding community. It is therefore connected with a presumed hereditary characteristic.

Modern hormone research has forced ornithologists to abandon other theories that they had clung to for a long time, including one that was supposed to explain the relation between gonads and plumage. One of its many variants, wrapped in a veil of natural philosophy by K. F. Burdach in 1830, ran to the effect that such phenomena as the growing of male plumage by female chickens "teach us that the life of the female in its early stages was hindered by the necessity of breeding from developing the full strength of its individuality, and especially from expressing it in peripheral attributes, those that define and individualize."[10] Modeste Bugdanov shared this view in 1868, and so did even Alexander Brandt in 1889, expressing it thus: "When the ovaries degenerate, a newly available excess of material previously used for breeding may re-evoke the latent efforts to produce exterior signs of differentiation."[11] The anatomist, Max Weber, who in 1890 described a gynandromorph chaffinch, thought the sex glands worked "through

*Advances in avian endocrinology in the last two decades have been so rapid that no adequate history can be provided here. For summaries see chapters 2, 3, and 4 in Donald S. Farner and James R. King, eds., *Avian Biology*, Vol. 3 (New York, 1973).

nervous influence on the channels for nutrition in the integument." But at that time the physiologists had already developed completely divergent theories. Soon after 1841, when Johannes Müller proposed the theory of the function of endocrine glands, or glands with internal secretion (like the thyroid, suprarenal glands, and thymus), A. H. Berthold (Müller's *Archiv für Anatomie, Physiologie und wissenschaftliche Medicin*, 1849), succeeded in proving, by transplanting the testes of a rooster, that the interactions between them and the other parts of the body were brought about not by the nervous system but by the productive activity of the testes through the circulation. The physiologist C. E. Brown-Séquard, working in France from 1869, further developed the theory of inner secretion. But this discovery bore fruit only after the turn of the century. In 1908 C. E. Walker described the influence of the testes on the secondary sexual characteristics of the chicken. In 1910 H. D. Goodale published the results of castrating ducks. A further milestone was M. B. Zavadovsky's book, *Sex and the Development of Sexual Characteristics*, published in Moscow in 1922. Combining the results of his own experiments with those of his predecessors, he explained that the males of chickens, pheasants, and ducks have asexual plumage, uninfluenced by sexual hormones, but that on the other hand the ovaries of these species produce a hormone that inhibits the pigmentation of the feathers, limiting it to a lower stage of development. The extension of such experiments to members of other bird groups later demonstrated that it is not simply this mechanism that universally accounts for sexual dimorphism in coloration, but that other means are employed as well.

After a period when only the sex hormones were studied, research was begun in several countries simultaneously in 1923 on the influence of thyroid-gland hormones on bird plumage. From 1927 on there followed experiments on the effect of the hormones of the anterior lobes of the hypophysis on the avian organism. These experiments produced notable results, especially after 1933, when rapid progress was achieved in isolating and fractionating the hormones.

At the moment [1951] it seems that in this very complicated process the endocrine organs have chiefly sexual and procreative functions. Regulated by the nervous system, on which they in turn react, they provide chemical control of metabolism, which must proceed irregularly according to biological demands. The motor of this system is the hypophysis, communicating directly (by means of channels in the vegetative nervous system) with the central nervous system, from which it receives its impulses, exciting the other endocrine organs with specific secretions and interfering purposively in their combined activity. In the sexual sphere the hormones influence the regulation of secondary sexual dimorphism, the breeding period, and the provision of metabolic products that are important for rearing the young; at the same time they govern instinctive behavior in the realms of sex and reproduction.

The physical changes evoked by hormones are coupled with corresponding changes in instinctive behavior. Castrated roosters lose the sexual urge and their entire pattern of domination, including crowing. They are unaffected by injections of the gonadotropic hormone from the hypophysis. If injected with androgen, a male sex hormone, they crow and display; if with gynogen, they behave like females and crouch in front of fellow members of the species to invite them to copulate. Noble and Wurm noted similar effects in night herons; injections of testosterone propionate (a male sex hormone) induced young as well as old unspayed females to behave like males.[12] Prolactin, a hormone for the care of nestlings, arouses among pigeons the instinctive behavior of incubating and rearing the young, and maintains it until the end of the breeding cycle.

Therefore, when the hormone mechanism is once set going, the medieval scholastic thesis holds true within wide limits: *animal non agit, sed agitur.* But the subject itself controls this machine, which runs with a minimal consumption of energy. It keeps a check on the machine's appropriate activity with the help of its system of associations, which connects up the external stimuli conveyed to it, and passes its commands on to the motor of the hormonal system by means of the (cortico?-)striato-hypothalamo-hypophyseal channel. The probability of mistakes is thus greatly reduced.

The task of discovering, in as many bird species as possible, what external stimuli start the breeding cycle (increasing daylight, weather, intervals of feeding, the presence of a mate, and others), was methodically undertaken in almost all climatic zones during the three decades 1920–1950. In order to disentangle the often complicated relations, one must have exact knowledge of the ecology of the species studied, courtship behavior, nidification, nest location, rearing of the young, social behavior, and so forth.

Noticeable progress has been made in this field since banding, originally intended for the study of migration, was employed for research into populations. It enabled S. Prentiss Baldwin to establish widely ramified genealogical trees for a breeding colony of house wrens (*Troglodytes aëdon*) in Cleveland, Ohio. In 1921 he published the results in the *Auk*, and was instrumental in attracting the attention of American banders (who in 1925 had started their own periodical, the *Bulletin* of the Northeastern Bird-Banding Association) to the study of individuals and population. The most magnificent undertaking of this kind, particularly because it greatly increased the knowledge of population dynamics, was started in 1922 by Oliver L. Austin and his coworkers in the tern colonies on Cape Cod. Up to 1944 almost 160,000 young birds and almost 40,000 adults were banded, with over 27,000 reports on birds found again — some of them in the same colony. It provided an enormous quantity of material, which Austin analyzed carefully in a series of publications, and used to arrive at various conclusions.

Further important insights into the structure and dynamics of societies were provided by the colored band, by which an individual bird can be recognized without being caught. J. P. Burkitt, in Ireland, was the first to think of this new technique, and with it was able to compile informative histories of individual robins (*Erithacus rubecula*), which he published in 1924.[13] Soon afterward the technique was further developed by Margaret M. Nice in her pioneering *Studies in the Life History of the Song Sparrow*,[14] where she recorded detailed accounts of scores of color-banded individuals in a population at Columbus, Ohio. For the first time it was possible to follow the fate of individual birds from birth to death, determine exactly their fertility rate, examine their relation to other members of the same population, and obtain much other information about which previously there had been only the vaguest notions. She was thus able to advance well into a new research area, where she was soon joined by other students, adding many new lines of investigation to those already pursued by the field ornithologist. In *A Guide to Bird Watching* (New York, 1943), by Joseph J. Hickey, an outline for the study of the life history of birds was included that brought together all the points to be noted by observers and banders. When Daudin did the same thing in his *Traité d'ornithologie* (1800), he thought of 35 questions; Hickey's total came to about 375!

It is true that, if one accustoms hand-reared nestlings so well to their surroundings and their keeper that they will return to their cage after every flight outdoors, the small details of instinctive and social behavior can be recorded more accurately than when banded birds are watched in the open. The domesticated birds regard the cage as their territory, build their nests there under the observer's eye, and rear their broods. This technique, recommended to the curious by Freiherr von Pernau, as early as the beginning of the eighteenth century, as an "agreeable way of passing the time," has been used again recently with astonishing success, first by Oskar Heinroth and then by Konrad Lorenz, who in 1927 established a colony of tame, color-banded jackdaws and has also kept other free-flying species of birds.

Together with behavioral study, which stimulated it and was in turn stimulated by it, ecological research became steadily more and more the focus of interest, assuming the position occupied for decades by historical zoogeography. The joining of many of its adherents with the geologists instead of the biologists harmed its reputation, the allied groups losing themselves in the boundless deeps of speculation. Darwin had early recognized the danger of keeping such company. On June 16, 1856, he wrote to Charles Lyell:

> My blood gets hot with passion and turns cold alternately at the geological strides, which many of your disciples are taking.
>
> Here, poor Forbes made a continent [extending] to North America and another (or the same) to the Gulf weed; Hooker makes one from New Zealand to South America and round the World to Kerguelen Land. Here is Wollaston speaking of Madeira and P. Santo "as the sure and certain witness of a former continent." Here is Woodward writes to me, if you grant a continent over 200 or 300 miles of ocean depths (as if that was nothing), why not extend a continent to every island in the Pacific and Atlantic Oceans? And all this within the existence of recent species! If you do not stop this, if there be a lower region for punishment of geologists, I believe, my great master, you will go there.[15]

Lyell was in danger of landing in this Tartarus with equally culpable zoologists, who had greatly oversimplified the serious zoogeographic problems in archipelagoes. But even when the distribution of animals on continents had to be explained, the zoogeographers went to work in a biased and superficial way. They allowed their judgment to be swayed by A. R. Wallace who, in his basic work *The Geographical Distribution of Animals* (London, 1876), had indeed mentioned the various factors that could help to determine distribution, but then, influenced by P. L. Sclater, had followed the pre-Darwinian cataloguing system, and had divided the earth into definite faunal regions, and the animal world into geographic categories. For decades afterward the zoogeographers (especially Russian ornithologists like Severzov, Menzbier, and Sarudny) tried to outdo him, and covered the map with a mosaic of smaller and smaller "zoogeographic units," although this in no way furthered research into causation. Finally it was admitted almost universally that these problems cannot be solved by statistical methods, because the history of distribution, and its limits, means, and speed, are hardly ever exactly the same for two species. So everything had to begin over again, and be restricted at first to the most accurate study of the ecology of individual species, exactly as the believers in the fixity of species had already recommended. One of them, Bernard Altum, had written in 1868, "Wild Nature is not a botanical or zoological garden, nor an area in which the animals have been introduced from outside; they belong very definitely to a landscape with peculiar conditions of geography, topography, soil, temperature, and climate. It is precisely to these that their organization, life, and all interests are adapted."[16] Gustav Jäger then demonstrated in his excellent book, *Deutschlands Thierwelt nach ihren Standorten zusammengestellt* [The animals of Germany classified by habitat] (Stuttgart, 1874), that this attitude could well be aligned with the historical-evolutionary view, and Carl Semper dedicated to it in 1879 his remarkable publication, already referred to, *Über die Aufgabe der modernen Thiergeographie*, wherein he wrote:

> Without accurate study of the most general comparative physiology we shall never be in a position to treat zoogeography really scientifically, that is, historically . . . Here is a whole great field open to research, but almost completely uncultivated, waiting for workers who will know how to make the virgin soil arable. Who will be the first? We do not know. Signs are growing that the pure zoologists are getting ready to take over, but on the other hand the physiologists do not seem reluctant to step in. If a battle should really start in the neutral zone between these two groups of scientists, who until very recently regarded each other almost as enemies, then the zoologists may easily come off worse, because the advantage of better preparation, such as the more abundant assistance provided by experiment in answering pertinent questions, undoubtedly lies with the physiologists.[17]

Actually years were to pass before the ornithologists, having begun to follow the path made by botanical ecology, seriously started to carry out this program.

The new era was introduced by "Ecological Relations of Bird Distribution," an article by S. E. Brook (1883–1918) in *British Birds* (1914). It contains the statement that "analysis of the environmental control is an indispensable aid towards a full conception of the phenomena and causes of geographical distribution." Brook did not restrict himself to such generalizations, however, but as a thoroughly experienced field ornithologist drew up a system of guidelines for analyzing the decisive environmental factors. At about the same time, in California, Joseph Grinnell published "Barriers to Distribution as Regards Birds and Mammals" in the *American Naturalist* (1914), but many years elapsed before the plan was put into practical use. Its first adoption in Europe was reported on by Ernst Mayr in his dissertation, "Die Ausbreitung des Girlitz" [The distribution of the serin] (*Journal für Ornithologie*, 1926). Further and considerably more detailed research on individual species of birds followed. In Finland, Pontus Palmgren created about 1928 a school of ornithologists working on ecology, whose methods steadily became more detailed and more subtle. While Palmgren added biological anatomy (and especially the physiology of locomotion and of the senses) to the auxiliary science of ecology, and American ecologists, particularly S. C. Kendeigh, were developing metabolic physiology, the English, led from 1933 by David Lack, related ecology to comparative animal psychology and attributed an important role in habitat selection to inheritance or the modifying "imprinting" of an individual (Lack, "Habitat Selection in Birds," *Journal of Animal Ecology*, 1933).

Subsequently, all important developments in ornithological research have tried to keep in close contact with the field of behavioral science; this is what welds together the framework of ornithology, which threatened to come apart at the seams. Today systematists, morphologists, zoogeographers, ecologists, and physiologists are all equally concerned to keep a close watch on the progress of animal psychology, in order to make use of it in their own work.

It was chiefly the research and deductions of Konrad Lorenz, the ethologist, that opened up new vistas after 1935.[18] He was interested in studying instinctive behavior in detail, first in birds and then in other vertebrates, and in working on instinct as a problem in evolution. Darwin and then Weismann (in 1882) had already attributed "composite instincts" (which had seemed to the teleologists in all periods, up to the present day, to be outside nature and not susceptible of causal explanation) to a "hereditary modification of the brain," and had supposed that its purposive increase of achievement and adaptation was the work of selection. In this century the mechanistic school of reflex physiologists, without paying special attention to the course of instinctive behavior, was content to explain it as a chain of unconditioned reflexes. Lorenz, on the other hand, proceeded from the discovery that the excitement leading to the specific reaction of instinctive behavior accumulates during the period when no reaction occurs, and is used up in the course of activity. This revealed a basic difference from the unconditioned reflex, which follows an "all-or-nothing law." Encouraged by the proof of spontaneous, automatic, and rhythmic production of stimuli in the central nervous system of invertebrates, obtained by the physiologist Erich von Holst, Lorenz attributed instinctive behavior as well to the production of stimuli by such endogenous automatisms, in which he conjectured the existence of complexes of multiple neural elements with different thresholds of sensitivity to stimulus. They are to be recognized as highly independent formations, as it were "fixed enclosures within the plastic confusion of mutual causal relations; they influence the form and the achievement of the whole, but they themselves are not at all or not significantly influenced by the whole." Characteristic of instinctive behavior is its inflexible course, not noticeably varying from individual to individual. It is provoked by definite combinations of a few stimuli, which represent the trigger adapted to the receptive correlative, and which Lorenz calls the releaser. As experiments with decoys have demonstrated, the inborn mechanisms respond in a specific manner only to their releaser (and not, with the help of a "memory-picture peculiar to the species," to the total appearance of an object or a situation). It is the performance of these two mechanisms, the transmitter and the receiver of stimuli (also called "a ready-made, innate system"), that enables the organism to react intelligently, without any experience or idea of a purpose, to the appearance of biologically significant and excitatory situations. The intensity of the process, and only that, can be determined by previous experience (fatigue, habituation); in addition, quantitative relations exist between the strength of a stimulus and the strength of reaction to it. Innate and individually acquired links usually are directly interlocked in a behavior chain ("interlocking of instinct and training") that functions singly and is aimed at a single purpose — to preserve the species. "Into gaps in the innate-behavior chain is inserted a capacity to acquire behavior patterns, instead of inborn instinctive behavior," which can be observed and experimentally proved in the nidification of birds, for example. Like Reimarus, who in 1760 supposed that "the blind but determined effort that accompanies the course of instinctive behavior is associated with pleasure and creates pleasure in the appropriate behavior,"[19] Lorenz concluded from his observations that the undisturbed running out of the behavior chain is clearly felt by the animal as satisfying. Because the pleasurable conclusion of the process begins, for example, only after usable nest materials have actually been brought, "birds learn surprisingly quickly to choose the biologically appropriate materials when they are only in the middle of the carrying reaction." This permits the conclusion that there are close relations between instinctive behavior and subjective manifestations. "In most cases we must attribute to animals much more widely differentiated feelings and passions than are known in man." Instinctive behavior can even be defined as a reflex process desired by its subject.

These analyses proceed in the direction indicated by Lloyd Morgan and Selous; on the other hand, in later research Lorenz chiefly continued the work of Heinroth, who was concerned to prove that distinctive coloration just as well as sounds could serve as signals to members of the same species. Lorenz demonstrated that complicated systems of releasers and innate schemata in many birds and other creatures form the basis of their entire sociology. All the behavior patterns relating to members of the same species of bird are inherited, and create a tapestry of innate action and reaction norms which, in the service of a common function — communication among members of a species — are still further differentiated, contemporaneously and parallel to one another. Such differentiation affects not only instinctive movements, but also colors and structures, which often reinforce the release function of the movements, and serve this purpose chiefly or exclusively.

From research into the genealogy of instinctive behavior, which Darwin had already designated as a task for the future, Lorenz promises important results for systematics. Every instinctive release pattern "can be used as a taxonomic characteristic just as well as the exterior form of some piece of skeleton or other organ," in fact, more reliably, because there is much less likelihood of convergence. Darwin had conceived the method when he recommended that we ought to find in the collateral lines of descent some evidence of the transitional gradations by which each complex instinct has been acquired.[20] Heinroth already knew that similar movements, constant in form, could serve to characterize genera, and even families and orders. Many of them have changed their meaning in the course of genealogical development, a phenomenon already worked on by Selous and Heinroth, which has recently attracted great attention. Many forms of movement that are effective releasers (especially in the sexual sphere) derive from movements to convey mood and intention that have been formalized into theatrical exaggerations.

Further research, especially by Otto Koehler and his school, concentrated on these components of behavior that are not provoked by endogenous automatisms, but are alterable by experience. They are the higher, equally innate, categories of thought process, whose function is to react intelligently to the many and various individual situations in the environment. Koehler has been able to analyze abstract thought in birds, and has stated that "abstract thought, suited to the environment and forming usable images of it," such as we encounter among birds, was the prerequisite for human speech.[21] From that arose the gift we prize as mankind's unique possession — the freely associative mind.

EPILOGUE

Materials for a History of American Ornithology

Ernst Mayr

The history of American ornithology has not yet been written, fascinating as such an account would be, although there is an excellent treatment of the earlier period (prior to Audubon) by Elsa G. Allen (1951). In his *Entwicklung der Ornithologie*, of which this volume is a translation, Stresemann dealt extensively with American ornithology and ornithologists, particularly in Chapters 2 (Brazil, Venezuela, Mexico, West Indies), 6 (Brazil), 9 (Bonaparte, Philadelphia, Wilson, Audubon), 13 (Cassin, Baird, Ridgway, Allen, Coues, Smithsonian Institution), and 15 (Davenport, Whitman, Beebe, Chapman, Mayr, Miller, and others).*

Stresemann's treatment, however, touches only the highlights and concentrates on the nineteenth century. It would be tempting to supplement this by providing a full treatment of American ornithology, particularly of the recent period, but this would require a separate volume, a task I am not prepared to undertake. Instead, I have brought together an informal, annotated bibliography of some of the more important developments in American ornithology. It will serve as a convenient takeoff point for any one who wants to probe deeper. My quick survey of the field has, of necessity, been extremely sketchy. The list of additional authors, publications, and institutions that could have been cited with equal justification is about as long as that which is actually pre-

*Page references will be found in the index and numerous additional references to American ornithologists in Chapters 18–21 and in the notes to this Epilogue.

sented. Furthermore, my account is unavoidably biased, owing to my special interests and degree of acquaintance with persons and institutions.

I hope, nevertheless, that the material I have gathered will prove useful as an introduction to a fascinating story and that it will spare others the unexpectedly time-consuming labor that was required to collect this information.

The first accounts of the birds of the Americas came from various chroniclers who followed the conquistadores into the New World colonies and from later colonial naturalists. Indeed, our knowledge of Mexico (Hernandez) and certain parts of South America (Marcgraf, Azara, and Molina), advanced in the beginning more rapidly than that of the birds of North America. But with Mark Catesby, William Bartram, Abbot, and many others, North American ornithology rapidly came into ascendancy, reaching a first climax in Alexander Wilson and John James Audubon.[1] Much of this is well reported by Stresemann and by Elsa G. Allen, and my own notes will concentrate on the developments of the last 100 years.

Perhaps the most striking difference between the classical ornithology of Europe and that of North America is the incomparably greater domination of European ornithology by a few outstanding personalities and by certain preeminent centers, like London and Tring in England, Paris in France, Leiden in the Netherlands, and Vienna in Austria. It was only in Germany that — in rivalry with Berlin — a whole series of minor centers developed, such as Dresden, Munich, Frankfurt, Bonn, Hamburg, Bremen, Brunswick, and Oldenburg. In the United States the early monopoly of Philadelphia and Washington was soon broken, and since then decentralization has proceeded at a steady pace and at many levels. It is reflected by the frequent establishment of new ornithological research centers at museums, by the large number of local and regional ornithological societies (in addition to three flourishing national ones), and of colleges and universities at which ornithology is taught and Ph. D. degrees in it are awarded. Furthermore, ever since the establishment of the U.S. Biological Survey (now U.S. Fish and Wildlife Service) in 1886, first the federal government and later many of the states have promoted avian research, particularly that of game species (waterfowl, gallinaceous birds, doves, and others). One can look at almost any other indicator of ornithological activity (such as the number of expeditions, ornithological journals, monograph series, or textbooks), and find additional evidence for the astonishingly broad basis of ornithology in America.

A good indication of the respect that ornithology commands within American science is provided by the fact that within the last thirty years eight ornithologists have been, or still are, members of the National Academy of Sciences: Frank Chapman, Herbert Friedmann, Daniel Lehrman, Robert MacArthur, Ernst Mayr, Alden Miller, S. Dillon Ripley, and Alexander Wetmore.

Museums and Ornithology

Most of the early bird collections in America were in private hands, just as in France, England, Germany, and the Netherlands. Even the first public museums, whether in Philadelphia, Cambridge, or New York, owed their existence to the enterprise and the public spirit of private citizens. When the American Ornithologists' Union was founded in 1883, more than a score of the founders had their own private collections, some of them with an ultimate size of 20–60,000 specimens. In the end all of these were incorporated in the collections of the public museums, contributing to their rapid growth.[2]

Philadelphia Academy. In the first half of the nineteenth century by far the most important American center of natural history studies was the Academy of Natural Sciences in Philadelphia.[3] It lost its predominance with the rise of the U.S. National Museum in Washington and other museums, but has continued to be a major center of ornithological interest. Among its recent curators, James Bond (b. 1900) has become the leading authority on the birds of the West Indies; he has also made major contributions to the knowledge of South American birds, particularly those of Peru and Bolivia. Rodolphe Meyer de Schauensee (b. 1901) concentrated at first on the birds of Thailand and southwest Africa but became in recent years one of the leading authorities on the birds of South America. After a number of preliminary reports on collections made by Sneidern, he published his *Birds of the Republic of Colombia* (1948–1951), followed by an illustrated guide, *The Birds of Colombia* (1964). This was a pioneering endeavor because until that time no adequate guide to the birds of any of the countries of tropical America had existed. Colombia having an extraordinarily rich bird fauna, it seemed logical to go on from there to South America as a whole, and this is indeed what Meyer de Schauensee undertook next, publishing *The Species of Birds of South America* in 1966 and *A Guide to the Birds of South America* in 1970. He and William Phelps, Jr., have since completed an illustrated guide to the birds of Venezuela and he is now working on a guide to the birds of China.

The U.S. National Museum. The history of the U.S. National Museum,[4] now the U.S. Natural History Museum, had some of its most brilliant moments almost immediately after the founding of the Smithsonian Institution in 1846. Stresemann has well described the period of Baird, Coues, and Ridgway (Chapter 13).[5] When Baird became assistant secretary, at the age of 27, he achieved wonders. Not only did the exploring expeditions in the West, which he organized, bring in enormous collections,[6] but his interests went far beyond the piling up of specimens. He and Coues were greatly interested in ornithology as a branch of scientific zoology, and they made pioneering contributions to the problems of geographic variation and the nature of species.

Robert Ridgway was America's foremost professional ornithologist during the period from about 1890 to the 1920s. In his work he set a standard of accuracy and reliability that served as a sound basis for the future. His monumental *Birds of North and Middle America*, Parts 1–8 (1901–1919) was never completed, even though his successor Herbert Friedmann (b. 1900) brought out three more parts. Ridgway and his associates (Oberholser, Richmond) had a great interest in the correct names of birds and were among the leading supporters of the principle of strict priority. As everywhere else in the world during this period, the description of new subspecies appeared to be the acme of ornithological activity. A strictly morphological definition of the genus led to the recognition of so many new genera that for Oberholser the average number of species per genus dropped to less than two.

Alexander Wetmore (b. 1886), who became director of the U.S. National Museum in 1925, had shown an early interest in the living birds, and his observations on the biology of ducks and on the birds of Argentina were long unsurpassed. In spite of his demanding duties, first as assistant secretary and later as secretary of the Smithsonian Institution, he steadily pursued ornithological studies, which culminated in his *Birds of the Republic of Panama* (1965– ; 3 volumes published).

The current secretary of the Smithsonian, S. Dillon Ripley (b. 1913), again an ornithologist, specializes in the birds of the tropical Old World, particularly India. With Salim Ali he is the author of the authoritative *Handbook of the Birds of India and Pakistan* (1968– ; 9 volumes published). The present staff of the Bird Department at the U.S. Natural History Museum is very diversified, with specialists on Eurasia and sea birds (George Watson), anatomy (Richard Zusi), and tropical America (Paul Slud). It is greatly strengthened by its association with the rich collections of the U.S. Fish and Wildlife Service.

The Museum of Comparative Zoology at Harvard University,[7] founded by Louis Agassiz in 1859, played an important role in the promotion of scientific ornithology. Its first curator of birds, J. A. Allen (1838–1921), was active in the founding of the Nuttall Ornithological

Club in 1873 and ten years later in the founding of the American Ornithologists' Union, of which he became the first president. The Cambridge area was blessed by the presence of a number of highly qualified amateurs, such as William Brewster, J. E. Thayer, C. F. Batchelder, Outram Bangs, A. C. Bent, and F. H. Kennard. Brewster's delightful natural history accounts, Bent's 23-volume *Life Histories of North American Birds* (1919–1968), and John C. Phillips's four-volume *Natural History of the Ducks* (1922–1926) are among the many significant contributions of this group. In 1931 James L. Peters (1890–1952) began the publication of a *Check-list of Birds of the World*, the first attempt since Sharpe's *Handlist* (1903–1910) to provide a complete listing of all species and subspecies of the birds of the world. Only seven volumes had been completed by the time of Peters's death in 1952, but James C. Greenway, Jr. (up to 1960), Raymond A. Paynter, Jr. (up to 1970), and Ernst Mayr took over the continuation. With the dedicated help of ornithologists from all over the world, this important 15-volume project is now nearing completion. Nothing like it exists for any other group of organisms.

The American Museum of Natural History. No other institution has had as profound an impact on the development of ornithology in America as the American Museum of Natural History in New York.[8] This is largely due to the influence of a single person, Frank M. Chapman (1864–1945). Although the American Museum is young in terms of ornithological history, having been founded only in 1869 by A. S. Bickmore (1839–1914), one of Louis Agassiz's pupils, it was favored by being situated in that energetic and wealthy metropolis, New York. As a result, it soon began to outdistance its sister institutions. J. A. Allen, famous for establishing Allen's Rule and other generalizations about the geographic variation, speciation, and geography of birds, became curator of both birds and mammals when he came from the Museum of Comparative Zoology in 1885. In 1888 Chapman was appointed as his assistant and when, in 1908, a separate department was established for the birds, Chapman was put in charge of it. Having left school at the age of 16 to become a bank clerk, he had had no college education. Birds, however, had been his major interest from boyhood on, and his excellent reports on bird migration brought him to the attention of the authorities of the American Museum. When he retired in 1942 at the age of 78, he had served the institution for 54 years.[9]

Chapman had his foibles like all of us. He was rather short and seemed a bit pompous to those meeting him for the first time. Yet he was an extraordinary personality. He introduced a new concept of exhibits into museums, that of habitat groups which accurately represent a definite spot in nature. In his *Camps and Cruises of an Ornithologist* (1908), he describes with what care he chose the sites for his bird groups in the American Museum's Hall of North American Birds and with what pains and accuracy all the accessories were collected. These groups and the later ones he supervised in the Birds of the World Hall and the Whitney Hall of Birds are monuments to Chapman's vision and artistic taste. He was a most understanding colleague, full of wisdom and friendliness. When, as a young man recently arrived from Germany, I reported to him and asked to be assigned a definite task, he said: "We engaged you as a specialist in South Sea birds, and I leave it entirely to your judgment what you would like to take up first. Come back with your first manuscript after it is completed." This certainly represented freedom of research in the best sense of the word. To one used to the more authoritarian customs of German science, this was an unforgettable experience.

Chapman's interest in young people was shown by his founding of the periodical *Bird-Lore* (in 1899), which stimulated generations of young ornithologists and had perhaps more to do with the spread of bird study and an appreciation of conservation in the United States than any other factor. Right up to his old age, Chapman kept an eye on *Bird-Lore*, until it was taken over by the National Audubon Society and eventually converted into *Audubon.**

Chapman deserves great credit for his furthering of bird art. He was a close personal friend of Louis Agassiz Fuertes,[10] to whom he gave every possible encouragement. He "discovered" Francis Lee Jaques[11] and engaged him to paint many of the magnificent backgrounds for the bird exhibits of the American Museum. Numerous other young bird artists received their first appreciation and encouragement from Chapman.

As a scientific ornithologist, Chapman had a much broader vision than some of his contemporaries at other museums, who seemed to be interested mainly in the "making" of new genera and subspecies. Even though he was not up to the technical perfection of a Ridgway or a

*Volume 1 of *Bird-Lore* was published in 1899. In 1941, after transfer to the National Audubon Society, it was renamed *Audubon Magazine*, and in 1953 simply *Audubon.* For a history see R. T. Peterson, "The Evolution of a Magazine," *Audubon*, 75, 1 (January 1973): 46–51. Another American ornithological journal was the *Osprey*, founded in 1896 and expiring with Vol. 7 in 1903. Best known for an outspoken editorial column ("Editorial Eyrie"), contributed by Elliott Coues, it never succeeded in its endeavor to become a broad-based journal. Chapman's *Bird-Lore* succeeded where the *Osprey* failed.

Chapman helped the young bird student in many other ways. His *Color Key to North American Birds* (1903, 2nd ed. 1912) was a quick identification guide serving much the same purpose as the later Peterson field guides. His *Handbook of Birds of Eastern North America* (1895, 2nd ed. 1912) was the most convenient reference work for generations of field ornithologists, and is still so useful that it was recently reprinted by Dover Publications.

Hellmayr, Chapman had vision that they did not have. His essays on the distribution and ecology of the birds of Colombia (1917) and Ecuador (1926) and his papers on the birds of the Venezuelan highlands (Roraima, Duida, 1929–1931) were pioneering contributions that not only (quite successfully) attempted to classify South American birds into ecological associations but also attempted to unravel the history of South American birds. Instead of accepting these faunas as something static, he asked how the bird faunas of the higher altitudes had evolved, what connections had existed formerly between the faunas of the temperate southern parts of the continent and those of the temperate zones at the higher altitudes of the Andes, and how the mountain faunas of isolated areas had evolved. With these questions he initiated a field of investigation that is still flourishing and still producing new answers.[12] His name will always be a shining one in the history of biogeography. Most remarkable, when Chapman was well in his sixties, he started an entirely new line of research, the detailed life history of tropical bird species. His studies of Wagler's oropendola (1928) and Gould's manakin (1935) were the beginning of an ever-increasing number of studies on the biology of tropical species, now appearing in the American literature.

Perhaps Chapman's most important achievement was to have brought together a remarkable staff of outstanding authorities to form the Bird Department of the American Museum: W. DeWitt Miller (1879–1929); Robert Cushman Murphy (1887–1973), an eloquent lecturer and the world's leading authority on seabirds; John T. Zimmer (1889–1954), trained by Hellmayr in the methods of the minute analysis of geographic races and whose "Studies of Peruvian Birds"[13] is the classic achievement in the study of subspecies of South American birds. Chapman had brought Zimmer from Chicago to New York to write a zoogeography and ecology of Peruvian birds equivalent to (and, it was hoped, even better than) his own essays on Ecuadorian and Colombian birds, but Zimmer lacked Chapman's vision and ability to generalize and never went beyond the production of generic revisions. When he died of cancer at a comparatively early age, Zimmer took with him his enormous knowledge of the distribution of South American birds. James P. Chapin (1889–1964), a specialist on African birds, had had an excellent training in modern biology (in T. H. Morgan's Zoology Department at Columbia University), such as Chapman would have loved to have himself. The introductory section of volume one (1932) of his *Birds of the Belgian Congo* was in its day — and in many ways still is — the finest treatment of a tropical bird fauna. Adaptation to special niches, seasonal variation including intratropical migrations, life histories, problems of speciation, and numerous other questions raised by tropical bird faunas were here treated virtually for the first time and for any continent and have served as a basis for the later studies of Reginald E. Moreau and others. I myself (from January 1931 on) had the unexcelled opportunity of working on the rich material of the Whitney South Sea Expedition (1920–1939), constantly helped and encouraged by these splendid colleagues. My studies culminated in *Systematics and the Origin of Species* (1942; see Chapter 15). There were (in the 1930s and 1940s) an intellectual excitement and a level of professional competence and ornithological universality at the American Museum that had nowhere existed previously and that perhaps can never again be duplicated. Much of ornithology has since become experimental and highly technical, so that its center of gravity has shifted from museums to experimental laboratories at the universities. The tradition of excellence is being maintained at the American Museum by the younger generation: E. Thomas Gilliard (1919–1965), Dean Amadon (b. 1912), Charles Vaurie (b. 1906; retired 1972), Wesley E. Lanyon (b. 1926), Lester L. Short (b. 1933), Eugene Eisenmann (b. 1906), and others.

One of Chapman's great triumphs — although the real credit should go to Dr. Leonard C. Sanford (1868–1950) — was the acquisition of the world-famous Rothschild collection for the American Museum (Chapter 14). The availability of this material greatly facilitated my work on New Guinea and South Sea island birds. Furthermore, after it had been incorporated into the already existing collections (which were newly classified under my supervision), it permitted a whole series of family revisions by Jean Delacour, E. Mayr, D. Amadon, and S. Dillon Ripley, and ultimately publication of Charles Vaurie's *Birds of the Palearctic Fauna* (1959, 1965), now the standard work for that region. Old World ornithologists have always been saddened by the fact that the havoc created by two world wars permitted the shift of the center of ornithological activity to the New World to such an extent that not only the *Check-list of Birds of the World* but also the standard work on the Eurasian fauna were written in America by American ornithologists.

The Field Museum of Natural History in Chicago[14] started its bird collection quite late (1893) and in 1933 possessed only about 100,000 bird skins. In recent years its collections have increased perhaps more rapidly than those of any other major museum, reaching 325,000 specimens in 1971. The institution acquired world fame in ornithological circles when in 1922 it asked Charles Hellmayr (1879–1943) to take over authorship of Cory's *Catalogue of Birds of the Americas.* This resulted in the publication of nine major volumes, the pinnacle of precise taxonomic scholarship. Since Hellmayr had personally examined virtually all type specimens of American birds (scattered as they were over scores of Old and New World museums), his *Catalogue* became the secure foundation of all future research on the systematics of New World birds. H. Boardman Conover (1892–1950), Austin Rand (b. 1905), Emmet R. Blake (b. 1909), and Melvin Traylor (b. 1915) took over when Hellmayr returned to Europe in the 1930s. Rand (retired 1970) worked on the birds of the New Guinea area, the Philippines, and Africa, whereas Traylor specialized in African birds. Rand, Blake, and Traylor have helped greatly in the continuation of Peters' *Check-list.* With Hellmayr one would not know from reading his publications whether he had ever seen a bird alive. His successors in Chicago are characteristic of the newer generation of systematic ornithologists by being interested in all sorts of aspects of the biology of birds, Rand for instance having made significant contributions to the study of bird behavior and Traylor to biogeography and the avian molt.

It is impossible in a cursory view of American ornithology to give detailed accounts of the numerous other museums where significant ornithological research is conducted. The list of such museums includes, for instance, the Carnegie Museum in Pittsburgh (Todd, Parkes), the Los Angeles County Museum (Hildegarde Howard, Stager), the California Academy of Sciences (Orr), and the museums of such universities as Michigan (Van Tyne, Storer, Tordoff), Yale (Ripley, Sibley), Louisiana (Lowery), Florida (Austin, Brodkorb), Kansas (Johnston, Mengel), Minnesota (Breckenridge, Tordoff), Utah (Behle), and Occidental College (Hardy). There are also two museums with major ornithological collections in Canada: the Royal Ontario Museum, Toronto, and the National Museum of Canada, Ottawa.

Other Centers of Ornithology

Ornithology at the Universities. Ornithology was essentially an amateur science until the first quarter of this century. The best-known ornithologists were clergymen, doctors, lawyers, or wealthy landowners, that is, they did not earn their living from ornithology. To be sure, some university professors, like C. O. Whitman (Chapter 15) studied the anatomy, physiology, or genetics of birds, but their research had virtually no contact with (and hence no influence on) the work published in the ornithological journals. Even some of the leading professional ornithologists, for instance, Ridgway and Chapman, had no university training in biology but had learned their trade as apprentices. There were a few exceptions to this generalization, but on the whole it was true. As a result, ornithology was not taken very seriously by the professional zoologists at the universities. This all changed in the twentieth century, for many reasons. One of these was, of course, the general professionalization of science, which finally caught up with ornithology. A second reason, no doubt, was that the work of the amateurs, enormously stimulated by the fight over the validity of Darwinism, produced so many challenging scientific problems connected with birds that certain aspects of bird study became quite attractive even to the professional zoologist. By 1933 some 36 colleges in the United States were giving courses in ornithology.[15] At that time 13 major universities in the United States and Canada permitted graduate students to write their dissertations on bird problems, but only Cornell, the University of California (Berkeley), Michigan, and Western Reserve (Cleveland) had truly active programs. Cornell was the first university (1915) to appoint a professor of ornithology, Arthur A. Allen. His courses, which included instruction in bird photography and other practical matters, were outstandingly popular, graduate courses were soon added, and many of his students earned a Ph. D.* A special Laboratory of Ornithology was founded in 1957, with its headquarters at a field station at Sapsucker Woods, near Ithaca.[16]

Joseph Grinnell (1877–1939) began in 1909 to give lectures on birds at the University of California (Berkeley)[17] and regular undergraduate courses in 1913 which very soon blossomed out into graduate teaching. Grinnell had taken his degree at Stanford under D. S. Jordan, who — although himself an ichthyologist — was liberal enough to accept an ornithological thesis. Similarly an occasional ornithological degree was awarded at a number of other universities. Josselyn Van Tyne (1902–1957) earned his at Michigan in 1928 and, after having become curator of the Museum of Zoology, he himself instituted a vigorous Ph.D. program, continued after his death by Robert W. Storer and H. B. Tordoff.

Now, 40 years after Allen's report, some 50 universities in the United States have Ph.D. programs dealing with birds, not including a considerable amount of avian research on wildlife biology conducted also in university departments (see below). Alden H. Miller (1906–1965), who after Grinnell's premature death took over from him at Berkeley, alone sponsored 31 Ph.D. candidates, many of whom subsequently assumed teaching positions in California or elsewhere, including such well-known ornithologists as W. L. Engels, Frank Richardson, H. I. Fisher, Albert Wolfson, A. S. Leopold, F. A. Pitelka, J. T. Marshall, C. G. Sibley, R. W.

*Apparently Allen started giving courses in ornithology soon after receiving his Ph.D. in 1911, but it was in 1915 that he was formally appointed assistant professor of ornithology. Some of Allen's better-known graduate students (Ph.D.'s) are Herbert Friedmann, Austin Rand, John T. Emlen, Dean Amadon, William C. Dilger, Francis Harper, P. P. Kellogg, Harrison F. Lewis, Ralph Palmer, Kenneth C. Parkes, Olin S. Pettingill, Allan R. Phillips, George M. Sutton, James T. Tanner, and Dwain W. Warner.

Storer, C. B. Koford, John Davis, T. R. Howell, K. L. Dixon, R. F. Johnston, R. K. Selander, R. I. Bowman, and N. K. Johnson. The studies of this large group of well-trained investigators have dealt with almost every attribute of birds, including anatomy, physiology, and behavior. In turn, many of them have established their own schools, in California, Utah, New Mexico, Texas, Kansas, and elsewhere.

In 1962, 160 graduate students in the United States were working on theses dealing with birds. Even though most of them became physiologists, anatomists, ecologists, or teachers of general biology (zoology), rather than professional ornithologists, these figures graphically illustrate the breadth of ornithological research in the United States.[18]

Departments of Wildlife Biology. An entirely independent trend of professionalization occurred in a different area. As the population of the United States increased, together with improved means of transportation and rising affluence, an increasing pressure developed on our game animals. After Aldo Leopold (1886–1948) had developed his new wildlife philosophy,[19] it became evident that game animals were an important renewable resource which could be harvested intelligently only if two conditions were met: (1) a flexible regulation of hunting pressure, and (2) preservation of the habitat, including, for migratory species, feeding and resting stations on their migration routes and in winter quarters. To meet this program required far more information than was available in the 1920s and 1930s, when these needs first became apparent. The result was the establishment of departments for wildlife research at many state universities. These sponsored research in wildlife species, some of it of first-rate quality and of far greater than purely applied significance. Leopold himself founded, at the University of Wisconsin, the first wildlife department (now the Department of Wildlife Ecology), which under his guidance and that of his successors J. J. Hickey and R. A. McCabe has been important in the education of highly qualified wildlife specialists and in significant research, particularly the effects of pesticides on bird populations. Frances Hamerstrom, the first woman Ph.D. in this field, got her degree at Iowa State. The *Journal of Wildlife Management* (founded in 1937) has published many investigations on bird biology and bird ecology that constitute major contributions.

Ornithological Research in Federal and State Departments. Not only at museums and academic institutions is ornithology cultivated, but also at the federal level through various agencies in the Department of Agriculture and the Department of the Interior (Fish and Wildlife Service). Most states now also have their own wildlife and conservation departments which employ specialists who received their education in zoology or wildlife departments. A considerable number of state bird faunas and of monographs of particular species, especially game birds, have been sponsored by state agencies. (See also Welker).[1]

Amateur Ornithology. Great Britain is perhaps the only other country in the world in which an interest in birds is as widespread as in the United States. This interest is evident at all levels, in the large number of popular magazines and books devoted to birds, in lectures and television programs, and in attractive museum exhibits. Pearson[20] provides some interesting figures on the sales of popular bird books up to July 1, 1933: John Burroughs (from 1880 on), 1,333,176 copies; Chapman's books (1895 ff), 230,000; Neltje Blanchan's *Bird Neighbors* (1897 ff), 150,000; Chester Reed's *Bird Guides* (1906 ff), 633,500 copies; E. T. Seton's books on mammals and birds (1880 ff), 2,425,715.* Of government pamphlets dealing with birds, their economic value, and protection, 27.5 million had been distributed. The National Audubon Society had published and supplied to the public about 270 million pages of printed information on birds and their protection.

No other living man has done so much for spreading an interest in birds as Roger Tory Peterson (b. 1908). His *Field Guides* are the constant companion of every bird watcher, giving him sound advice on the identification of difficult species. His other books and his photographs, motion pictures, and lectures have all given great enjoyment to literally millions of people.

A great deal of the popular interest in birds is largely recreational and much of it is now correlated with conservation efforts. Yet leading amateurs have always made major contributions to ornithology and continue to do so. For instance, most of the founders of the national ornithological societies were non-professionals. Although the trend toward professionalization is undeniable, amateurs have continued to do splendid work. One might mention, for instance, the work of Laurence Walkinshaw (a dentist) on cranes, of Harold Mayfield (a glass manufacturer) on Kirtland's warbler, of William Schorger (an engineer) on the passenger pigeon, of H. Roy Ivor (an engineer) on anting, and of Crawford Greenewalt (a chemist) on bird flight and bird vocalization. Migration watching and bird banding have particularly appealed to amateurs. Many of the regional ornithological societies, such as the Nuttall Ornithological Club (Cambridge) or the Linnaean Society of New York, are now almost entirely run by spare-time ornithologists.

Women have played a far greater role in ornithology in America than in other parts of the world. There are several women among the most

*Pearson's report was written before the Peterson guides, and many other bird guides have appeared since. With every bird watcher having one or several of these guides, their total sales must be many millions. Bird study has become the major avocation of an appreciable percentage of all Americans.

successful bird banders, but women have made even more significant contributions to behavioral studies. In addition to Margaret Morse Nice, one might mention Louise de K. Lawrence and Josephine Michener (mockingbirds), Winifred Sabine (emberizines), Frances Hamerstrom (grouse, hawks), Doris Hauser (sunbathing), Marguerite Heydweiller Baumgartner and Helen Hays (duck behavior). Women have also contributed importantly in other areas, such as paleontology (Hildegarde Howard), faunistic studies (Elsie Binger Naumburg, on South American birds), and avian anatomy (Mary Heimerdinger Clench).

Field Research

Ornithological Expeditions. The Lewis and Clark Expedition (1804–1806) was the first major scientific expedition the young republic was able to organize. It was the beginning of a great period of exploration of the West, particularly in connection with the boundary and railroad surveys (Chapter 13).[6] Most of the surveys had primarily geographic aims, but owing to Baird's insistence many of them were accompanied by zoological collectors and greatly enriched the U.S. National Museum. Very active at the same period were a number of army surgeons who made collections in out-of-the-way places and discovered numerous new species of birds.*

Soon the activities were expanded into Central America, particularly Mexico, and various parts of South America. Expeditions to other continents followed in quick order, such as that of Herbert Lang (with Chapin) to Africa, the Archbold Expeditions to Madagascar and New Guinea, the Whitney South Sea Expedition (1920–1939), and others too numerous to mention. The earliest American voyage (with ornithological results) was probably that of the U.S. Exploring Expedition to the Pacific (see Chapter 12), which was soon followed by others to the Arctic and Antarctic. Several expeditions specialized in the study of the sea birds, such as the Brewster Expedition (Rollo H. Beck) to South America and the Blossom Expedition to the islands of the Atlantic.

The grand expeditions of former periods have now been largely replaced by the individual investigator who — aided by the improved means of modern transportation — quickly gets to the desired location

*E. E. Hume has provided a mine of information in his *Ornithologists of the United States Army Medical Corps* (Baltimore: Johns Hopkins Press, 1942; 608 pp.). The 36 army surgeons for whom biographies are provided include such well-known ornithologists as Charles Bendire (1836–1897), James G. Cooper (1830–1902), Elliott Coues (1842–1899), A. L. Heermann (1827–1865), Edgar A. Mearns (1856–1916), Robert W. Shufeldt (1850–1934), Casey A. Wood (1856–1942), Leonard Wood (1860–1927), and John Xantus (1825–1894).

and does his work there until he has obtained the specimens and the information he needs.

Field Stations. When the taxonomic-faunistic researches began to suffer from the law of diminishing returns, it became evident that a field station is a far more effective tool for ecological and behavioral research than is an expedition. Many of the museums and university departments have therefore established their own field stations. To mention only a few of the better known: the American Museum staffs the Archbold Station (Lake Placid, Florida), the Southwestern Research Station (Portal, Arizona), and the Kalbfleisch Station (Huntington, Long Island); the Museum of Vertebrate Zoology (Berkeley) established the Hastings Reservation; the Smithsonian Institution, the Tropical Research Institute in Panama; and the Zoological Museum of the University of Michigan, the George Reserve. For water birds there are the Delta Waterfowl Research Station (Delta, Manitoba), and the Weller Wildlife Station (Sinton, Texas). A good deal of very substantial work has also been done in some of the national parks.

Communication among Ornithologists

Ornithological Societies. Perhaps nowhere else in the world have ornithological societies played such an important role in stimulating ornithology as in North America. The first of these societies to be founded, in 1873, was the Nuttall Ornithological Club[21] in Cambridge, Massachusetts. It initiated almost at once a vigorous program of publication and attracted not only local members but also leading ornithologists from New York and elsewhere in the United States. Only 10 years later, in 1883, three of its members — J. A. Allen, Elliott Coues, and William Brewster — sent a letter to 48 leading American ornithologists inviting them to attend a meeting for the purpose of "founding an American Ornithologists' Union." In response to the call, 21 persons met in the library of the American Museum of Natural History in New York on September 26, 1883, and chose Coues as temporary chairman. After a constitution had been adopted, Allen was elected the first president and C. Hart Merriam the first secretary. Most officers served for lengthy terms, and this "has been largely responsible for maintaining the conservative and consistent policy of the Union," as T. S. Palmer said in 1933 with considerable pride. Others found this conservative atmosphere rather stifling, and it eventually led, soon after 1933, to the adoption of various changes in the constitution of the A.O.U. that gave the younger members greater voice in the affairs of the Union.

The publication problem of the newly founded A.O.U. was solved ingeniously: "Thanks to the generosity of the Nuttall Ornithological

Club, an offer was made of the 'Bulletin' of the Club together with its editor, circulation, and good will. This offer was promptly accepted, and thus the 'Bulletin' for October, 1883, which contained an account of the New York meeting, became the last issue of the Nuttall Club series. The next issue which appeared in January, 1884, came out as the initial number of a new series known as 'The Auk,' but the general style and make-up were essentially the same, and Dr. J. A. Allen was continued as Editor of the new journal."[22]

The American Ornithologists' Union had its greatest importance in the first decades after its founding. Some of its committees, such as those on migration, on geographic distribution, and on the English sparrow, speedily outgrew the resources of the Union and within two years were turned over to the U.S. Department of Agriculture, forming the nucleus of what is now the Fish and Wildlife Service (originally the Biological Survey, later the Bureau of Sport Fisheries and Wildlife).* The Committee on Bird Protection played an important role in the founding of the National Audubon Society.† The Committee on Classification and Nomenclature of North American Birds issued the important series of *Check-lists of North American Birds*[23] of which the earlier ones in particular were pioneering in popularizing trinomials and setting the stage for the adoption of polytypic species, in other words, spearheading the new systematics (Chapter 13).

From the very beginning, Canadians‡ were active in the A.O.U. and the border between the two countries has been of little influence in the

*The role the A.O.U. played in the establishment of the Biological Survey is well portrayed in Chapman and Palmer, (ref. 1). Detailed accounts are given in in the chapters by Palmer, Rowan (bird migration), Lincoln (bird banding) and McAtee (economic ornithology). The only comprehensive history of the Biological Survey is Jenks Cameron, *The Bureau of Biological Survey, Its History, Activities and Organization* (Baltimore: Johns Hopkins Press, 1929; 349 pp.). A more recent report on the ornithological activities of the agency are given by D. L. Leedy, "Some Federal Contributions to Bird Conservation during the Period 1885–1960," *Auk*, 78 (1961): 167–175. It must be emphasized that the interests of the Bureau go far beyond game species.

†American ornithologists have consistently played a major role in the conservation movement. The most active officers of the National Audubon Society as well as those of the state Audubon Societies and more local organizations have had their major interest in birds. The A. O. U., the Wilson Ornithological Society, and most other ornithological societies have their own conservation committees, which have been instrumental in the protection of certain species, in preservation of their habitat, and in public education. The history of the American conservation movement has not yet been written, but when this is done, the leadership of ornithologists will be apparent. (See also Welker, ref. 1.)

‡Thomas McIlraith of Hamilton, Ontario, was one of the founders of the A.O.U. His *Birds of Ontario* (1886) was an influential publication (see Palmer et al., ref. 1, pp. 386–387). There is no published history of Canadian ornithology, but there are several manuscript histories of certain aspects of conservation and natural history and other relevant material in the archives of the Canadian Wildlife Service, of several museums, and of the Hudson's Bay Company. The Canadian Wildlife Service began with the appointment of Hoyes Lloyd in 1917 as "Ornithologist," charged with administering the Migratory Bird Treaty in Canada.

election of officers or the choice of the location for the annual meeting. However, the United States and Canada together cover an entire huge continent, and in the days of railroad travel attendance at annual meetings required much time. The founding of additional regional ornithological societies was a natural further development.

The first of these, the Wilson Ornithological Club (1888),* centered in the Middle West. It soon established (1894) its own journal, the *Wilson Bulletin*, and developed more and more into a second national society. Indeed, the *Wilson Bulletin* was perhaps the leading ornithological journal of the country during the period (1939–1948) when Josselyn Van Tyne was its editor.[24]

The West Coast soon followed suit through the founding in 1893 of the Cooper Ornithological Club, which became the Cooper Ornithological Society in 1952.[25] Again, in the beginning it was strictly a regional society, chiefly in California and adjacent states, with northern and southern divisions which had actually been founded independently. Its journal, the *Condor* (founded in 1898), in due time reached a level of excellence that attracted members from all over the United States and Canada and indeed from all over the world. None of the other societies was as clearly dominated by a single person as was for many years the Cooper Club by Joseph Grinnell (1877–1939). Even after his death his most eminent student, Alden H. Miller, continued the Grinnellian tradition, ably supported by Jean Linsdale and other Grinnell students and associates. In addition to publishing the *Condor*, the Cooper Ornithological Society made a major impact by the publication of the *Pacific Coast Avifaunas* (see below).

These are the three societies that achieved national and international importance. Additional societies were founded in due time in many states of the Union and Canadian provinces, many of them issuing their own monthly or quarterly publications.† Only a few of these have the ambition of being truly scientific societies; most of them rather concentrate on local records, migration data, and at best the publication of

*Founded in December 1888, as the "Wilson Ornithological Chapter of the Agassiz Association" (Aaron Bagg, MS., Library of the Wilson Ornithological Society, Museum of Zoology, University of Michigan), it became the Wilson Ornithological Club in 1903 and Society in 1955.

†New ornithological societies were founded almost annually in the 20 years following the founding of the Nuttall Club in 1873: 1878, Linnaean Society of New York; 1879, Agassiz Association; 1883, A.O.U. and Ridgway Ornithological Club (Chicago); 1884, Bristol County (Mass.) Ornithological Club; 1886, Audubon Society (now National Audubon Society), Colorado State Ornithological Association, and Young Ornithologists' Association (forerunner of the W.O.S.); 1888, Wilson Ornithological Club and New Haven (Conn.) Ornithological Club; 1889, Audubon Ornithological Club (Chicago) and California Ornithological Club; 1890, Delaware Valley Ornithological Club; 1891, Western Pennsylvania Ornithological Association; 1892, Potomac Valley Ornithological Club; 1893, Cooper Ornithological Club.

local bird faunas. However, many a young ornithologist received his first stimulation and encouragement in one of these local societies. The Linnaean Society of New York, in spite of its name essentially an ornithological society, acquired wide fame through its publication of Margaret Morse Nice's *Life History of the Song Sparrow* (1937–1943) and other monographs.

The spread of bird banding eventually resulted in the founding of bird-banding societies (see below).

Monograph Series. Stresemann (Chapter 21) has described the pioneering work of Margaret Morse Nice on individually marked populations of the song sparrow (*Passerella melodia*). The trail blazed by her has been followed by scores of other ornithologists, who have studied the life histories of particular bird species, often with color-banded individuals. Many of these studies produced such rich results that it was impossible to publish them in short papers in the standard journals. The need arose for monograph series, and several of these were established in the United States. The Linnaean Society of New York initiated this new policy by publishing volumes 1 and 2 of Margaret M. Nice's *Life History of the Song Sparrow*, as just noted, and Tinbergen's *Behavior of the Snow Bunting in Spring* (1939). The National Audubon Society published monographs on the roseate spoonbill (1942) by R. P. Allen, the whooping crane (1952) by R. P. Allen, the ivory-billed woodpecker (1942) by J. Tanner, and the California condor (1953) by C. B. Koford. The Nuttall Ornithological Club in Cambridge established in 1957 a monograph series, *Publications of the Nuttall Ornithological Club*, of which 15 volumes have so far been published, and a similar series was initiated in 1964 by the American Ornithologists' Union (16 volumes published). Not all ornithological monographs are devoted to life histories; some are largely faunistic or devoted to bird anatomy. For instance, the Cooper Ornithological Society publishes a series, *Pacific Coast Avifauna* (1900–) of which 36 volumes have appeared, most of them faunistic.

General Bird Literature. The flourishing state of the science of ornithology is also documented by the large number of textbooks of ornithology among which those of Welty (1962) and Van Tyne-Berger (1959)[26] are perhaps the best known. The richness of research results made Marshall's two-volume *Biology and Comparative Physiology of Birds* (1960–1961) obsolete within 10 years and it has now been replaced by a four-volume work, *Avian Biology* (1971–1974) edited by D. S. Farner and J. R. King.[27] *The New Dictionary of Birds* (1964), edited by Sir A. Landsborough Thomson, is an enormously successful summary of the most important ornithological information, in which many Americans have participated.

Ornithological Libraries. America has many fine ornithological libraries. Among the best are those of the Philadelphia Academy, the American Museum, the Museum of Comparative Zoology, the Smithsonian Institution, and various other museums and universities. The excellence of these library facilities is reflected in the publication of a number of outstanding ornithological bibliographies.[28]

The Recent History of Ornithological Research in America

Birds, in many ways, are ideal material for biological investigations. Their taxonomy is exhaustively known, most of them are diurnal and easily observed, and they make themselves conspicuous by vocal and visual displays. Ornithologists, as a result, became leaders in various branches of biology, ranging from the new systematics and speciation research to endocrinology and behavior biology.* It would be of interest to the historian to take a closer look at the progress in various areas of ornithological research, but space permits only a rather unbalanced reference to certain key areas.

Avian Relationships. Even though birds are better known at the species level than any other group of animals, when it comes to the relationships of families and orders of birds, there is still great uncertainty and controversy. The reasons for this are beginning to become evident. It appears that birds acquired their major adaptations and characteristics soon after their origin in the Jurassic, and the major groups of birds were apparently established during the Cretaceous. The phylogenetic "tree" of birds consequently resembles more the structure of a broad bush than that of a tree, and the conventional labels of "higher" and "lower", usually employed in the comparison of major taxa, are quite meaningless. Since there is a necessity for a linear sequence of taxa in a printed work, it becomes obvious that in bird taxonomy such a sequence can be established only arbitrarily. A sequence for the nonpasserines proposed by Wetmore was adopted by Peters in his *Check-list of Birds of the World* and is therefore now followed rather widely. Some major groupings are also apparent within the passerine birds, such as the New World nine-primaried oscines, but there is still disagreement as to the best sequence for the songbird families. Early in the nineteenth century the elaboration of the central nervous system

*See references 18 and 27 for summaries of important biological work on birds, often with extensive historical notes.

was accepted (following Cuvier) as a criterion of advanced taxonomic status. The Corvidae, among which the raven (*Corvus corax*) has the relatively largest forebrain (Portmann), were therefore widely accepted as the "highest" group of songbirds, that is, the one to be listed last in the sequence. This sequence was the one chosen by the Old World ornithologists at the Basel International Ornithological Congress (1954) as that most widely adopted in different parts of the world and most likely to lead to universal acceptance.[29] Wetmore (1930) had suggested placing the finches last, as representing a recent development with the potential of further evolution. This is the sequence adopted by most New World ornithologists.

Although the argument about the best linear sequence of avian families can probably never be settled, because no twig of a bush is "higher" or "more advanced" than any other twig, there is nevertheless a legitimate interest among ornithologists in the question: Which family is closest to what other family? Inasmuch as conventional morphological analysis seemed to be up against a stone wall, or, at best, was making only very slow progress, increasing interest developed in new approaches to the comparative anatomy of birds and in the employment of additional taxonomic characters. Beecher, Tordoff, Berger, Hudson, H. I. Fisher, and others made various contributions, some of them of monographic extent, to the muscle systems (and associated bones) of specified groups of birds. It became obvious that mere description would always give only ambiguous results, and for this reason there has been an increasing stress on the functional interpretation of differences between different groups (see Chapter 15). Walter Bock, [30] in a series of analyses, has attempted to investigate carefully the evolutionary significance of anatomical differences and to develop a new methodology for these researches.

In spite of all the work that has already been accomplished, one gains the impression that the field is still in an early stage of development and that further analysis may well produce more significant results than the work of the past. Perhaps birds are particularly apt to respond to specific selection pressures by convergent developments, which means that the evidence from many different character systems will have to be studied in order to determine which similarities are the result of relationship and which reflect convergent adaptation.

Avian paleontology. The study of fossil birds has made rapid strides in recent decades, even though the number of workers actively engaged in this research is still comparatively small; in the United States, Loye Miller (1874–1970), Hildegarde Howard (b. 1901), Alexander Wetmore (b. 1886), and Pierce Bordkorb (b. 1908) are particularly noteworthy. The number of known fossil species is increasing annually, but, what is perhaps more important, the number of known deposits of fossil birds is also increasing, so that the geographic and chronological distribution of species has become increasingly better known.[31] It is now possible, at least for the later Tertiary and the Pleistocene, to describe the distribution of whole avian faunas and to draw conclusions on past climates and the distribution of habitats. Many misidentifications of the past have been corrected in recent years and new lines of relationship established. This work has made it quite clear that the origin of the avian orders, and perhaps even that of many of the families, precedes the Tertiary. The phylogenetic connections will have to be searched for in the Cretaceous. Unfortunately, the fossil record of birds from the Cretaceous is very meager, for, alas, birds do not have teeth or other parts that would fossilize easily, and avian paleontologists, much to their chagrin, have so far not yet been able to find fossils that would connect entire orders of birds with each other. Those who would like to establish the actual line of descent of the avian orders may have to search for the evidence in altogether different areas.

Molecular Phylogeny. One of the most interesting, and in some ways unexpected, discoveries of recent times is that fact that the major molecules of organisms show a structural evolution just like muscles, bones, or any other product of the genotype. This was first indicated by serological evidence, although serology never succeeded in proving anything but the obvious. The demonstration by serologists that the chicken is more closely related to the turkey than to the duck did not surprise anyone. Within the last dozen years, however, entirely new techniques have been developed by biochemists, which permit the dissection of protein molecules into their individual amino acids, and this may ultimately give us unequivocal evidence on the relationship of families and orders. Charles G. Sibley (b. 1917) has conducted avian protein analysis with great energy for some 15 years and has made many suggestive findings. The solution of traditional problems of avian phylogeny will quite likely be assisted by these new techniques before the end of this decade.[32]

Bird Migration and Orientation. There is, perhaps, no other continent in which the manifestations of bird migration are as spectacular as in North America, with its huge temperate and subarctic area and its attenuation toward the subtropical belt. As Stresemann reported (Chapter 19), networks of observers were organized as early as the 1880s, and migration studies were for many decades the major interest of most American bird watchers. William Rowan[33] (1933) has presented an admirable report of the organization of migration watching in the United States. Wetmore's *Migration of Birds* (1926) and Frederick

Lincoln's book (1939) of the same title ably summarized these researches. There has been considerable interest in recent years in quantitative studies of bird migration and in the application of new techniques, such as the tracing of migrating birds by radar and their nocturnal passage across the moon as observed by telescopes.[34]

From an early period on, American bird students were particularly interested in the causation of migration. At first (for instance, Chapman[35]) the discussion was purely speculative, but following William Rowan's discovery of the effect of light on testis growth (Chapter 21), a new period of physiological research began (see below), which is still producing exciting results.

One aspect of migration is that of orientation. Research in orientation and homing has been extremely active in the United States ever since Werner Rüppell and Gustav Kramer revitalized this area of research in Europe. Frank Bellrose, Stephen Emlen, Donald Griffin, W. J. Hamilton III, H. B. Hitchcock, W. T. Keeton, Martin Michener, Franz Sauer, and Charles Walcott are among those who have been particularly active in the last two or three decades in North America.

Avian Physiology. The physiology of birds was long a distressingly neglected area. The work on food utilization carried out in poultry departments did not lead to biologically significant generalizations. For a long time, the most interesting work was that devoted to explaining the physiology of sexual dimorphism and of seasonal plumages. That the male plumage of the domestic fowl is genetically determined and is retained even after the removal of the testes has been known at least since the Middle Ages, through the observation of capons (castrated roosters), which were considered special delicacies. Berthold showed in 1849 that a product of the ovary was responsible for hen feathering (Chapter 21), and the details of this process were worked out by many laboratories, particularly that of F. R. Lillie in Chicago, who showed that it is a female steroid hormone (estrogen) that induces feather germs to produce henny feathers.[36] Much active research, particularly by Emil Witschi (1890–1971) and his students, has shown that in other groups of birds sexual dimorphism is not at all affected by either male or female gonadal hormone, or else that it is affected by a pituitary hormone, as in certain weaver birds.

The most important breakthrough in avian physiology came when William Rowan (1925) of Edmonton, Alberta, showed that the increase in testis size in juncos in the spring was not caused by a rise in temperature but by an increase in day length (Chapter 21). This discovery of photoperiodicity led to the development of a flourishing field of avian physiology, which revealed how complex the control of the annual cycle is in birds. Migratory restlessness, deposition of fat, molt, and many other aspects of the physiology of birds that vary seasonally are under the control of a delicate interaction between internal rhythms (internal clocks) and external stimuli. A fascinating account of these researches can be found in summarizing volumes edited by Wolfson (1955), Marshall (1960, 1961) and Farner and King (1971–1974).[27] Albert Wolfson, T. H. Bissonette, A. H. Miller, Donald S. Farner, James R. King, and Carl Helms are among the American ornithologists who have been particularly active in this field. Most recently Michael Menaker was able to pinpoint the phase in the circadian rhythm when a short illumination was most effective. He also provided definitive proof that birds (house sparrows) without eyes responded to light stimulation as well as normal birds.

Ornithologists are also in the forefront of research in what is often called "environmental physiology." This term refers to the physiological specializations that adapt an individual bird to the changing temperatures and food conditions of the seasons and to regional climatic conditions. The pioneering publication in this field was Baldwin and Kendeigh's *Physiology of the Temperature of Birds* (1932), and S. C. Kendeigh (b. 1904) and his students have continued to make contributions to the study of heat regulation and energy metabolism in birds. With their high body temperature, birds have special problems in regulating their temperature and this results in a selective premium for economizing metabolic expenditure. One of the great discoveries in this field is that hummingbirds, swifts, and poorwills (Caprimulgidae) can drop their body temperature quite drastically for a night or, in the case of the poorwills, for longer periods ("hibernation").[27]

In more recent years interest has shifted to the investigation of the adaptation of birds to special climates. For instance, many birds in the southwestern United States are adapted to the extreme heat and aridity of the deserts. Bartholomew, Dawson, Lasiewski, and Schmidt-Nielsen have studied the water economy of these birds. Schmidt-Nielsen discovered that the nasal glands of marine birds are able to excrete salt, and this explains how oceanic and shore birds can regulate their salt balance. Certain desert birds are likewise able to excrete salt in this manner, adding to the repertory of desert adaptations. The study of physiological adaptations is one of the most rapidly advancing branches of avian biology.

Behavior. In his twentieth chapter Stresemann has described the revolution in the study of animal behavior brought about by Lloyd Morgan, Edmund Selous, Eliot Howard, and Oskar Heinroth. Here he gives full credit to F. H. Herrick for his pioneering writings about "the home life of wild birds." In addition to many studies by amateurs (most of them not quite able to escape the anecdotal approach), there was also some interest in bird behavior among professional biologists, particularly shown in the work of Watson and Lashley (Chapter 21) on the terns of the Dry Tortugas. A new era was ushered in by G. K. Noble (1894–1940),[37] who carried out a number of experiments on wild birds making use of hormone injections, alterations of plumage marks, and dummies. Nothing, however, had a greater impact on American behavior studies than the development in Europe of the new behavior school, ethology.

Nikolas Tinbergen visited America in 1947 when, at the American Museum of Natural History in New York and at Columbia University, he gave a series of lectures, which were subsequently published as *The Study of Instinct* (1951). This, as well as the ethological papers in the European ornithological literature (Chapter 21), attracted many American ornithologists into ethology. Martin Moynihan (b. 1928) went to Oxford to take his degree under Tinbergen and carried out a notable series of ethological analyses on members of the gull family, beginning with the blackheaded gull, extended later to other groups of birds, particularly in the tropics.[38] Andrew Meyerriecks, taking his degree at Harvard, made a very thorough analysis of the courtship displays of the green heron (*Butorides virescens*) in comparison with the previously described display of the European gray heron and certain other American herons.[39] At Cornell, at Berkeley, at Ann Arbor, and at other universities, the frequency of theses devoted to bird behavior increased steadily. Several authors analyzed the display of various species of ducks, expanding upon the previous work of Heinroth, Delacour, and Lorenz. Much of this work (including his own) was summarized by Paul Johnsgard.[40] Others undertook similar studies for storks, Pelecaniformes, woodpeckers, grouse, and many other orders and families of birds. Dilger's careful analysis of the life history of American thrushes[41] revealed the interesting fact that behavior elements indicate quite clearly that the relationship of the woodthrush (*Hylocichla*) to the four other North American species of thrushes is more remote than previously suspected and that these other species should be combined with the genus *Catharus.* This is only one of many instances in which behavioral characters have clarified a taxonomic situation. (I have mentioned here only the authors of the earliest of these studies. The number of more recent students of bird behavior is legion, including Colin Beer, Nicholas Collias, J. T. Emlen, R. W. Ficken, Jack Hailman, P. H. Klopfer, Frank McKinney, and G. H. Orians.)

In the original ethological research the main emphasis was on "ultimate causes," that is, on genetically programmed behavior and its evolution; hence the interest in comparative studies and in the construction of behavior phylogenies. Another group of students of behavior concentrated on "proximate causes," that is, the developmental aspects of behavior. They studied physiological mechanisms and the nature of external and internal stimuli responsible for specific behaviors and behavior changes. Oscar Riddle, the discoverer of prolactin, was a pioneer in this field; in recent years Daniel Lehrman (1919–1972) and his associates at Rutgers University have greatly advanced our understanding of the respective roles of various hormones in the reproductive cycle of birds, in particular that of the ring dove (*Streptopelia risoria*).[42]

Bird Song. Albert Brand (1889–1940), a middle-aged businessman,[43] introduced himself one day in 1929 to Frank M. Chapman, saying that he was tired of dedicating his life to making money and that he would rather make some useful contribution in the broad field of natural history. Chapman advised him to go to Cornell and study birds with Arthur Allen. Since Brand (who never before had been a bird watcher) had trouble remembering the innumberable bird songs and bird calls — such a knowledge being almost indispensable for correct identification — he decided to make a collection of recordings, which would help both himself and others in a similar predicament. Recording equipment at that time was still utterly primitive, since this was before the age of electronics. With great energy and ingenuity, and recruiting all the help attainable from physicists and engineers, Brand became a pioneer in the out-of-doors recording of bird songs. In due time this activity was taken up by many others, both in the United States and abroad, and, because the recording equipment has steadily improved, superb recordings are now available from most parts of the world. However, Brand was not satisfied merely to make recordings; he was interested in the scientific study of bird song. He analyzed the sound tracings on his tapes under the microscope and was beginning to make real headway when he was struck by his final illness. At first no one continued his work, because it was evident that Brand's method of analysis was altogether too inefficient. D. J. Borror (b. 1907) at Ohio State University deserves credit for being the first to apply systematically a new device, the Kay sonagraph, for the translation of the sound recordings into a visible picture accessible to rapid interpretation.[44] He was able to demonstrate how useful — indeed, how indispensable — this method is for the analysis of bird songs. The study of avian vocalization, which is as old as bird study itself, acquired a new lease on life by being subjected to this precise, quantitative analysis.

The question of the learning of song was investigated in depth by W. H. Thorpe and associates (R. A. Hinde, Peter Marler, R. J. Andrew) in Cambridge, England. The new methods made it possible to ask and answer entirely new questions, such as the effect of environmental context on variations in bird song. W. John Smith (then at Harvard University, now at the University of Pennsylvania) analyzed for the eastern kingbird with great precision which songs were given under what conditions and how greatly any of these songs became modified under the influence of different emotional stresses (1966).[45] W. E. Lanyon showed that an analysis of call notes does not support the often-made claim of widespread hybridization between eastern and western meadowlarks (1957).[46] Applying his methods of analysis to the crested flycatcher (*Myiarchus*) group, he has shown that several Central American and West Indian forms had been associated with the wrong species.

A beginning has now been made for a far deeper analysis of bird vocalization. In the new studies vocalization is considered a method of communication, a system of exchanged signals, in which the messages as well as the answers depend on a constellation of external conditions and internal physiological states. This raises a host of new questions. For instance, why do some species seem to have a far richer vocabulary than others? Is there an inverse correlation between richness of visual display and variety of vocalization? The questions that the analysis of avian vocalization have raised are also applicable to the vocalization of mammals and perhaps even to that of small children. Again, it seems that bird study has opened up a highly rewarding new field.

But these are only some of the problems of bird vocalization that have been studied in recent years. Crawford H. Greenewalt, building on the foundation laid by Werner Rüppell, has produced a fine monograph on the acoustics and physiology of bird song (1968).[47] His most important conclusion is that many songs are composed of the independent contributions of the two tracheal membranes. Other authors (Masakasu Konishi, Fernando Nottebohm) have been concerned particularly with the interaction between hearing and vocalization.

Ecology. The occurrence of birds in different habitats has been of keen interest to ornithologists for centuries (Chapter 21). C. Hart Merriam's life-zone concept was largely based on the distribution of birds. It found its most useful application in western North America, where life zones are best developed owing to the vertical structuring of the country. Grinnell at Berkeley and his associates (Tracy Storer, A. H. Miller) were particularly active in testing the life-zone concept and showing how in different parts of California the local faunas could be classified into Sonoran, Transitional, Canadian, and other elements.* Even though it was subsequently shown by Kendeigh and others that the precise temperature correlation postulated by Merriam does not

*Since the life-zone concept played such an important role in the study of western American birds, it is worthwhile to list a few of the local faunas (including some on other vertebrates) in which the life-zone concept was featured.[48] After 1951 the concept was no longer applied in detailed ecological analyses, even though one might still loosely refer to a Canadian or a Sonoran element.

exist, the fact remains that both vegetation and associated animal life show definite zonation and correlation with climate and other physiographic factors. In this approach the emphasis is on the presence and absence of species and on the association of species.

A somewhat different approach is that of quantitative censuses. The National Audubon Society, at the initiative of William Vogt, started in 1935 a project of breeding-bird censuses. There was nothing new in this method, since breeding-bird censuses had already been conducted by economic ornithologists some 30 years earlier.[49] However, this new project had a national basis and involved scores of young ornithologists, all of them becoming aware of the importance of territory size, the fluctuations of population density from year to year, the influence of slight changes in vegetation and landscape on the composition of their bird fauna, and so forth. Problems of competition, territorial exclusion among closely related species, and indeed the whole complex of the factors controlling species diversity in a given area began to arouse the interest of bird ecologists.

In the 1950s, G. Evelyn Hutchinson at Yale University, a limnologist by profession but indeed a very broad biologist, encouraged his student, Robert MacArthur (1930–1972), to analyze avian diversity, with the help of sophisticated mathematical models, far more precisely than had been done previously. This led to new questions: for example, if two pieces of forest greatly differ in species diversity, is this caused by the greater variety of species of trees in one of the forests or by the richer stratification of vegetation in its substage and canopy? Inevitably, such questions led to a comparison of tropical- and temperate-zone bird faunas and to a new study of the old question why tropical faunas are so much richer in species than those of the temperate zones. Also, to pose some other questions: As far as islands are concerned, is the number of species on an island determined by its area, by its vertical elevation, by the distance from other islands and from the nearest mainland, by its latitude, rainfall, vegetation, or by what other factors? What is the proportional contribution made by each of these factors? Soon ecology began to pose questions that were of as much interest to the biogeographer as to the student of environmental biology.[50] The regulation of the diversity of bird species is a fascinating scientific problem to which, in addition to those of MacArthur, valuable contributions have been made by such other American biologists as M. L. Cody, K. K. Crowell, P. R. Grant, Jared Diamond, N. K. Johnson, John Terborgh, T. H. Hamilton, G. H. Orians, F. W. Preston, and T. W. Schoener.

Cycles in the abundance of birds, particularly arctic birds, have always been of interest to ornithologists. It has long been realized that the flights of snowy owls, for instance, are the direct result of crashes in Arctic rodent (lemming) populations. Frank Pitelka (b. 1916) showed, however, that predatory birds, for instance jaegers, have a noticeable, if not a decisive, impact on these cycles.[51]

Another aspect of the study of the relation between bird and environment is the study of adaptive radiations. The pioneering achievements in this area was David Lack's *Darwin's Finches* (1947). This was followed by Paul H. Baldwin's "Annual Cycle, Environment and Evolution in the Hawaiian Honeycreepers" (1953), by another analysis of the Galapagos finches (Robert I. Bowman, 1961), and by various other papers on niche utilization by birds and its effect on morphological adaptations.

Techniques of Investigation

Bird Photography. Bird photography has played an important part in the history of American ornithology.[52] Although occasional bird photographs had been taken by others before him, it was Frank M. Chapman who introduced bird photography to this country. He lectured extensively in his younger years and had himself to provide the material for his slides. With a self-designed reflex camera and a novel type of blind, he took (from 1888 on) a series of what were for his time superb photographs. His *Bird Studies with a Camera* (1900) and *Camps and Cruises of an Ornithologist* (1908) stirred a large group of followers, many of them also stimulated by the earlier publications (1898-99) of the Keartons in England. F. H. Herrick, W. L. Finley, Norman McClintock, H. H. Cleaves, H. K. Job, and many others published splendid avian photographs and entertained enthusiastic audiences on their illustrated lecture tours. While waiting in the blinds, these photographers had the opportunity for detailed observations, but F. H. Herrick and A. O. Gross were among the few who took advantage of this opportunity. Almost 90 percent of the photographs were pictures of incubating birds or birds feeding their young. In more recent years, it has been particularly Arthur A. Allen of Cornell and his students who have raised the standard of bird photography.

With modern telephoto lenses, highly sensitive films, synchronized-flash equipment, and color photography, photographs of exquisite beauty and highest fidelity have been achieved. Crawford Greenewalt's hummingbird pictures are examples of outstanding bird portraiture. They also show how important high-speed photography is for the study of bird flight.

Nothing, perhaps, benefited the study of the living bird more than the introduction of the motion-picture camera. Chapman took his first motion pictures of birds in 1908, and the study of film segments is now a standard component in the analysis of avian courtship, bird flight, and other aspects of avian behavior. It is no longer necessary to rely on memory for an accurate perception of a courtship action that occurs in a fraction of a second.

Bird Banding. After some pioneering experiments beginning in 1902 by various individuals (Paul Bartsch, L. J. Cole, P. A. Taverner), the American Bird-Banding Association was formed in 1909 (at a meeting of the American Ornithologists' Union in New York). Banding, first of nestlings but later also of trapped adults, was ever more actively pursued from this point on. In the beginning entirely handled by volunteers, the banding program was taken over on March 1, 1920, by the U.S. Biological Survey.[53] Since that time many millions of birds have been banded and new uses for the banding method are continually being found. Although migration routes and the location of winter quarters were at first the major interest of the banders, banding (particularly with colored bands) increasingly became the most important method for the study of bird individuals. The information derived from banding returns has become absolutely vital in the regulation of the hunting pressure on game birds (grouse, ducks, doves). The introduction of the Japanese mist nets into America by Dr. Oliver Austin, Jr., was a significant methodological advance and greatly facilitated the capture of certain previously elusive species.* Banding is now usually supplemented by a detailed study of the captured bird, and records are taken of its weight, stage in the molt, presence of parasites and pathological conditions, and so forth.[54]

There are a number of bird-banding associations in different parts of the country and most of the results are published in a quarterly, *Bird-Banding*, founded in 1930 by the Northeastern Bird-Banding Association. Recoveries of banded game birds are often published in special publications of the U.S. Fish and Wildlife Service.

Regional Ornithology

The United States and Canada cover such an enormous area that the bird life is very different in Florida, New England, the Prairie Provinces, Labrador, or the arid Southwest. It is not surprising then that from the very beginning there has been great interest in America in regional studies. Such studies may be restricted to the vicinity of a major city (New York City region), a natural physiographic district (Delaware River

*Italian bird nets were regularly used at the Austin Ornithological Research Station on Cape Cod from 1930 until 1950, when Austin brought mist nets with him on his return from Japan (he had previously mailed some to Seth Low at Patuxent, Maryland). The first comprehensive report was published by Seth Low: "Banding with Mist Nets," *Bird-Banding*, 48 (1957), 115–128.

Valley, Great Basin), a county (Essex County, Dutchess County), or an entire state. Only the *Handbook of North American Birds*, ed. Ralph Palmer (vol. 1, 1962), covers all of North America.

State Bird Books. Ornithological interest was greatly stimulated by the so-called state bird books. Many of them have placed special emphasis on colored illustrations, and artists like Louis Agassiz Fuertes, Allan Brooks, Francis Lee Jaques, George M. Sutton, Walter A. Weber, and many others added in this way to their reputations. *The Birds of New York* (by E. H. Eaton, 1910–1914) and the *Birds of California* (by W. L. Dawson, 1923) were among the earlier well-known state faunas. Forbush's *Birds of Massachusetts and Other New England States* (1925–1929) set completely new standards, by recording a great deal about the living bird, its food, and its breeding habits. Other, more recent, outstanding local faunas include Roberts (1932) *Birds of Minnesota*, Todd (1940) *Birds of Western Pennsylvania*, Howell (1932) *Florida Bird Life*, Mengel (1965) *Birds of Kentucky*, Florence Merriam Bailey (1928) *Birds of New Mexico*, Jewett *et al.* (1953) *Birds of Washington State*, Phillips *et al.* (1964) *Birds of Arizona*, Gabrielson and Lincoln (1959) *Birds of Alaska*, and Stewart and Robbins (1958) *Birds of Maryland and the District of Columbia.*

State books have often been criticized for the artificiality of their geographic limitations, but most other attempts to provide for localized delimitations are not much of an improvement. State borders certainly have the advantage of precision and convenience. [55]

Some local faunas have adopted a more limited scope. Stone's *Bird Studies at Old Cape May* (1937) is an excellent example. Others are provided by the series of ecological surveys published by Grinnell and his school.[48]

Canada. Bird study has had a long tradition in Canada. P. A. Taverner was the author of earlier volumes on the birds of Canada and western Canada, and an excellent treatment of the *Birds of Canada* (1966) by Godfrey is now a reliable source of information.[56]

The major centers of ornithological research in Canada are the National Museum (Ottawa), the Royal Ontario Museum (Toronto), McGill University (Redpath Library), the Arctic Institute, and the biology departments of several of the universities.

Canada has long led in the study of game birds and excellent research on ducks has been conducted for many years at the Delta Research Station in Manitoba, resulting in publications by Albert Hochbaum and others.

Latin America. Ornithological research south of the U.S. border was in the hands of European naturalists during the earlier part of the nineteenth century, as Stresemann has shown. Unfortunately, no comprehensive account of the history of Mexican ornithology seems to exist. However, as the birds of Texas, New Mexico, Arizona, and California became better known, collecting trips by American authors into Mexico were undertaken ever more frequently. The faunistic-taxonomic results of these activities were summarized in the *Distributional Check-List of the Birds of Mexico* by Herbert Friedmann, Ludlow Griscom, R. T. Moore, and A. H. Miller (1950–1957), in various more localized faunas (birds of Sonora, Yucatán, etc.), and in a series of field guides to the birds of Mexico. These activities were soon extended to countries south of Mexico. Bird faunas were provided for Guatemala (Griscom, Land), British Honduras (Russell), Honduras (Monroe), El Salvador (Dickey and Van Rossem), and Costa Rica (Carriker, Slud). It is planned to include all of Central America in the next A.O.U. *Check-list.*

Tropical America. One of the most fascinating developments in American ornithology in recent years has been the enormous interest in tropical ornithology. At first this merely consisted in sending expeditions to poorly known areas in Central and South America, resulting in the long series of volumes by Chapman, Zimmer, Carriker, Todd, Griscom, Wetmore, Meyer de Schauensee, and others, but Chapman's studies of the biology of Barro Colorado (Panamá) birds initiated a new chapter in tropical American ornithology.

Even earlier, William Beebe and the New York Zoological Society had established a research station in Trinidad, where Beebe and later David Snow (now at the British Museum) conducted a whole series of fascinating studies on the life history and courtship displays of tropical American birds. The Trinidad station was the last of a series of temporary research stations that Beebe had in various parts of tropical America. This was soon followed by Alexander Skutch's rich series of life-history studies of Costa Rican birds, many of which brought us the first information on characteristically tropical genera and families. This trend was accelerated when in 1957 Martin Moynihan took over the directorship of the Smithsonian Tropical Research Institute in the Canal Zone (including Barro Colorado). He not only carried out his own research but attracted a very capable staff and was most encouraging to temporary graduate students and postdoctorals. The work of W. John Smith on Tyrannidae, of Neil Smith on oropendolas and cowbirds, and of E. O. Willis on antbirds are examples of these researches. Recently the Smithsonian Institution organized also a station near Belem (Brazil) and the University of California (Berkeley) one near Cali (Colombia).

Most gratifyingly, many South American countries have developed ornithological centers of their own. The late Dr. William H. Phelps, Sr. (1875–1965),[57] and his son, William H. Phelps, Jr., in Caracas, made a thorough survey of the birds of Venezuela, particularly of the isolated sandstone plateaus (tepuis) and gathered an outstanding national collection of birds in the Phelps Museum. Paul Schwartz (Caracas) has built up a fine collection of avian vocalizations and has become the foremost authority on the life of Venezuelan birds. For Surinam there is an excellent avifauna by Frans Haverschmidt (1968). In Brazil there are a number of ornithologists (such as Pinto), continuing a long tradition in part described by Stresemann (Chapters 2, 6, 10). Helmut Sick has contributed not only to our knowledge of the poorly known areas in the interior but also to the life histories of various rare species. Augusto Ruschi has an outstanding knowledge of hummingbirds and has rediscovered certain highly localized species. He has helped Crawford Greenewalt by supplying some of the material for his splendid volume on the hummingbirds. In Peru, Maria Koepcke, who so tragically lost her life in an air disaster on Christmas Eve, 1971, did a great deal for the knowledge of Peruvian birds, building on the foundation laid by Taczanowski and others.

Temperate South America made very early contributions to the history of ornithology: Azara in Paraguay (Chapter 16) and Molina in Chile.[58] Argentina has long been a center of ornithological activity, Hudson's writings, of course, being best known. The Ornithological Society of Argentina publishes its own journal, *El Hornero*, and C. C. Olrog's *Lista y distribución de las aves argentinas* (1963) presents a competent faunal survey of that country.

The birds of the oceans, although few in number of species, are of particular interest, owing to the unique specializations in their life cycles. Major orders of birds are entirely (penguins, albatrosses, shearwaters, petrels) or largely (Pelecaniformes, Lari) restricted to the seas. R. C. Murphy's *Oceanic Birds of South America* (1936) has become the classic in this field. Seabirds, particularly penguins and Tubinares, have also been a major component in the U.S. Antarctic Research Program.

Outlook

Sketchy and necessarily one-sided as this survey had to be, it brings out a number of points that are worth emphasizing. First, there has never been a time when the study of birds has been more active and more diversified than the present. Quite remarkable also is the constant flux in the field. Institutions have periods of flowering and periods of decline. Certain subjects suddenly become fashionable and almost as suddenly are eclipsed by new interests. A third point is the remarkable professionalization of ornithology in the last 30 to 50 years, which — in turn — has resulted in the leadership of ornithologists in various fields of biology. This has not been altogether a blessing. As more and more professional societies were formed, often under the leadership of ornithologists dealing with such special fields as systematic zoology, evolution, and behavior, leading ornithologists tended to attend the meetings of these professional groups as well as the national meetings of the American Institute of Biological Science or of the American Association for the Advancement of Science, rather than the meetings of the ornithological societies, leaving the latter increasingly to the amateurs. There are signs that this trend is now being reversed.

Birds are among the best indicators of a healthy environment. Ornithologists were among the earliest conservationists (in the 1880s and 1890s) and ornithologists have continued to be in the forefront of the fight for a better environment. Although the study of birds is primarily a hobby and a source of recreation to most people, it is at the same time an activity of considerable importance for our understanding of nature and quite likely even for a better planning of Man's future. In all these aspects, the history of ornithology is an important component of the history of biology as a whole.*

*I received great help from G. W. Cottrell, Raymond Paynter, and Jack Hailman in the planning of this Epilogue and in the organization of the material. For information on various matters I am much indebted to John T. Aldrich, Dean Amadon, Oliver L. Austin, Jr., Emmet R. Blake, Walter Bock, John Davis, William C. Dilger, Jocelyn Crane Griffin, Edward Gruson, Jack Hailman, Joseph J. Hickey, Hoyes Lloyd, R. Meyer de Schauensee, Robert Cushman Murphy, Thomas D. Nicholson, Jane Oppenheimer, Roger T. Peterson, O. Sewall Pettingill, Jr., S. Dillon Ripley, H. Radclyffe Roberts, Robert W. Storer, and Melvin A. Traylor.

REFERENCES

INDEX

References

Chapter 1. From Classical Times to the Renaissance

1. *The Works of Aristotle*, ed. J. A. Smith and W. D. Ross, IV, *Historia animalium*, trans. D'Arcy W. Thompson (Oxford, 1910), book 6, chap. 8.
2. *Ibid.*, book 8, chap. 28.
3. Aristotle, *De partibus animalium*, trans. William Ogle, book 1, chap. 5, 645a, in Smith and Ross, *The Works of Aristotle*, V (Oxford, 1912).
4. *Ibid.*, book 4, chap. 13, 697b, 11, 14–15, 23–24.
5. Fritz Schultze, *Philosophie der Naturwissenschaft*, I (Leipzig, 1881), p. 119.
6. See Max Wellman, "Alexander von Myndos," *Hermes*, 26 (1891): 481–566.
7. See Max Wellmann, "Der Physiologos: Eine religions-geschichtlich — naturwissenschaftliche Untersuchung," *Philologus*, Supplementband 22, Heft 1 (1930): 1–116.
8. Konrad von Megenberg, *Puch der Natur*, modernized by Hugo Schulz (Greifswald, 1897).
9. Ernst Kantorowicz, *Kaiser Friedrich der Zweite* (Berlin, 1936), p. 331.
10. Frederick II, *Reliqua librorum Friderici II imperatoris de arte venandi cum avibus cum Manfredi regis additionibus*, ed. Johann Gottlieb Schneider, 2 vols. (Leipzig, 1788–1789).
11. Blasius Merrem, *Versuch eines Grundrisses zur allgemeinen Geschichte und natürlichen Eintheilung der Vögel* (Leipzig, 1788), pp. 22–23. A useful history of Frederick's work, with a survey of contents, manuscripts, and editions, is provided by Michael Henss in *Journal für Ornithologie*, 111 (1970): 456–481, in the course of a review of C. A. Willemsen's facsimile edition of the Vatican manuscript (Codex Platinus Latinus 1071; Frankfurt a. M., 1970).
12. Frederick II, *Reliqua librorum Friderici II. Imperatoris de arte venandi cum avibus* (Augsburg, 1596), p. 155; see note 10 above.
13. Konrad Lorenz, "Beobachtetes über das Fliegen der Vögel und über die Beziehungen der Flügel- und Steuerform zur Art des Fluges [Observations on bird flight and on the relation of wing and tail shape to the form of flight]," *Journal für Ornithologie*, 81 (1933): 107–236; Lorenz, "Fliegen mit dem Wind und gegen den Wind [Flight with the wind and against the wind]," *Journal für Ornithologie*, 81 (1933): 596–607.
14. See Heinrich Balss, *Albertus Magnus als Biologe* (Stuttgart, 1947).
15. See W. H. Mullens, "William Turner," *British Birds*, 2 (1908): 5–13.
16. As Henry Scherren demonstrated (*Proceedings of the Fourth International Ornithological Congress*, London, 1905, pp. 183–190), Turner also compiled the list of "Volucrum appellationes" which the humanists Eber and Peucer added to their book, *Vocabula rei nummerariae . . . collecta* (Wittenberg, 1549).
17. Merrem, *Versuch eines Grundrisses*, p. 25.
18. *Turner on Birds*, ed. and trans. A. H. Evans (Cambridge, 1903), pp. 119–121.
19. Paul Delaunay, *L'Aventureuse existence de Pierre Belon du Mans* (Paris, 1926; reprinted from *Revue du XVIe Siècle*, 9–12, 1922–1925).

20. Pierre Belon, *L'Histoire de la nature des oyseaux* (Paris, 1555), p. 8.
21. It is sufficient to cite here the appreciation given by Ernst H. F. Meyer in his *Geschichte der Botanik*, IV (Königsberg, 1857), pp. 322–334.
22. See *Intorno alla vita e alle opere di Ulisse Aldrovandi: Studi di A. Baldacci* [*et al*]. (Bologna, 1907).
23. Juan Eusebio Nieremberg (1595–1658), *Historia naturae, maxime peregrinae, libris XVI. distincta* (Antwerp, 1635).

Chapter 2. The Beginnings of Exotic Ornithology

1. S. Killermann, "Zur Geschichte der Einführungs der Papageien," *Naturwissenschaftliche Wochenschrift*, n.s., 20 (1921): 545–550.
2. In Konrad Gesner, *Historia animalium*, III (Zürich, 1555), p. 613.
3. Francisco Lopez de Gomara, *La istoria de las Indias* (Saragossa, 1552), p. 564. See also Erwin Stresemann, "Was wussten die Schriftsteller des XVI. Jahrhunderts von den Paradiesvögel?", *Novitates Zoologicae*, 21 (1914): 13–24.
4. Pierre Belon, *L'Histoire de la nature des oyseaux* (Paris, 1555), p. 8.
5. *Ibid.*, p. 355.
6. *Ibid.*, p. 319.
7. See, further, Erwin Stresemann, "Die Anfänge ornithologischen Sammlungen," *Journal für Ornithologie*, 71 (1923): 112–127, and for living collections Gustave Loisel, *Histoire de ménageries de l'antiquité à nos jours*, 3 vols. (Paris, 1912).
8. Michel de Montaigne, *Journal de voyage en Italie, par la Suisse et l'Allemagne*, ed. Charles Dédéyan (Paris, 1946), p. 187.
9. *Ibid.*, p. 140.
10. Carolus Clusius, *Exoticorum libri decem* (Leiden, 1605), p. 359.
11. E. H. F. Meyer, *Geschichte der Botanik*, IV (1857), p. 353.
12. Clusius, *Exoticorum libri decem*, foreword.
13. Baron Georges Cuvier, *Histoire des sciences naturelles, depuis leur origine jusqu'à nos jours*, ed. Magdeleine de Saint-Agy, II (Paris, 1841), p. 134.
14. Hinrich Lichtenstein, "Die Werke von Marcgrave und Piso über die Naturgeschichte Brasiliens, erläutert aus den wieder aufgefundenen Original-Abbildungen," *Abhandlungen Kgl. Akad. Wiss. Berlin, Phys. Kl.* (1818–1826); Adolf Schneider, "Die Vogelbilder zur Historia naturalis Brasiliae des Georg Marcgrave," *Journal für Ornithologie*, 86 (1938): 74–106.
15. Cuvier, *Histoire des sciences naturelles*, II, p. 143.
16. Georg Marcgraf, *Historia naturalis Brasiliae* (Leiden, 1648), p. 190.
17. Peter Simon Pallas, *Elenchus zoophytorum* (The Hague, 1766), praefatio.
18. See J. E. Heeres, "Rumphius' Levensloop," in *Rumphius Gedenkboek 1702–1902* (Koloniaal Museum te Haarlem, 1902), pp. 1–16; H. C. D. de Wit, "Georgius Everhardus Rumphius," in *Rumphius Memorial Volume* (Baarn, 1959), pp. 1–26.
19. *Rumphius Gedenkboek 1702–1902*, p. 6.
20. *Ibid.*, pp. 174–175.
21. Facsimile edition by W. Junk (Berlin, 1923); quotations pp. iii and 1.

Chapter 3. The Influence of Methodology

1. See Charles E. Raven, *John Ray Naturalist: His Life and Works*, 2nd ed. (Cambridge, 1950).
2. William Charleton, *Onomasticon zoicon* (1668).
3. Supplement to *Musei Petiveriana centuria secunda et tertia* (London, 1698).

4. Frederik Ruysch, *Thesaurus animalium* (Amsterdam, 1710).
5. Quoted by C. Mylius, "Johann Leonhard Frischens Vorstellung der Vögel," *Hamburgisches Magazin*, 4 (1749): 412.
6. Biography in Wilfrid Blunt, *The Compleat Naturalist: A Life of Linnaeus* (London, 1971).
7. See H. Engel, "The Life of Albert Seba," *Svenska Linné-Sällskapets Årsskrift*, 20 (1937): 75–100.
8. Peter Artedi, *Philosophia ichthyologica* (Leiden, 1738), p. 2.
9. G.-L. Leclerc, Comte de Buffon, *Histoire naturelle*, I (Paris, 1749), introduction.
10. M. J. Brisson, *Le règne animal* (Paris, 1756), preface.
11. C. Linnaeus, *Systema naturae*, 10th ed., I (Stockholm, 1758), p. 2.
12. Buffon, *Histoire naturelle des oiseaux* (Paris, 1770), p. 75.
13. Pierre Sonnerat, *Voyage à la Nouvelle Guinée* (Paris, 1776), p. 75.
14. Letter from Réaumur to Séguier, May 35, 1749.
15. Buffon, *Histoire naturelle*, I (1749), p. 38.
16. First part of German text by F. H. W. Martini, continued by B. C. Otto, 35 vols. (Berlin, 1772–1809).

Chapter 4. Scientific Expeditions Between 1767 and 1795

1. Letter to Brünnich at Leiden, November 2, 1765, *Bref och scrifvelser af och till Carl von Linné*, II, 1 (Upsala, 1916), p. 392.
2. Thomas Pennant, *A Tour on the Continent (1765)*, ed. B. R. de Beer (London, 1948), p. 162.
3. *The Literary Life of the Late Thomas Pennant, Esq., by Himself* (London, 1793), p. 9.
4. P. S. Pallas, *Naturgeschichte merkwürdiger Thiere, Fünfte Sammlung* (Berlin and Stralsund, 1776), pp. 13, 15. The Latin edition, *Spicilegia zoologica, fasciculus quintus*, had appeared in Berlin in 1769.
5. K. A. Rudolphi, "Peter Simon Pallas. Ein biographischer Versuch," in Rudolphi, *Beyträge zur Anthropologie und allgemeinen Naturgeschichte* (Berlin, 1812), pp. 1–64.
6. Pallas, *Reise durch verschiedene Provinzen des Russischen Reiches, Dritter Theil* (St. Petersburg, 1776), p. 690.
7. Erwin Stresemann, "Dr. C. H. Mercks ornithologische Aufzeichnungen während der Billingschen Expedition von Ochotsk nach Alaska (1787–1791) [Dr. C. H. Merck's ornithological notes during the Billings Expedition from Okhotsk to Alaska (1787–1791)]," *Zoologische Jahrbücher* (Abteil. Syst.), 78 (1948): 97–132.
8. Pallas, *Zoographia Rosso-Asiatica*, I (St. Petersburg, 1811), p. v.
9. Johann Reinhold Forster, *Enchiridion historiae naturali inserviens* (Halle, 1788), p. 3.
10. Georg Forster, *A Voyage round the World . . . in the Years 1772, 3, 4, and 5*, I (London, 1777), p. v; reprinted in Georg Forster, *Werke*, ed. Gerhard Steiner, I (Berlin, 1968), p. 10.
11. Erwin Stresemann, "Birds Collected during Capt. James Cook's Last Expedition (1776–1780)," *Auk*, 67 (1950): 66–88.
12. See W. H. Mullens, "The Leverian Museum," *Museums Journal*, 15 (1915): 123–129.
13. Heinrich Friedrich Link, *Travels in Portugal* (London, 1801), p. 27; translation of Part 1 of *Bemerkungen auf einer Reise durch Frankreich, Spanien, und vorzüglich Portugal* (3 pts., Kiel, 1801–1804).
14. For Pennant see references 2 and 3 above, and W. H. Mullens, "Thomas Pennant," *British Birds*, 2 (1909): 259–266.
15. D. Ferguson, "Joan Gideon Loten, F.R.S., the Naturalist Governor of

Ceylon (1752–1757), and the Ceylonese Artist de Bavere," *Journal of the Ceylon Branch of the Royal Asiatic Society,* 19 (1907): 217–271.

16. J. R. Forster, *A Catalogue of the Animals of North America* (London, 1771), p. 40.

17. Pierre Sonnerat, *Voyage aux Indes orientales et à la Chine, fait par ordre de Louis XVI., 1774–1781* (Paris, 1782).

18. In Part 2, 1786, pp. 84–96 ("Specimen zoologicum"); reprinted in *Scopoli's Ornithological Papers,* ed. Alfred Newton (London: Willughby Society, 1882).

Chapter 5. François Levaillant (1753–1824)

1. F. M. Daudin, *Traité élémentaire et complet d'ornithologie,* I (Paris, 1800), p. 459.

2. I. Geoffroy Saint-Hilaire, *Catalogue méthodique de la collection . . . des oiseaux . . . du Muséum d'Histoire Naturelle de Paris* (Paris, 1851).

3. *Göttingische Anzeigen von gelehrten Sachen,* 3 April 1790, p. 537.

4. Hinrich Lichtenstein, "Johannes Centurius Graf von Hoffmannsegg," *Dresdner Album,* ed. Elfriede von Mülenfels, 2nd ed. (Berlin, 1856), pp. 24–44.

5. H. F. Link, *Travels in Portugal* (London, 1801), p. 27 (see Chapter 4, reference 13).

6. K. A. Rudolphi, *Bemerkungen aus dem Gebiet der Naturgeschichte, Medicin und Thierarzneykunde, auf einer Reise durch einen Theil von Deutschland, Holland und Frankreich,* 2 parts (Berlin, 1804–1805).

7. Andrew Smith, "A Description of the Birds Inhabiting the South of Africa," *South African Quarterly Journal,* 1 (1830): 9; reprinted in Sir Andrew Smith's *Miscellaneous Ornithological Papers* (London: Willughby Society, 1880).

8. Georg Forster, *Allgemeine Literatur-Zeitung,* No. 98, 31 March 1791, cols. 780–781.

9. François Levaillant, *Voyage,* I (1790), p. 359.

10. Levaillant, *Oiseaux d'Afrique,* I (1796), p. 97.

11. *Ibid.,* II (1799), p. 139.

12. Levaillant, *Voyage,* I, xxiii.

13. Levaillant, *Oiseaux d'Afrique,* II (1799), p. 73.

14. Johann Wagler, "Monographia psittacorum," *Abhandlungen der Math.-Phys. Classe der K. Bayerischen Akademie der Wissenschaften,* Munich, 1 (1832): 467.

15. Carl Jacob Sundevall, "Kritisk framställning af fogelarterna uti äldre ornithologiska arbeten" [Critical examination of bird species in older ornithological works], *K. Svenska Vetenskaps-Akademiens Handlingar,* n.s., 2 (1857); portion concerning Levaillant translated by Léon Olphe-Galliard as "Les oiseaux d'Afrique de Levaillant, critique de cet ouvrage par Carl J. Sundevall," *Revue et Magazin de Zoologie,* 2nd ser., 17–18 (1865–1866). See also *François Le Vaillant Traveller in South Africa and His Collection of 165 Water-colour Paintings 1781–1784,* 2 vols. (Cape Town: Library of Parliament, 1973).

16. Levaillant, *Histoire naturelle des oiseaux de paradis,* I, p. 106.

17. J. C. F. von Schiller, "Xenien," *Musen Almanach für das Jahr 1797* (Tübingen, 1796), p. 214.

18. Levaillant, *Second voyage,* II (1795), pp. 290–292.

Chapter 6. Carl Illiger (1775–1813)

1. Ernst Hartert, "Eine bedeutende Vogelsammlung des 18. Jahrhunderts," *Ornithologische Monatsberichte,* 31 (1923): 73–75. Also, Wilhelm Petry, "Eine bedeutende Vogelsammlung des 18. Jahrhunderts," *Ornithologische Monatsberichte,* 45 (1937): 157–162.

2. See Erwin Stresemann, "Die brasilianischen Vogelsammlungen des Grafen von Hoffmannsegg ans den Jahren 1800 bis 1812," *Bonner Zoologische Beiträge*, 1 (1950): 43–51, 126–143.

3. H. F. Link, *Travels in Portugal* (1801), p. 229 (see Chapter 4, reference 13).

4. Hinrich Lichtenstein, "Graf von Hoffmannsegg," *Dresdner Album* (1856), pp. 24–44. For the Berlin Museum see A. Brauer, "Das zoologische Museum," in Max Lenz, *Geschichte der . . . Universität zu Berlin*, III (Halle a. d. S., 1910), pp. 372–389; Erwin Stresemann, "Die Entwicklung der Vogelsammlung des Berliner Museums unter Illiger und Lichtenstein," *Journal für Ornithologie*, 70 (1922): 498–503.

5. For Lichtenstein see Erwin Stresemann, "Hinrich Lichtenstein: Lebensbild des ersten Zoologen der Berliner Universität," *Forsche und Wirken: Festschrift zur 150-Jahr-Feier der Humboldt-Universität zu Berlin 1810–1960*, I (Berlin, 1960), pp. 73–96.

6. Manuscript notes by Rammelsberg in the archives of the Berlin Zoological Museum.

7. Lichtenstein, "Ehrendenkmal des Herrn J. C. W. Illiger," *Abhandlungen der Königlichen Akademie der Wissenschaften in Berlin . . . 1814–1815* (1818): 62.

8. Lichtenstein, *Dresdner Album* (1856), pp. 24–44.

9. Lichtenstein, "Ehrendenkmal," p. 53.

10. Archives of the Berlin Zoological Museum.

11. Carl Illiger, *Abhandlungen der Königlichen Akademie der Wissenschaften in Berlin . . . 1812–1813* (1816), Phys. Kl., pp. 221–236 and 20 leaves of tables.

12. Blasius Merrem, *Versuch eines Verzeichnisses zur allgemeiner Geschichte . . . der Vögel* (Leipzig, 1788), p. 4.

13. Included as a separate essay in Illiger's *Versuch einer . . . Terminologie . . .* (1800); English translation in Ernst Mayr, "Illiger and the Biological Species Concept," *Journal of the History of Biology*, 1 (1968): 163–178.

Chapter 7. Coenraad Jacob Temminck (1778–1858)

1. For this chapter see J. A. Susanna, "Levensschets van [Biographical sketch of] Coenraad Jacob Temminck," *Handelingen [Levensberigten] . . . Maatschappij der Nederlandsche Letterkunde te Leiden* (1858), pp. 47–78; Agatha Gijzen, *'s Rijks Museum van Natuurlijke Historie 1820–1915* (Rotterdam, 1938).

2. Thomas Pennant, *Tour on the Continent, 1765*, ed. G. R. de Beer (London, 1948), p. 155.

3. See H. Engel, "Alphabetical List of Dutch Zoological Cabinets and Menageries," *Bijdragen tot de Dierkunde*, 27 (1939): 247–346.

4. Georg Forster, *Ansichten vom Niederrhein* (Berlin, 1791); *Werke*, ed. Gerhard Steiner, IX (1958), p. 314.

5. F. Levaillant, *Oiseaux d'Afrique*, I (1796), p. 36.

6. *Ibid.* Auction catalogue, 1787, made by Dr. F. C. Meuschen (Musculus).

7. Forster, *Ansichten vom Niederrhein* (1791); *Werke*, ed. Steiner, IX (1958), p. 295.

8. François Levaillant, *Oiseaux d'Afrique*, I, pp. 34–35. See also C. J. Temminck, *Histoire naturelle générale des pigeons et des gallinacés* (Amsterdam), II (1813), p. 372; III (1815), p. 35.

9. Temminck, *Histoire . . . des pigeons et des gallinacés*, II, p. 458.

10. Susanna, "Levensschets van . . . Temminck," pp. 49–50.

11. Peter Thomsen, ed., "Aus dem Briefwechsel von Johann Friedrich Naumann mit Coenraad Jacob Temminck, 1816–1820," in manuscript; the original letters are in the archives of the Naumann Museum at Köthen.

12. "Etwas ueber naturhistorische Museen des Auslandes. Von einem Reisenden," Oken's *Isis*, II (1821), Literarischer Anzeiger, col. 338.

13. *Ibid.*, col. 337.

14. See Erwin Stresemann, "Histoire des origines des 'Planches coloriées' de Temminck et Laugier," *Oiseau*, 21 (1951): 33–47. The work was continued until

the 102nd part (1839), and contained 600 colored plates with accompanying text, portraying about 800 species.

15. See R. P. Lesson, *Bulletin des Sciences Naturelles et de Géologie*, 25 (1831): 365–366; Paul H. Oehser, "Louis Jean Pierre Vieillot," *Auk*, 65 (1948): 568–576.

16. See Agatha Gijzen, *'s Rijks Museum van Natuurlijke Historie, 1820–1915* (Rotterdam, 1938).

Chapter 8. The Natural History Commission of the Dutch Indies

1. For this chapter see J. A. Susanna, "Levensschets van . . . Temminck" (see Chapter 7, reference 1); Gustav Schlegel, "Levensschets van Hermann Schlegel," *Jaarboek van de Koninklijke Akademie van Wetenschappen . . . Amsterdam voor 1884*, pp. 1–85 (largely autobiography by Hermann).

2. J. W. von Goethe, *Tagebücher*, Gesamtausgabe, XII (Stuttgart: Cotta, 1957), p. 380.

3. Goethe, *Ueber Kunst und Alterthum im den Rhein und Mayn Gegenden*, Erstes Heft (Stuttgart, 1816); Gesamtausgabe, X (Stuttgart: Cotta, 1954), p. 678.

4. This letter and others from Kuhl and van Hasselt were published in Oken's *Isis* for 1822, cols. 472–476, 893–904.

5. Th. van Swinderen, "Bijdragen tot eene Schets van het Leven, het Karakter en de Verdiensten van wijlen Dr. H. Kuhl," *Groninger Almanak voor 1823*, 11 (1822): 1–79.

6. Letter in the archives of the Berlin Zoological Museum.

7. H. J. Veth, *Overzicht van hetgeen . . . gedaan is voor de Kennis der Fauna van Nederlandsch Indië* [Survey of what . . . has been accomplished toward a knowledge of the fauna of the Netherlands Indies] (Leiden, 1879).

8. Letter in the archives of the Berlin Zoological Museum.

9. Partly published in Oken's *Isis* for 1828, cols. 1025–1035.

10. See J. A. Susanna, *Levensschets van* [Biographical sketch of] *Hendrik Boie* (Amsterdam, 1834).

11. Lady Sophia Raffles, *Memoir of the Life and Public Services of Sir Thomas Stamford Raffles* (London, 1830), pp. 569, 568.

12. See Sophia Raffles, *Memoir*, and Henry Scherren, *The Zoological Society of London: A Sketch of Its Foundation and Development* (London, 1905), chapter 1.

13. P. Lemoine, "Le Muséum National d'Histoire Naturelle: Son histoire, son état actuel," *Archives du Muséum d'Histoire Naturelle*, (6) 12 (1935): 1–79.

14. William Swainson, *On the Natural History and Classification of Birds*, I (London, 1836), p. 222.

15. Letter in the archives of the Berlin Zoological Museum.

16. Gustav Schlegel, "Levensschets van Hermann Schlegel," pp. 37–38.

Chapter 9. Charles Lucien Bonaparte (1803–1857)

1. C. L. Bonaparte, *Journal of the Academy of Natural Sciences in Philadelphia*, 3, 2 (1824): 227–233.

2. Quoted in Witmer Stone, "Some Philadelphia Ornithological Collections and Collectors, 1784–1850," *Auk*, 16 (1899): 166–177.

3. See F. L. Burns, "Alexander Wilson. V. The Completion of the American Ornithology," *Wilson Bulletin*, 21 (1909): 16–35.

4. Letter in the archives of the Berlin Zoological Museum.

5. *Audubon and His Journals*, ed. Maria R. Audubon, I (New York, 1897), pp. 256–257.

6. *Fauna Boreali-Americana, Vol. II: Birds* (London, 1931), p. lx.
7. C. L. Bonaparte, *American Ornithology*, I (Philadelphia, 1825), p. 52.
8. Bonaparte, *Conspectus generum avium*, I (Leiden, 1850-1851) p. 413.
9. Gustav Hartlaub, "Bericht über die Leistungen in der Naturgeschichte der Vögel (während des Jahres 1840)," *Archiv für Naturgeschichte*, 17, 2 (1851): 35.
10. Baron Johann Wilhelm von Müller, "Biographische Notiz über Carl Lucian Bonaparte," *Naumannia*, 2, 1 (1852): 93.
11. Bonaparte, *Journal für Ornithologie*, 4 (1856): 259, 258.
12. Bonaparte, "Excursions dans les divers Musées d'Allemagne, de Hollande et de Belgique," *Comptes Rendus de l'Académie des Sciences*, Pàris, 43 (1856): 410-411, source also for the comments on collections here quoted.
13. Gustav Schlegel, "Levensschets van [Biographical sketch of] Hermann Schlegel," *Jaarboek van de Koninklijke Akademie van Wetenschappen . . . Amsterdam voor 1884*, pp. 45-47.
14. Elliott Coues, "Third Instalment of American Ornithological Biography," *Bulletin U.S. Geological and Geographical Survey*, 5 (1880): 677.
15. Bonaparte, *American Ornithology*, II (1828), p. 5.
16. C. L. Bonaparte, "Nouvelles especes ornithologiques . . . Première partie: Perroquets," *Comptes Rendus de l'Académie des Sciences*, Paris, 20 (1850): 132.

Chapter 10. The Effect of Natural Philosophy

1. For this chapter see the two works of Henri Daudin, *De Linné à Jussieu: méthodes de la classification et idée de série en botanique et en zoologie (1740-1790)* and *Cuvier et Lamarck: les classes zoologiques et l'idée de série animale (1790-1830)*, both Paris, 1926.
2. See Arthur O. Lovejoy, *The Great Chain of Being* (Cambridge, Mass., 1936).
3. August Thienemann, "Die Stufenfolge des Dinge, der Versuch eines natürlichen Systems der Naturkörper aus dem achtzehnten Jahrhundert [The chain of being, the attempt at a natural system of creation in the eighteenth century]," *Zoologische Annalen*, 3 (1910): 234.
4. Fritz Schultze, *Philosophie der Naturwissenschaft*, I (Leipzig, 1881), p. 346.
5. Immanuel Kant, *Critique of Pure Reason*, trans. N. K. Smith (London, 1929), p. 548.
6. Charles Bonnet, *Contemplation de la nature*, I (Amsterdam, 1764), p. iv (preface).
7. *Ibid.*, pp. 66-68.
8. Kassel, 1784; reprinted in Georg Forster, *Werke*, ed. Gerhard Steiner, II (Frankfurt am Main, 1969), pp. 44-45.
9. Vicq d'Azyr, *Traité d'anatomie et de physiologie* (Paris, 1786); *Oeuvres*, ed. J. L. Moreau, IV (Paris, 1805), p. 199.
10. K. A. Rudolphi, *Beyträge zur Anthropologie und allgemeinen Naturgeschichte* (Berlin, 1812).
11. Baron Justus von Liebig, *Eigne biographischen Aufzeichnungen* [Autobiographical notes], ed. K. Esselborn (Giessen, 1926).
12. N. A. Vigors, *Transactions of the Linnean Society of London*, XIV (1825): 395-517.
13. William Swainson, *On the Natural History and Classification of Birds*, I (London, 1826), pp. 195, 193. The succeeding presentation, including tables and figures, is drawn from both volumes of this work (1836-1837).
14. William Swainson, *Flycatchers* (Edinburgh, 1838), pp. 160-172, 188-192, 231, 250 (*The Naturalist's Library: Ornithology*, Vol. 10).
15. Henry Scherren, *The Zoological Society of London* (London, 1905), chap. 1.
16. Keith A. Hindwood, "A Note on William Swainson," *Emu*, 49 (1950): 208-210.
17. T. C. Jerdon, *The Birds of India*, I (Calcutta, 1862), p. xxxiii.

18. Ludwig Reichenbach, *Das natürliche System der Vögel* (Dresden, 1852), pp. 2–11.

19. Reichenbach, "Ueber den Begriff der Art in der Ornithologie," *Journal für Ornithologie*, 1 (1853): 6–7.

20. Reichenbach, "Aufzählung der Colibris oder Trochilideen in ihrer wahren natürlichen Verwandtschaft [Enumeration of the hummingbirds in their true natural relationship]," *Journal für Ornithologie*, 1, Extraheft (1854): 3.

21. *Ibid.*

22. J. J. Kaup, "Einige Worte über die systematische Stellung der Familie der Raben: Corvidae," *Journal für Ornithologie*, 2 (1854): Erinnerungschrift, pp. xlvii–lvi and pl. 10, fig. 10; "Monograph of the Strigidae," *Transactions of the Zoological Society of London*, 4 (1862): 201–260 (read 1852).

23. Gustav Hartlaub, "Bericht über die Leistungen in der Naturgeschichte der Vögel während des Jahres 1851," *Archiv für Naturgeschichte*, 18, 2 (1852): 17.

24. J. F. Brandt, "Beiträge zur Kenntniss der Naturgeschichte der Vögel," *Mémoires de l'Académie Impériale des Sciences de Saint-Pétersbourg*, 6. série, Sciences Math., Phys. et Naturelles, vol. 5, pt. 2, Sciences Naturelles, vol. 3, livr. 1–2 (1839), p. 152 (read May 19, 1837).

25. H. E. Strickland, *Report of the Fourteenth Meeting of the British Association for the Advancement of Science . . . 1844* (London, 1845), p. 177.

26. Immanuel Kant, *Critique of Teleological Judgement*, trans. J. C. Meredith (Oxford, 1928), pp. 79, 80. See also J. Brock, "Die Stellung Kant's zur Deszendenztheorie [The position of Kant regarding the theory of descent]," *Biologisches Centralblatt*, 8 (1889): 641–648.

27. Johann Friedrich Blumenbach, *Beyträge zur Naturgeschichte* (Göttingen, 1790), p. 3.

28. Carl Illiger, *Versuch einer systematischen vollständigen Terminologie für das Thierreich und Pflanzenreich* (Helmstedt, 1800), pp. xxxvii–xxxviii.

29. J. H. Blasius, *Bericht über die XIII. Versammlung der Deutschen Ornithologen-Gesellschaft zu Stuttgart . . . 1860* (Stuttgart, 1861?), pp. 44–45.

Chapter 11. Hermann Schlegel (1804–1884)

1. For this chapter see Gustav Schlegel, "Levensschets van Hermann Schlegel" (see Chapter 8, reference 1); H. J. Veth, *Overzicht van hetgeen . . . gedaan is voor de Kennis der Fauna van Nederlandsch Indië* (Leiden, 1879); Agatha Gijzen, *'s Rijks Museum van Natuurlijke Historie 1820–1915* (Rotterdam, 1938).

2. Incorporated in Gustav Schlegel, "Levensschets van Hermann Schlegel."

3. Schlegel, "Levensschets," pp. 8–9. For Pastor Brehm, see Chapter 17, below.

4. Bijleveld, *Verloren Glorie. Jhr. Dr. Ph. F. B. von Siebold en zijne Buitenplaats Nippon* (Leiden, 1923).

5. Claus Nissen, *Die illustrierten Vogelbücher* (Stuttgart, 1953), pp. 63–64. See also A. H. Palmer, *The Life of Joseph Wolf* (London, 1895).

6. Nissen, *Die illustrierten Vogelbücher*.

7. Hermann Schlegel, "Notice sur le sous-genre *Tanysiptera*," *Nederlandsch Tijdschrift voor de Dierkunde*, 3 (1866): 276–277.

8. Immanuel Kant, *Critique of Pure Reason*, trans. N. K. Smith (1929), p. 541.

9. C. L. Brehm, "Ueber Species und Subspecies," *Naumannia*, 3 (1853): 10–18.

10. C. F. Bruch, "Ornithologische Beytrage," Oken's *Isis*, 21 (1828), col. 725.

11. See A. R. Wallace, *My Life: A Record of Events and Opinions*, 2 vols. (London, 1905).

12. P. L. Sclater, "On the General Geographical Distribution of the Members of the Class Aves," *Journal of the Proceedings of the Linnean Society, Zoology*, 2 (1858): 131.

13. Wallace, *Annals and Magazine of Natural History*, 2nd ser., 16 (1855): 184–196.

14. Hermann Schlegel, "Einige Worte über die Schwarz-Kakatu's und über die Paradiesvögel," *Journal für Ornithologie*, 9 (1861): 382.

15. H. A. Bernstein, "Over de zoogenoemde eetbare Vogelnesten en den Nestbouw van eenige andere Javasche Vogels [On the so-called edible bird nests and the nest building of certain Javanese birds]," *Verhandelingen der Natuurkundige Vereeniging in Nederlandsch Indië*, 3 (Batavia, 1857–1858), 34 pages; Bernstein, "Ueber das Vorkommen eines einzigen Hodens bei [On the occurrence of a single testis] in *Centropus medius* Müll. und *Centropus affinis* Horsf.," *Archiv für Anatomie, Physiologie und Wissenschaftliche Medicin* (1860): 161–168.

16. A. Gijzen, *'s Rijks Museum van Natuurlijke Historie 1820–1915* (Rotterdam, 1938), pp. 60–61.

17. Published in 14 volumes, Leiden, 1862–1908.

18. Eduard Baldamus, "Ornithologisage Reiseskizzen vam Jahre 1860," *Journal für Ornithologie*, 9 (1861): 61–64.

19. P. L. Martin, *Dermoplastik und Museologie* (Weimar, 1870), introduction.

20. Baron Ferdinand Droste-Hülsoff, "Das Reichsmuseum zu Leyden," *Journal für Ornithologie*, 15 (1867): 352–355.

21. J. H. Blasius, *Bericht über die XIII. Versammlung der Deutschen Ornithologen-Gesellschaft zu Stuttgart . . . 1860* (Stuttgart, 1861?), p. 41.

22. Jean Cabanis, [presentation of birds from Peru collected by Konstantin Jelski], *Journal für Ornithologie*, 21 (1873): 317.

23. Blasius, *Bericht*, p. 46.

24. Hermann Schlegel, "Notice sur le sous-genre *Tanysiptera*," *Nederlandsch Tijdschrift voor de Dierkunde*, 3 (1866): 277.

25. Charles Darwin, *On the Origin of Species* (London, 1859; facsimile edition, Cambridge, Mass., 1964), p. 484.

26. Jean Cabanis, "Uebersicht der im Berliner Museum befindlichen Vögel von Costa Rica," *Journal für Ornithologie*, 9 (1861): 244.

27. Darwin, *On the Origin of Species*, pp. 484–485.

28. Schlegel, "Levensschets," pp. 26, 85.

Chapter 12. Otto Finsch (1839–1917)

1. Including Capitonidae and Bucconidae, published in Schlegel's *Revue Méthodique et Critique de la Collection des Oiseaux* (Muséum d'Histoire Naturelle des Pays-Bas), 2nd and 3rd livraisons (1863).

2. Otto Finsch, "Zur Versöhnung zweier toten Meister [Toward reconciliation of two dead masters] (Hartlaub-Petényi)," *Journal für Ornithologie*, 50 (1902): 353.

3. Sir William Jardine, *Memoirs of Hugh Edwin Strickland, M.A.* (London, 1858), pp. ccxxxi–ccxxxii.

4. Paul Leverkühn, "Zur Erinnerung an Dr. Gustav Hartlaub," *Journal für Ornithologie*, 49 (1901): 347.

5. Gustav Hartlaub, *System der Ornithologie Westafrica's* (Bremen, 1857), p. vi.

6. Jules Sebastien César Dumont d'Urville (1790–1842), *Voyage au Pole Sud et dans l'Océanie sur les corvettes l'Astrolabe et la Zélée . . . pendant les anneés 1837–1838–1839–1840* (Paris, 1842–1854), 23 vols. and atlas of 7 vols.

7. Gustav Hartlaub, *Journal für Ornithologie*, 2 (1854): 160–171.

8. Hans Johansen, "Die Vögel Westsibiriens," *Journal für Ornithologie*, 91 (1943): 78.

9. Original edition, Dunedin, 1865; Finsch's translation in *Journal für Ornithologie*, 15 (1867): 305–347.

10. Otto Finsch, *Journal für Ornithologie*, 20 (1872): 81–112, 161–188, 241, 274; 22 (1874): 167–224.

11. Otto Finsch, "Wie ich Kaiser Wilhelms-Land erwarb," *Deutsche Monatschrift für das gesamte Leben der Gegenwart*, 1 (1902), 4 installments; A. Wichmann, "Entdeckungsgeschichte von Neu-Guinea [History of the discovery of New Gui-

nea]," in *Nova-Guinea. Uitkomsten* [Results] *der Nederlandsche Nieuw-Guinea Expeditie in 1903*, vols. 1-2 (Leiden, 1909–1912).

12. Otto Finsch and A. B. Meyer, "Vögel von Neu Guinea . . . I. Paradiseidae," *Zeitschrift für die gesammte Ornithologie*, 2 (1885): 369–391 (quotation pp. 369–370); English translation as "On Some New Paradise-Birds," *Ibis* (5) 4 (1886): 237–258.

13. *Ibid.*

14. An annotated catalogue of Finsch's travels, publications, honors, and so forth, appears in O. Finsch, *Systematische Uebersicht der Ergebnisse seiner Reisen und Schriftstellertätigkeit (1859–1899)* (Berlin, 1899).

15. Published in *Notes from the Leyden Museum*, vols. 22–23 (1900–1901).

16. Otto Finsch, *Journal für Ornithologie*, 58 (1910); 520–521.

17. Arnold Jacobi, "Nachruf," *Dresdner Anzeiger*, February 18, 1917.

18. H. Schalow, meeting of Deutsche Ornithologische Gesellschaft, Berlin, February 5, 1917, in *Journal für Ornithologie*, 65 (1917): 236.

19. Letter to the editors, *Ibis*, (8) 4, (1904): 660.

Chapter 13. The Effect of the Theory of Evolution

1. William MacGillivray, *A History of British Birds* (London, 1837), p. 7.

2. Hugh E. Strickland, *Report of British Association*, 1845.

3. Johannes Müller, "Über die bisher unbekannten typischen Verschiedenheiten der Stimmorgane der Passerinen," *Abhandlungen der Königlichen Akademie der Wissenschaften zu Berlin* . . . 1845 (1847): 325.

4. Max Fürbringer, *Untersuchungen zur Morphologie und Systematik der Vögel* (Bijdragen tot de Dierkunde, vol. 15, Amsterdam, 1888), p. 1121.

5. Anton Reichenow, "System und Genealogie," *Ornithologische Monatsberichte*, 1 (1893): 117.

6. Léon Olphe-Gaillard, *Naumannia*, 7 (1857): 151–177.

7. J. H. Blasius, *Bericht über die XIII. Versammlung der Deutschen Ornithologen-Gesellschaft zu Stuttgart . . . 1860* (Stuttgart, 1861?), p. 46.

8. Wilhelm von Nathusius, "Nachweis des Speciesunterschiedes von [Proof of the specific difference between] *Corvus corone* und *Corvus cornix*," *Journal für Ornithologie*, 22 (1874): 23.

9. Friederich Kutter, *Journal für Ornithologie*, 25 (1877): 396–423; 26 (1878): 300–348.

10. Nathusius, "Betrachtungen über die Selectionstheorie vom Standpunkt der Oologie aus [Reflections on the theory of natural selection in the light of oology]," *Journal für Ornithologie*, 27 (1879): 228.

11. Nathusius, "Nachweis," p. 22.

12. Kutter, "Betrachtungen," *Journal für Ornithologie*, 25 (1877): 416.

13. Nicolai Severzov, "Allgemeine Uebersicht der aralo-tianschanischen Ornis," *Journal für Ornithologie*, 21 (1873): 324.

14. J. A. Palmén, *Om foglarnes flyttningsvägar* [Concerning the migration routes of birds] (Helsingfors, 1874).

15. S. F. Baird, "Notes on a Collection of Birds Made by Mr. John Xantus, at Cape St. Lucas, Lower California," *Proceedings of the Academy of Natural Sciences of Philadelphia. 1859* (1860): 300.

16. S. F. Baird, "The Distribution and Migrations of North American Birds," *American Journal of Science and Arts*, 2nd ser., 41 (1866): 190–191.

17. C. L. Bonaparte, *Journal für Ornithologie*, 4 (1856): 258.

18. Robert Ridgway, "Notices of Certain Obscurely Known Species of American Birds," *Proceedings of the Academy of Natural Sciences of Philadelphia. 1869* (1869): 126.

19. J. A. Allen, *Bulletin of the Museum of Comparative Zoology*, 2 (1871): 161–450 (quotations pp. 239, 246).

20. Allen, *Radical Review*, 1 (1877): 108-140.
21. Elliott Coues, "Progress of American Ornithology," *American Naturalist*, 5 (1871): 371, a review of J. A. Allen, "On the Mammals and Winter Birds of East Florida" (reference 19, above).
22. Ridgway, *Nomenclature of North American Birds Chiefly Contained in the United States National Museum* (Washington, D.C., 1881), pp. 9-10.
23. Cf. Leonhard Stejneger, "On the Use of Trinomials in American Ornithology," *Proceedings of the U.S. National Museum*, 7 (1885): 70-80.
24. Henry Seebohm, *The Geographical Distribution of the Family Charadriidae* (London, 1887), p. 64.

Chapter 14. Ernst Hartert (1859-1933)

1. "Zoological Nomenclature," *Nature*, 30 (1884): 257-259, 277-279.
2. "Bericht über die (IX.) Jahresversammlung," *Journal für Ornithologie*, 33 (1885): 15.
3. Berlepsch in a letter to Hartert, May 5, 1890. The letters of ornithologists to Hartert during the period from 1885 to 1892 are in the archive of the Deutsche Ornithologen-Gesellschaft.
4. "Bericht über die (XV.) Jahresversammlung," *Journal für Ornithologie*, 39 (1891): 11.
5. "Regeln für die zoologische Nomenclature angenommen . . . auf der XVI. Jahresversammlung," *Journal für Ornithologie*, 39 (1891): 322.
6. Sharpe's own account of the expansion will be found in R. B. Sharpe, *The History of the Collections Contained in the Natural History Department of the British Museum*, vol. 2, *Birds* (London, 1906).
7. Sharpe, *A Review of Recent Attempts to Classify Birds* (Budapest, 1891), p. 90 (address before the Second International Ornithological Congress).
8. [P. L. Sclater,] "Sharpe's Catalogue of Birds," *Nature*, 16 (1877): 541.
9. See reference 7.
10. Baron Rothschild of Tring from 1915. See Karl Jordan, "In Memory of Lord Rothschild," *Novitates Zoologicae*, 41 (1938): 1-41.
11. On the Brehm collection, see Ernst Hartert, "The Brehm Collection," *Novitates Zoologicae*, 8 (1901): 38-48; Günther Niethammer, "Die Brehm-Sammlung, ihr Schicksal und ihre Bedentung," *Verhandlungen der Deutschen Zoologischen Gesellschaft 1960* (1961): 505-512; Ludwig Baege, "Dokumentarisches zur Geschichte der Brehm-Sammlung," *Südthüringer Forschungen*, 2/66 (1967): 69-119.
12. Hartert, *Katalog der Vogelsammlung im Museum der Senckenbergischen Naturforschenden Gesellschaft* (1891), p. xiii.
13. Otto Kleinschmidt, "Was ist die Subspecies?" *Ornithologische Monatsberichte*, 5 (1897): 74-76.
14. Reichenow, "Über Begriff und Benennung von Subspecies [On the concept and naming of subspecies]," *Ornithologische Monatsberichte*, 9 (1901): 148.
15. Otto Kleinschmidt, "Beiträge zur Ornis des Grossherzogtums Hessen . . . IV. *Parus salicarius* C. L. Brehm und die ähnlichen Sumpfmeisen (mit einigen Bemerkungen über die Nomenklatur-Frage)," *Journal für Ornithologie*, 45 (1897): 112-137.
16. Carlo Freiherr von Erlanger, "Beiträge zur Avifauna Tunisiens: Über die Galerita Arten im Allgemeinen," *Journal für Ornithologie*, 47 (1899): 324-352.
17. P. L. Sclater, review, *Ibis*, (8) 4 (1904): 293.
18. R. B. Sharpe, *A Hand-list of the Genera and Species of Birds*, V (London, 1909), p. v. This was the concluding volume of a work that began publication in 1899.
19. See Chapters 12 and 13.
20. Published as "Report of a Committe Appointed 'to Consider of the Rules by which the Nomenclature of Zoology May Be Established on a Uniform and

Permanent Basis,' " *Report of the Twelfth Meeting of the British Association for the Advancement of Science . . . 1842* (London, 1843), pp. 105–121.

21. Johann Wagler, *Systema avium: pars prima* (Stuttgart, 1827), p. [26].

22. H. E. Strickland, "On the Arbitrary Alteration of Established Terms in Natural History," *Magazine of Natural History*, ed. J. C. Loudon, 8 (1835): 36–40.

23. Jean Cabanis, "Ornithologische Notizen," *Archiv für Naturgeschichte* 13 (1847): 312–313.

24. Cabanis, *Museum Heineanum*, I (Halberstadt, 1850–1851): p. vii (preface dated October 1851).

25. A. O. Hume, "Die Papageien," *Stray Feathers*, 3 (1874): 2.

26. Printed in *Second Congrès Ornithologique International, Budapest 1891, Compte-Rendu* (Budapest, 1892), p. 182.

27. American Ornithologists' Union, *The Code of Nomenclature and Check-list of North American Birds* (New York, 1886), p. 21 (Principle V).

28. Hartert, presidential address, "Über die Entwicklung und die Fortschritte der Ornithologie seit 1910 [Concerning the development and the advances of ornithology since 1910]," *Verhandlungen des VI. Internationalen Ornithologen-Kongresses in Kopenhagen 1926* (Berlin, 1929), p. 36.

29. See Friedrich Steinbacher and Erwin Stresemann, "Zum Dedächtnis Ernst Harterts," *Journal für Ornithologie*, 82 (1934): 169–183; Lord Rothschild, "Ernst Johann Otto Hartert (1859–1933): An Appreciation," *Ibis* (13) 4 (1934): 350–377 (bibliography pp. 357–377); Stresemann, "Rückblick und Ausblick zu Ernst Harterts 100. Geburtstag [Retrospect and prospect on Ernst Hartert's 100th birthday]," *Journal für Ornithologie*, 100 (1959): 433–438. On the Tring bird collection see Stresemann, "Die Vogelsammlung des Tring-Museums, ihr Aufbau und ihr Ende," *Ornithologische Monatsberichte*, 40 (1932): 65–73; H. F. Witherby, "The Tring Collection of Birds," *British Birds*, 26 (1932): 17–21; R. C. Murphy, "Moving a Museum," *Natural History*, 32 (1932): 497–511; Murphy, "Leland Cutter Sanford," *Auk*, 68 (1951): 409–410; D. W. Snow, "Robert Cushman Murphy and His 'Journal of the Tring Trip,' " *Ibis*, 115 (1973): 607–611.

Chapter 15. The Effect of the Theory of Mutation

1. Otto Kleinschmidt, "Arten oder Formenkreise?", *Journal für Ornithologie*, 48 (1900): 136.

2. Oskar Heinroth, "Ornithologische Ergebnisse der 'I. Deutschen Südsee Expedition' von Br. Mencke," *Journal für Ornithologie*, 51 (1903): 103–104.

3. William Beebe, "Geographical Variation in Birds with Especial Reference to the Effects of Humidity," *Zoologica*, 1 (1907): 12.

4. *Ibid.*, p. 15.

5. A. Adlerspaare, "*Poëphila gouldiae: Poëphila mirabilis.* En korsning mellan australiska rafvarfåglar," *Fauna och Flora*, 13 (1918): 193–208.

6. Erwin Stresemann, "Übersicht über die 'Mutationsstudien' I–XXIV und ihre wichtigsten Ergebnisse," *Journal für Ornithologie*, 74 (1926): 384.

7. Bernhard Rensch, *Das Prinzip geographischer Rassenkreise und das Problem der Artbildung* (Berlin, 1929).

8. *University of California Publications in Zoology*, No. 44.

9. Letter from Charles Darwin to Joseph Hooker, January 11, 1844, in *The Life and Letters of Charles Darwin*, ed. Francis Darwin, II (London, 1887), p. 23.

10. *Evolution*, 2 (1948): 238–265.

Chapter 16. Early Theories and Collections of Material

1. See Erwin Stresemann, "Beiträge zu einer Geschichte der deutschen Ornithologie," *Journal für Ornithologie*, 73 (1925): 603–621; Stresemann, "Baron von Pernau, Pioneer Student of Bird Behavior," *Auk*, 64 (1947): 36–52.

2. Stresemann, "Beiträge," pp. 621–628.

3. J. A. H. Reimarus in the fourth edition of his father H. S. Reimarus's *Allgemeine Betrachtungen über die Triebe der Thiere* (Hamburg, 1798; first edition, 1760).

4. See W. H. Mullens, "Synopsis of the Life of Gilbert White," *Bulletin of the British Ornithologists' Club*, 33 (1913): 9–27; Anthony Rye, *Gilbert White and His Selborne* (London, 1970).

5. White, letter to Barrington, August 8, 1778.

6. Alfred Newton, "Gilbert White," *Dictionary of National Biography*, 61 (1900).

7. White, letter to Pennant, April 18, 1768.

Chapter 17. The Naumann Period

1. For Brehm see Hugo Hildebrandt, "Christian Ludwig Brehm als Ornithologe," *Mitteilungen aus dem Osterlande*, n.s., 20 (1929): 23–38 (translation in *Ibis*, (13) 2, 1932: 308–316); Gerhard Buchda, "Zur Lebensgeschichte und zum wissenschaftlichen Werk des Pfarrers und Ornithologe Christian Ludwig Brehm," *Wissenschaftliche Zeitschrift der Fr.-Schiller-Universität*, 3 (1954), Math.-Wiss. Reihe: 459–466.

2. Published as the introduction to *Abhandlungen der Naturforschenden Gesellschaft zu Görlitz*, 1 (1827): 1–21.

3. E. F. von Homeyer, *Ornithologische Briefe* (Berlin, 1881), p. iv.

4. For Faber see H. Helms, "Frederik Faber," *Naturens Verden*, 12 (1928): 145–160.

5. For the Naumanns see Paul Leverkühn, *Biographisches über die drei Naumanns und Bibliographisches über ihre Werke* Gera-Untermhaus, 1904); Peter Thomsen and Erwin Stresemann, *Johann Friedrich Naumann der Altmeister der deutschen Vogelkunde* (Leipzig, 1957).

6. J. A. Naumann, autobiographical notes, first published in his *Ausführliche Beschreibung aller Wald-, Feld- und Wasser- Vögel*, Vol. II (Köthen, 1797), afterword; reprinted in Vol. I of the *Naturgeschichte der Vögel* (1820, 1905), and in Leverkühn, *Biographisches über die drei Naumanns* (1904).

7. *Ibid.*

8. Published in Erwin Stresemann and Peter Thomsen, "Ornithologen-Briefe aus den Jahren 1816 bis 1820 gewechselt zwischen J. F. Naumann und C. J. Temminck," *Centaurus*, 2 (1952): 102–103.

9. C. L. Brehm, *Ornis*, 1 (1827): 134–135.

10. "Etwas über Brehms neue Vogelarten von Brehm," Oken's *Isis*, 18 (1826): col. 193.

11. *Ibid.*, col. 191.

12. Frederik Faber, "Einige Bemerkungen über Hn. Brehms neue Arten der hochnordischen Schwimmvögel," Oken's *Isis*, 18 (1826), cols. 319, 318.

13. Oskar Heinroth, "Joh. Friedr. Naumann im Lichte der heutigen Forschung," *Journal für Ornithologie*, 65 (1917): 120.

14. For a summary of the rather involved history of the two general German ornithological societies and their related journals, see Erwin Stresemann, "Einhundert Jahre Deutsche Ornithologen-Gesellschaft," *Vogelwarte*, 15 (1950): 209–213.

15. Letter of December 20, 1854, published in Thomsen and Stresemann, *Johann Friedrich Naumann* (1957), p. 165.

Chapter 18. From Fixed to Evolving Species

1. [The foregoing presentation may well be complemented by the following references: Francis H. Herrick, *Audubon the Naturalist*, 2nd ed., 2 vols. in 1

(New York, 1938); Robert C. Murphy, "John James Audubon (1785–1851): An Evaluation of the Man and His Work," *New-York Historical Society Quarterly*, 40 (1956): 315–350; Alexander B. Adams, *John James Audubon* (New York, 1966); *The Original Water-color Paintings by John James Audubon for the Birds of America. Reproduced in Color for the First Time*, 2 vols. (New York, 1966); Waldemar H. Fries, *The Double Elephant Folio* (Chicago, 1973).]

2. Louis Agassiz, "Geographical Distribution of Animals," *Christian Examiner and Religious Miscellany*, 48 (1850): 181-204.

3. Frederik Faber, *Das Leben der hochnordischen Vögel* (1825).

4. René Primevère Lesson, *Compléments de Buffon*, II, *Oiseaux* (Paris, 1838), revising ideas first promulgated by Linnaeus.

5. K. A. Rudolphi, *Beyträge zur Anthropologie und allgemeinen Naturgeschichte* (Berlin, 1812).

6. Burmeister, *Geschichte der Schöpfung* [History of creation] (Leipzig, 1843).

7. L. K. Schmarda, *Die geographische Verbreitung der Thiere* [The geographic distribution of animals] (Vienna, 1853).

8. Louis Agassiz, "The Natural Relations between Animals and the Elements in Which They Live," *American Journal of Science and Arts*, 2nd ser., 9 (1850): 369–394.

9. P. L. Sclater, *Journal of the Proceedings of the Linnean Society, Zoology*, 2 (1858): 130.

10. H. S. Reimarus, *Allgemeine Betrachtungen über die Triebe der Thiere*, 4th ed., ed. J. A. H. Reimarus (Hamburg, 1798).

11. Thomas Huxley, in *The Life and Letters of Charles Darwin*, ed. Francis Darwin, II (London, 1887), p. 197.

12. *Ibid.*, 312.

13. Charles Darwin, *On the Origin of Species*, p. 175.

14. *Ibid.*, pp. 209–210.

15. J. A. H. Reimarus, ed., in H. S. Reimarus, *Allgemeine Betrachtungen.*

16. C. W. L. Gloger, "Über die Farben der Eier der Vögel. Ein teleologischer Versuch [Concerning the colors of birds' eggs. A teleological investigation]," *Verhandlungen der Gesellschaft Naturforschender Freunde, Berlin*, 1 (1829): 332–347.

17. Bernard Altum, "Allgemeines und Spezielles zur Färbung der Vogeleier [General and particular information concerning the coloration of birds' eggs]," *Bericht über die XIII. Versammlung der Deutschen Ornithologen-Gesellschaft zu Stuttgart . . . 1860* (Stuttgart, 1861?), p. 34.

18. Gustav Seidlitz, *Die Bildungsgesetze der Vogeleier und das Transmutationsgesetz der Organismen* [The laws of growth of birds' eggs and the laws of transmutation of organisms] (Leipzig, 1869).

19. A. R. Wallace, *Westminster Review* (July 1867), p. 3.

20. *Ibid.*, p. 34.

21. All quotations from Wallace, "The Philosophy of Birds' Nests," *Intellectual Observer*, 11 (1867): 418–420.

22. Wallace, *The Geographical Distribution of Animals, with a Study of The Relations of Living and Extinct Faunas as Elucidating the Past Changes of the Earth's Surface*, I (London, 1876), pp. 25-27.

23. Leipzig, 1876; a translation, with revision of a dissertation published in Swedish (Helsingfors, 1874); see Chapter 13, reference 14.

24. E. F. von Homeyer, "Die Heerstrassen und Stationen der Vögel, mit Ruchsicht auf die ererbten Gewohnheiten [The highways and way stations of birds, with regard to inherited habits]," *Journal für Ornithologie*, 26 (1878): 23, 26.

25. Wallace, *Tropical Nature and Other Essays* (London, 1878), p. 209.

26. Wilhelm von Reichenau, *Die Nester und Eier der Vögel in ihren natürlichen Beziehungen betrachtet* (Leipzig, 1880).

27. Darwin to Moritz Wagner, October 13, 1876, in Francis Darwin, ed., *Life and Letters of Charles Darwin*, III (1887), p. 159.

28. Henry Seebohm, *The Geographical Distribution of the Family Charadriidae* (London, 1887), p. 17.

Chapter 19. The Return to Empiricism

1. See Herman Schalow, "Bernard Altum 1824–1900," in Schalow, *Beiträge zur Vogelfauna der Mark Brandenburg* (Berlin, 1919), pp. 555–569.
2. Bernard Altum, "Die Nahrung unserer Eulen," *Journal für Ornithologie,* 11 (1863): 41–46.
3. Adolf and Karl Müller, "Eine Entgegnung auf die Schrift: '*Der Vogel und sein Leben,* ' " *Journal für Ornithologie,* 16 (1868): 265–284, 340–352.
4. See also Chapter 13.
5. Edited by Rudolf Blasius; English edition, slightly revised, *Heligoland as an Ornithological Observatory,* trans. Rudolph Rosenstock (Edinburgh, 1895); ensuing quotations are from the English edition, pp. 148, 42, 148, 132.
6. W. W. Cooke, *Report,* edited and revised by Merriam (U.S. Department of Agriculture, Division of Economic Ornithology, Bulletin No. 2), p. 5.
7. J. A. Naumann, *Naturgeschichte der Land- und Wasservögel,* IV (Köthen, 1803), part 2.
8. For Mortensen, see *Studies in Bird Migration: Being the Collected Papers of H. Chr. C. Mortensen,* ed. Paul Jespersen and A. V. Tåning (Copenhagen, 1950).
9. Johannes Thienemann, "Die Zug des weissen Storches *Ciconia ciconia,*" *Zoologisches Jahrbücher,* Supplement 12 (1910): 665–686.

Chapter 20. The Reform of the Theory of Behavior

1. C. Lloyd Morgan, *Habit and Instinct* (London, 1896), p. 305.
2. *Ibid.,* p. 315.
3. *Ibid.,* p. 320.
4. *Ibid.,* p. 255.
5. *Ibid.,* p. 212.
6. Edmund Selous, *The Bird Watcher in the Shetlands* (London, 1905), p. 323.
7. Selous, *Bird Life Glimpses* (London, 1905), p. 210.
8. *Ibid.,* p. 184.
9. Selous, *The Bird Watcher in the Shetlands,* p. 200.
10. Selous, *Bird Life Glimpses,* p. 199.
11. Selous, *The Bird Watcher in the Shetlands,* pp. 199–201, 280–283.
12. Biography and full list of publications in Katharina Heinroth, *Oskar Heinroth, Vater der Verhaltensforschung* (Stuttgart, 1971).
13. Oskar Heinroth, *Sitzungsberichte der Gesellschaft naturforschender Freunde* (Berlin, 1898), pp. 95–118.
14. Heinroth, "Ornithologische Ergebnisse der 'I. Deutschen Südsee Expedition' von Br. Mencke," *Journal für Ornithologie,* 50 (1902): 393.
15. Heinroth, "Beiträge zur Biologie, namentlich Ethologie und Psychologie der Anatiden," *Verhandlungen des V. Internationalen Ornithologen-Kongresses in Berlin . . . 1910* (Berlin, 1911): p. 589.
16. Heinroth, "Beiträge," p. 701.
17. Charles Darwin, *On the Origin of Species* (1859), p. 135.
18. Carl Semper, *Über die Aufgabe der modernen Thiergeographie* (Berlin, 1879), pp. 13–14 (Heft 322 in *Sammlung gemeinverständlicher wissenschaftlicher Vorträge,* ed. R. Virchow and Fr. von Holtzendorff, XIV. Series).

Chapter 21. Ramification and Interconnection

1. R. Demoll, *Zeitschrift für Biologie,* 90 (1930): 199–230.
2. Max Stolpe and Karl Zimmer, "Der Schwirrflug des Kolibri im Zeitlupenfilm

[The whirring flight of hummingbirds in slow-motion films]," *Journal für Ornithologie,* 87 (1939): 136-155.

3. Karl Zimmer, "Der Flug des Nektarvogels [The flight of the sunbird] (*Cinnyris*)," *Journal für Ornithologie,* 91 (1943): 371-387.

4. J. E. Silberschlag, "Vom dem Fluge der Vögel [On the flight of birds]," *Schriften der Berlinischen Gesellschaft Naturforschender Freunde,* 2 (1781): 214-270.

5. R. R. Graham, "Safety Devices in Wings of Birds," *British Birds,* 24 (1930): 2-21, 34-47, 58-65.

6. Konrad Lorenz, *Journal für Ornithologie,* 81 (1933): 107-236.

7. Werner Rüppell, "Versuche über Heimfinden ziehender Nebelkrähe nach Verfrachtung [Experiments with home-seeking hooded crows following displacement]," *Journal für Ornithologie,* 92 (1944): 123.

8. Gustave Kramer and Ursula von St. Paul, "Ein wesentlicher Bestandteil der Orientierung der Reisetaube: Die Richtungsdressur [An essential component in the orientation of the carrier pigeon: directional training]," *Zeitschrift für Tierpsychologie,* 7 (1950): 620-631.

9. For subsequent developments see G. V. T. Matthews, *Bird Navigation,* 2nd ed. (Cambridge, England, 1968); Ernst Schütz, *Grundriss der Vogelzugskunde* [Outline of the science of bird migration], 2nd ed. (Berlin, 1971).

10. K. F. Burdach, *Die Physiologie als Erfahrungswissenschaft* [Physiology as an empirical science], III (Leipzig, 1830), p. 399.

11. Alexander Brandt, "Anatomisches und Allgemeines über die sogenannte Hahnenfedrigkeit und über anderweitige Geschlechtsanomalien bei Vögeln [Anatomical and general considerations on so-called cock feathering and other sexual anomalies in birds]," *Zeitschrift für wissenschaftliche Zoologie,* 48 (1889): 101-190.

12. G. K. Noble and K. Wurm, "The Effect of Testosterone Propionate on the Black-crowned Night Heron," *Endocrinology,* 26 (1940): 837-850.

13. J. P. Burkitt, "A Study of the Robin by Means of Marked Birds," *British Birds,* 17 (1924): 294-303, 18 (1924-25): 97-103, 294-303, 19 (1925): 120-124, 20 (1926): 91-101.

14. M. M. Nice, *Transactions of the Linnean Society of New York,* 4 (1937) and 6 (1943).

15. Francis Darwin, ed., *The Life and Letters of Charles Darwin,* II (1887), p. 7.

16. Bernard Altum, *Der Vogel und sein Leben* (Münster, 1868), p. 54.

17. Carl Semper, *Über die Aufgabe der modernen Thiergeographie* (Berlin, 1879), pp. 25, 27.

18. Konrad Lorenz, "Die angeborenen Formen möglicher Erfahrung [The innate programs for possible experience]," *Zeitschrift* für Tierpsychologie, 5 (1943): 235-409; Lorenz, "Ganzheit und Teil in der tierischen und menschlichen Gemeinschaft [Whole and part in animal and human society]," *Studium Generale,* 3 (1950): 455-499; Lorenz, *Über tierisches und menschliches Verhalten,* 2 vols. (Munich, 1965), translated by Robert Martin as *Studies in Animal and Human Behaviour,* 2 vols. (Cambridge, Mass., 1970-1971).

19. H. S. Reimarus, *Allgemeine Betrachtungen über die Triebe der Thiere* (Hamburg, 1760), paragraph 41.

20. Charles Darwin, *On the Origin of Species* (1959), p. 210.

21. Otto Koehler, "Vorsprachliches Denken und 'Zählen' der Vögel," [Preverbal thinking and 'counting' in birds], in *Ornithologie als biologische Wissenschaft,* ed. Ernst Mayr and Ernst Schüz (Festschrift Stresemann; Heidelberg, 1949), p. 145.

Epilogue. Materials for a History of American Ornithology

1. For general accounts: C. G. Sibley, "Ornithology," in *A Century of Progress in the Natural Sciences, 1853-1953* (San Francisco: California Academy of

Sciences, 1955), pp. 629–659. R. H. Welker, *Birds and Men* (Cambridge, Mass.: Harvard University Press, 1955), 240 pp.; chapters dealing with Alexander Wilson, Audubon, bird art, bird poetry, education, and conservation; with bibliography. F. M. Chapman and T. S. Palmer, eds., *Fifty Years' Progress of American Ornithology 1883–1933* (Lancaster, Pa.: American Ornithologists' Union, 1933), 249 pp.; referred to in subsequent notes as "Chapman and Palmer." T. S. Palmer *et al.*, *Biographies of Members of the American Ornithologists' Union* (Washington, D. C.: no publisher given, 1954), 630 pp.

For the early period (pre-Audubon) particularly: Elsa G. Allen, "The History of American Ornithology before Audubon," *Trans. Amer. Phil. Soc.*, n.s., 41 (1951): 385–591. G. R. Frick and R. P. Stearns, *Mark Catesby: The Colonial Audubon* (Urbana: University of Illinois Press, 1961), 148 pp. Joseph Ewan, *William Bartram: Botanical and Zoological Drawings, 1756–1788* (Philadelphia: *American Philosophical Society*, 1968), 190 pp. Francis Harper, *The Travels of William Bartram* (Naturalist's edition; New Haven: Yale University Press, 1958), 788 pp.

2. J. L. Peters, "Collections of Birds in the United States and Canada: Study Collections," in Chapman and Palmer (ref. 1), pp. 131–141. R. C. Banks, M. H. Clench, and J. C. Barlow, "Bird Collections in the United States and Canada," *Auk*, 90 (1973): 136–170; report on 283 collections with 4 million specimens.

3. See Chapters 9 and 13; E. J. Nolan, *A Short History of the Academy of Natural Sciences of Philadelphia* (Philadelphia: Academy of Natural Sciences, 1909), 38 pp. Witmer Stone, [History of bird collection,] *Proc. Acad. Nat. Sci. Phil. 1899* (1900), 5–11.

4. G. B. Goode, *The Smithsonian Institution 1846–1896* (Washington, D. C.: Smithsonian Institution, 1897), 856 pp. Walter Karp, *The Smithsonian Institution* (Washington, D. C.: Smithsonian Institution and *American Heritage Magazine*, 1965), 125 pp. C. W. Richmond, The Division of Birds and Its Collections (Unpublished MS in the U.S. Natural History Museum, 1919), 13 pp.

5. For Baird: W. H. Dall, *Spencer Fullerton Baird: A Biography* (Philadelphia: Lippincott, 1915), 478 pp. G. B. Goode, *Bibliographies of American Naturalists I: The Published Writings of Spencer Fullerton Baird, 1843–1882; Bull. U.S. Nat. Mus.*, No. 20 (1883), 393 pp. J. S. Billings, "Memoir of Spencer Fullerton Baird, 1823-1887," *Nat. Acad. Sci., Biog. Mem.*, 3 (1895): 141–160.

For Coues: J. A. Allen, "Biographical Memoir of Elliot Coues, 1842–1899," *Nat. Acad. Sci., Biog. Mem.*, 6 (1909): 397–446.

For Ridgway: Henry Harris, "Robert Ridgway, with a Bibliography of his Published Writings and Fifty Illustrations," *Condor*, 30 (1928): 5–118.

6. For the history of railroad and boundary surveys: W. H. Goetzmann, *Army Exploration in the American West 1803–1863* (New Haven: Yale University Press, 1959), 509 pp.; idem, *Exploration and Empire* (New York: Knopf, 1966), 678 pp.

7. G. R. Agassiz, "The History of the M.C.Z.," *Notes Concerning the History and Contents of the Museum of Comparative Zoology by Members of the Staff* (Cambridge, Mass.; Museum of Comparative Zoology, 1936). Edward Lurie, *Louis Agassiz: A Life in Science* (Chicago: University of Chicago Press, 1960), particularly Chapter 6. C. F. Batchelder, "A Bibliography of the Published Writings of William Brewster," *Mem. Nuttall Ornith. Club*, 20 (1951), 54 pp.

8. Histories of the American Museum and of its Bird Department exist only in manuscript: the Museum, by Jean Le Corbeiller, in the hands of Dr. Thomas Nicholson, director, and the Bird Department, by Robert C. Murphy, in the library of the Department. See also F. M. Chapman, "Joel Asaph Allen." *Nat. Acad. Sci., Biog. Mem.*, 11 (1927): 1–20; *Mem. Nat. Acad. Sci.*, 21, First Memoir.

9. F. M. Chapman, *Autobiography of a Bird Lover* (New York: Appleton-Century, 1933), 436 pp. R. C. Murphy, "Frank Michler Chapman, 1864–1945," *Auk*, 67 (1950): 307–315. W. K. Gregory, "Biographical Memoir of Frank Michler Chapman, 1864–1945," *Nat. Acad. Sci., Biog. Mem.*, 25 (1949), 111–145.

10. Mary Fuertes Boynton, *Louis Agassiz Fuertes* (New York: Oxford University Press, 1956), 335 pp. F. G. Marcham, ed. *Louis Agassiz Fuertes and the Singular*

Beauty of Birds: Paintings, Drawings, Letters (New York: Harper & Row, 1971), 232 pp.

11. Florence P. Jaques, *Francis Lee Jaques: Artist of the Wilderness World* (Garden City, N. Y.: Doubleday, 1973), 392 pp.

12. Some recent papers on the ornithogeography of South America, particularly of the Andes: Jürgen Haffer, "Speciation in Colombian Forest Birds West of the Andes," *Amer. Mus. Novit.*, No. 2294 (1967), 57 pp.; idem, "Speciation in Amazonian Forest Birds," *Science*, 165 (1969), 131–137; idem, *Avian Speciation in Tropical America* (Cambridge, Mass.: Nuttall Ornithological Club, Publ. no. 14, 1974), 398 pp. Ernst Mayr, "Inferences Concerning the Tertiary American Bird Faunas," *Proc. Nat. Acad. Sci.*, 51 (1964), 280–288. Ernst Mayr and W. H. Phelps, Jr., "The Origin of the Bird Fauna of the South Venezuelan Highlands," *Bull. Amer. Mus. Nat. Hist.*, 136 (1967), 269–328. R. A. Paynter, Jr., "Biology and Evolution of the *Atlapetes schistaceus* Species-Group (Aves: Emberizinae)," *Bull. Mus. Comp. Zool.*, 143 (1972), 297–320. François Vuilleumier, "Population Structure of the *Asthenes flammulata* Super-species (Aves: Furnariidae)," *Breviora*, No. 297 (1968), 21 pp.; idem, "Insular Biogeography in Continental Regions. I. The Northern Andes of South America," *Amer. Nat.*, 104 (1970), 373–388.

13. J. T. Zimmer, "Studies of Peruvian Birds," *Amer. Mus. Novit.* (1931–1955), Nos. 1–66.

14. Alan Solem, "The World Is Our Study" [An outline history of the Zoology Department], *Field Mus. Nat. Hist. Bull.*, 43 (7) (1972), 13–16. J. T. Zimmer, "In Memoriam — Carl Eduard Hellmayr," *Auk*, 61 (1944), 616–622.

15. A. A. Allen, "Ornithological Education in America," in Chapman and Palmer (ref. 1), pp. 215–229.

16. A. A. Allen, "Cornell's Laboratory of Ornithology," *Living Bird*, 1 (1962), 7–36; see also Chapman and Palmer (ref. 1), pp. 167–179.

17. Hilda W. Grinnell, "Joseph Grinnell: 1877–1939," *Condor*, 42 (1940), 3–34. A. H. Miller, ed., *Joseph Grinnell's Philosophy of Nature: Selected Writings* (Los Angeles: University of California Press, 1943), 242 pp. Ernst Mayr, "Alden Holmes Miller," in *Biographical Memoirs, National Academy of Sciences* (New York: Columbia University Press, 1973), pp. 177–214. N. K. Johnson, "The Museum of Vertebrate Zoology, University of California, Berkeley," *American Birds*, 25 (1971): 537–538.

18. Ernst Mayr, "The Role of Ornithological Research in Biology," *Proc. XIII Int. Ornith. Congress . . . 1962* (American Ornithologists' Union, 1963), pp. 27–38.

19. Aldo Leopold, *Game Management* (New York: Scribner's, 1933), 502 pp.; idem, *A Sand County Almanac* (New York: Oxford University Press, 1949), 226 pp.; P. L. Errington, "In Appreciation of Aldo Leopold," *Journ. Wildlife Management*, 12 (1948): 341–350.

20. T. G. Pearson, "Fifty Years of Bird Protection in the United States," in Chapman and Palmer (ref. 1), pp. 199–213.

21. Charles Foster Batchelder, "An Account of the Nuttall Ornithological Club, 1873–1919," *Mem. Nuttall Ornith. Club*, 8 (1937), 109 pp.

22. T. S. Palmer, "A Brief History of the American Ornithologists' Union," in Chapman and Palmer (ref. 1), pp. 7–27.

23. American Ornithologists' Union, *Check-list of North American Birds.* 1st ed. (1886), 400 pp.; 2nd ed. (1895), 384 pp.; 3rd ed. (1910), 430 pp.; 4th ed. (1931), 546 pp.; 5th ed. (1957), 705 pp.

24. See Lynds Jones, *Wilson Bull.*, 26 (1914): 23–27; R. M. Strong, *Wilson Bull.*, 51 (1939): 3–16. Also memorials for T. C. Stephens, *Auk*, 66 (1949): 396, and for Lynds Jones, *Auk*, 69 (1952): 258–265.

25. H. S. Swarth, *A Systematic Study of the Cooper Ornithological Club* (San Francisco: no publisher given, 1929, 78 pp.) See also ref. 17.

26. J. C. Welty, *The Life of Birds* (Philadelphia: Saunders, 1962), 560 pp. Josselyn Van Tyne and A. J. Berger, *Fundamentals of Ornithology* (New York: Wiley, 1959), 636 pp.

27. Albert Wolfson, ed., *Recent Studies in Avian Biology* (Urbana: University of Illinois Press, 1955), 479 pp. A. J. Marshall, ed., *Biology and Comparative Physiol-*

ogy (2 vols., New York: Academic Press, 1960–1961). D. S. Farner and J. R. King, eds., *Avian Biology* (4 vols., New York: Academic Press, 1971–1974).

28. J. T. Zimmer, *Catalogue of the Edward E. Ayer Ornithological Library* (2 vols., Chicago: Field Museum of Natural History, 1926; Zoological Series, vol. 16). C. A. Wood, *An Introduction to the Literature of Vertebrate Zoology* (London: Oxford University Press, 1931), 664 pp.; based chiefly on the libraries of McGill University, Montreal. R. M. Strong, *A Bibliography of Birds* (4 vols., Chicago: Field Museum of Natural History, 1939–1959; Zoological Series, vol. 25, pts. 1–4). S. D. Ripley and L. L. Scribner, *Ornithological Books in the Yale University Library* (New Haven: Yale University Press, 1961), 338 pp. Robert Mengel, *A Catalogue of the Ellis Collection of Ornithological Books in the University of Kansas Libraries* (Lawrence: University of Kansas Press, vol. 1, A-B, 1972), 288 pp.

29. Alexander Wetmore, "A Systematic Classification for the Birds of the World," *Proc. U.S. Nat. Mus.*, 76 (24) (1930): 1–8; see also *Auk*, 43 (1926): 337–346. Herbert Friedmann, "Recent Revisions in Classification and Their Biological Significance," in Albert Wolfson, ed., *Recent Studies* (ref. 27), pp. 23–43. Ernst Mayr and J. C. Greenway, Jr., "Sequence of Passerine Families (Aves)," *Breviora*, No. 58 (1956), 11 pp. Adolf Portmann, "Études sur la cérébralisation chez les oiseaux," *Alauda*, 14 (1946): 2–20; 15 (1947): 1–15, 161–176. Adolf Portmann and Werner Stingelin, "Cerebralization and Related Problems," in A. J. Marshall, ed., *Biology and Comparative Physiology of Birds* (ref. 27), vol. 2, pp. 29–35.

30. To indicate the scope of the modern approach to bird anatomy the titles of some of Bock's publications may be listed: "Preadaptation and Multiple Evolutionary Pathways," *Evolution*, 13 (1959): 194–211; with W. de W. Miller, "The Scansorial Foot of the Woodpeckers, with Comments on the Evolution of Perching and Climbing Feet in Birds," *Amer. Mus. Novit.*, No. 1931 (1959), 45 pp.; "Secondary Articulation of the Avian Mandible." *Auk*, 77 (1960): 19–55; "The Palatine Process of the Premaxilla in the Passeres. A Study of the Variation, Function, Evolution and Taxonomic Value of a Single Character throughout an Avian Order," *Bull. Mus. Comp. Zool.*, 122 (1960): 361–488; "Evolution and Phylogeny in Morphologically Uniform Groups," *Amer. Nat.*, 97 (1963): 265–285; "Kinetics of the Avian Skull." *Journ. Morph.*, 114 (1964): 1–42; "The Role of Adaptive Mechanisms in the Origin of Higher Levels of Organization," *Syst. Zool.*, 14 (1965): 272–287; with Gerd von Wahlert, "Adaptation and the Form-Function Complex," *Evolution*, 19 (1965): 269–299; "Comparative Morphology in Systematics," in *Systematic Biology* (Washington, D. C.: National Academy of Sciences, 1969), pp. 411–448; "Microevolutionary Sequences as a Fundamental Concept in Macroevolutionary Models," *Evolution*, 24 (1970): 704–722.

31. Alexander Wetmore, "Paleontology," in Albert Wolfson ed., *Recent Studies* (ref. 27), pp. 44–56; Pierce Brodkorb, "Catalogue of Fossil Birds," *Bull. Florida State Mus.*, 7 (1963): 179–293; 8 (1964): 195–335; 11 (1967): 99–220; 15: (1971): 163–266.

32. C. G. Sibley, "A Comparative Study of the Egg-White Proteins of Passerine Birds,". *Bull. Peabody Mus. Nat. Hist.*, 32 (1970), 165 pp.; C. G. Sibley and J. E. Ahlquest, "A Comparative Study of the Egg-White Proteins of Non-Passerine Birds," *Bull. Peabody Mus. Nat. Hist.*, 39 (1972), 322 pp.

33. William Rowan, "Fifty Years of Bird Migration," in Chapman and Palmer (ref. 1), pp. 51–63.

34. G. H. Lowery, Jr., "A Quantitative Study of the Nocturnal Migration of Birds," *Univ. Kansas Publ. Mus. Nat. Hist.*, 3 (1951), 361–472; G. H. Lowery and R. J. Newman, "Direct Studies of Nocturnal Bird Migration," in Albert Wolfson, ed., *Recent Studies* (ref. 27), pp. 238–263; W. H. Drury, Jr., and J. A. Keith, "Radar Studies of Songbird Migration in Coastal New England," *Ibis*, 104 (1962), 449–489; I. C. T. Nisbet and W. H. Drury, Jr., "Short-term Effects of Weather on Bird Migration: A Field Study Using Multivariate Statistics," *Animal Behaviour*, 16 (1968), 496–530.

35. F. M. Chapman, "Remarks on the Origin of Bird Migration," *Auk*, 11 (1894), 12–17.

36. Emil Witschi, "Sex and Secondary Sexual Characters." in A. J. Marshall,

ed., *Biology and Comparative Physiology of Birds* (ref. 27), vol. 2, pp. 115–168.

37. For a summary of his researches and a bibliography: Ruth Noble, *The Nature of the Beast* (Garden City, N.Y.: Doubleday, Doran, 1945), 237 pp.

38. Martin Moynihan, "Some Aspects of Reproductive Behavior in the Black-headed Gull (*Larus ridibundus ridibundus L.*) and Related Species," *Behaviour*, Suppl. 4 (1955), 220 pp.; idem, "A Revision of the Family Laridae (Aves)," *Amer. Mus. Novit.*, No. 1928 (1959), 42 pp.; idem, "Hostile and Sexual Behavior Patterns of South American and Pacific Laridae, *Behaviour, Suppl.* 8 (1962), 365 pp.; idem, *The Organization and Probable Evolution of Some Mixed Species Flocks of Neotropical Birds* (Washington, D. C.: Smithsonian Miscellaneous Collections, 143, 7, 1962), 140 pp.

39. A. J. Meyerriecks, *Comparative Breeding Behavior of Four Species of North American Herons* (Cambridge, Mass.: Nuttall Ornithological Club, Publ. no. 2, 1960), 158 pp.

40. P. A. Johnsgard, *Handbook of Waterfowl Behavior* (Ithaca, N. Y.: Cornell University Press, 1965), 394 pp.

41. W. C. Dilger, "Hostile Behavior and Reproductive Isolating Mechanisms in the Avian Genera *Catharus* and *Hylocichla*," *Auk*, 73 (1956), 313–353; idem, "Adaptive Modifications and Ecological Isolating Mechanisms in the Thrush Genera *Catharus* and *Hylocichla*," *Wilson Bull.*, 68 (1956), 171–199.

42. D. S. Lehrman, "Control of Behavior Cycles in Reproduction," in William Etkin, ed., *Social Behavior and Organization among Vertebrates* (Chicago: University of Chicago Press, 1964), pp. 143–166; idem, "Interaction between Internal and External Environments in the Regulation of the Reproductive Cycle of the Ring Dove," in F. A. Beach, ed., *Sex and Behavior* (New York: Wiley, 1965), pp. 355–380; idem, "Semantic and Conceptual Issues in the Nature-Nurture Problem," in Lester Aronson, *et al.*, eds., *Development and Evolution of Behavior: Essays in Memory of T. C. Schneirla* (San Francisco: Freeman, 1970), pp. 17–52.

43. A. A. Allen, "Albert Rich Brand," *Auk*, 58 (1941): 444–448.

44. For a history of the introduction of graphic methods into the study of bird song: R. K. Potter, G. A. Kopp, and H. C. Green, *Visible Speech* (New York: Van Nostrand, 1947), 441 pp. (sonagrams on p. 441); N. E. Collias, "The Development of Social Behavior in Birds," *Auk*, 69 (1952): 127–159 (see Fig. 1, p. 131, and Fig. 3, p. 135); D. J. Borror and C. R. Reese, "The Analysis of Bird Songs by Means of a Vibralyzer," *Wilson Bull.*, 65 (1953), 271–276; idem, "Vocal Gymnastics in Wood Thrush Songs," *Ohio J. Sci.*, 56 (1956), 177–182 (10 figures).

45. W. J. Smith, *Communication and Relationships in the Genus* Tyrannus, (Cambridge, Mass.: Nuttall Ornithological Club, Publ. no. 6, 1966), 262 pp.

46. W. E. Lanyon, *The Comparative Biology of the Meadowlarks* (Sturnella) *in Wisconsin* (Cambridge, Mass.: Nuttall Ornithological Club, Publ. no. 1, 1957), 108 pp.

47. C. H. Greenewalt, *Bird Song: Acoustics and Physiology* (Washington, D. C.: Smithsonian Institution Press, 1968), 202 pp.

48. Joseph Grinnell, "The Biota of the San Bernardino Mountains," *Univ. Calif. Publ. Zool.*, 5 (1908): 1–170; idem, "An Account of the Mammals and Birds of the Lower Colorado Valley," *Univ. Calif. Publ. Zool.*, 12 (1914): 51–294; idem, "A Distributional List of the Birds of California," *Pacific Coast Avifauna*, No. 11 (1915), 217 pp.; idem, "Observations upon the Bird Life of Death Valley," *Proc. Calif. Acad. Sci.*, Ser. 4, 13 (1923), 43–109; idem, "A Distributional Summation of the Ornithology of Lower California," *Univ. Calif. Publ. Zool.*, 32 (1928), 1–300; Joseph Grinnell and H. S. Swarth, "An Account of the Birds and Mammals of the San Jacinto Area of Southern California," *Univ. Calif. Publ. Zool.*, 10 (1913), 197–406; Joseph Grinnell and T. I. Storer, *Animal Life in the Yosemite* (Berkeley: University of California Press, 1924), 770 pp.; Joseph Grinnell, Joseph Dixon, and J. M. Linsdale, "Vertebrate Natural History of Northern California through the Lassen Peak Region," *Univ. Calif. Publ.* Zool., 35 (1930), 1–594; Joseph Grinnell and J. M. Linsdale, *Vertebrate Animals of Point Lobos Reserve, 1934–1935* (Washington, D. C.: Carnegie Institution, 1936), 165 pp.; Joseph Grinnell and

A. H. Miller, "The Distribution of the Birds of California," *Pacific Coast Avifauna*, No. 27 (1944), 608 pp.; E. R. Hall, *Mammals of Nevada* (Berkeley: University of California Press, 1946), 721 pp.; E. R. Hall and Joseph Grinnell, "Life-Zone Indicators in California," *Proc. Calif. Acad. Sci.*, Ser. 4, 9 (1919), 37–67; A. H. Miller, "An Analysis of the Distribution of the Birds of California," *Univ. Calif. Publ. Zool.*, 50 (1951), 531–644; H. S. Swarth, "A Distributional List of the Birds of Arizona, *Pacific Coast Avifauna*, No. 10 (1914), 133 pp.

49. Literature on bird censuses: R. R. Graber and J. W. Graber, "A Comparative Study of Bird Populations in Illinois, 1906–1909 and 1956–1958," *Bull. Ill. Nat. Hist. Survey*, 28 (1963), 383–528; S. C. Kendeigh, "Measurement of Bird Populations ," *Ecol. Monographs*, 14 (1944), 67–106.

50. R. H. MacArthur and E. O. Wilson, *The Theory of Island Biogeography* (Princeton: Princeton University Press 1967), 215 pp.; R. H. MacArthur, *Geographical Ecology* (New York: Harper & Row, 1972), 287 pp.

51. F. A. Pitelka, P. Q. Tomich, and G. W. Triechel, "Ecological Relations of Jaegers and Owls as Lemming Predators near Barrow, Alaska," *Ecol. Monographs*, 25 (1955), 85–117; F. A. Pitelka, "Some Aspects of Population Structure in the Short-Term Cycle of the Brown Lemming in Northern Alaska," *Cold Spring Harbor Symp. Quant. Biol.* (1957), 22 (1958), 237–251.

52. A. O. Gross, "History and Progress of Bird Photography in America," in Chapman and Palmer (ref. 1), pp. 159–180. For the use of motion pictures in avian research see, for instance, W. C. Dilger, "Methods and Objectives of Ethology," *Living Bird*, 1 (1962), 83–92; J. H. Storer, *The Flight of Birds Analyzed through Slow-Motion Photography* (Bloomfield Hills, Mich.: Cranbrook Institute of Science, Bulletin No. 28, 1948), 110 pp.; C. H. Greenewalt, "The Wings of Insects and Birds as Mechanical Oscillators," *Proc. Amer. Phil. Soc.*, 104 (1960), 605–611.

53. F. C. Lincoln, "Bird Banding," in Chapman and Palmer (ref. 1), pp. 65–87.

54. Hans Bub, *Vogelfang und Vogelberingung. Teil II. Fang mit grossen Reusen, Fangkäfigen, Stellnetzen und Decknetzen* (2nd ed.; Wittenberg: Ziemsen, 1972), 212 pp. (Neue Brehm-Bücherei 377).

55. For further information on faunistic literature: J. J. Hickey, *A Guide to Bird Watching* (New York: Oxford University Press, 1943), 276 pp.; O. S. Pettingill, Jr., *A Guide to Bird Finding East of the Mississippi* (New York: Oxford University Press, 1951), 682 pp.; idem, *A Guide to Bird Finding West of the Mississippi* (New York: Oxford University Press, 1953), 734 pp.; idem, *Ornithology in Laboratory and Field* (4th ed.; Minneapolis: Burgess, 1970), 540 pp.; R. G. Welker, *Birds and Men* (ref. 1), Chap. 13, "Federal and State Bird Books."

56. W. E. Godfrey, *The Birds of Canada* (Ottawa: Queen's Printer, 1966), 432 pp. (valuable faunistic bibliography, pp. 410–414).

57. R. C. Murphy, "In Memoriam: William Henry Phelps," *Auk*, 87 (1970), 419–424; William H. Phelps and W. H. Phelps, Jr., *Lista de las aves de Venezuela con su distribucion* (2 vols., Caracas: Editorial Sucre, 1958–1963).

58. For a history of the ornithological exploration of Chile: C. E. Hellmayr, *The Birds of Chile* (Chicago: Field Mus. Nat. Hist., Zoological Series, vol. 19, 1932), pp. 6–12. There is now an excellent avifauna of Chile: A. W. Johnson, *The Birds of Chile and Adjacent Regions of Argentina, Bolivia, and Peru* (2 vols., Buenos Aires: Platt, 1965–1967; Supplement, 1972, 118 pp.).

Index